NEW HORIZONS IN ASTRONOMY
Frank N. Bash Symposium 2005

COVER ILLUSTRATION:

Graphic from the conference poster designed by J. Umbarger. Images from NASA, ESA, and J. Hester (Arizona State University).

ASTRONOMICAL SOCIETY OF THE PACIFIC
CONFERENCE SERIES

A SERIES OF BOOKS ON RECENT DEVELOPMENTS IN ASTRONOMY AND ASTROPHYSICS

Volume 352

EDITORIAL STAFF

Managing Editor: J. W. Moody
Publication Manager: Enid L. Livingston
Technical Specialist: Lisa B. Roper
Technical Consultant: Jared Bellows
E-book Specialist: Jeremy Roper

PO Box 4666, Room N221– ESC, Brigham Young University, Provo, Utah, 84602-4666
Phone: 801-422-2111   Fax: 801-422-0553
E-mail: aspcs@byu.edu   E-book site: http://www.aspbooks.org

$L^A T_E X$ *Consultant:* T. J. Mahoney (Spain) – tjm@iac.es

Beginning 2004, ASP-CS Volumes may be found as e-books with color images at:
http://www.aspbooks.org

A listing of all IAU Volumes published
and the ASP Conference Series published during the past one year
may be found at the back of this volume.

For all ASP-CS Volumes published see: http://www.astrosociety.org/pubs.html

ASTRONOMICAL SOCIETY OF THE PACIFIC
CONFERENCE SERIES

Volume 352

NEW HORIZONS IN ASTRONOMY
Frank N. Bash Symposium 2005

Proceedings of a meeting held at
The University of Texas, Austin, Texas, USA
16-18 October 2005

Edited by

Sheila J. Kannappan, Seth Redfield,
Jacqueline E. Kessler-Silacci,
Martin Landriau, and Niv Drory
Department of Astronomy, University of Texas, Austin, Texas, USA

SAN FRANCISCO

ASTRONOMICAL SOCIETY OF THE PACIFIC
390 Ashton Avenue
San Francisco, California, 94112-1722, USA

Phone: 415-337-1100
Fax: 415-337-5205
E-mail: service@astrosociety.org
Web Site: www.astrosociety.org
E-books: www.aspbooks.org

ISBN: 1-58381-220-2

Library of Congress (LOC) Cataloging in Publication (CIP) Data:
Main entry under title
Library of Congress Control Number (LCCN): 2006929346

Printed in United States of America by Sheridan Books, Ann Arbor, Michigan

Contents

Part 2. Research Highlights

Contents   vii

Preface

The novel suggestion of an astronomy symposium devoted entirely to the work and ideas of young researchers first arose in 2003 as a way to honor Frank N. Bash, retiring Director of McDonald Observatory. Recognizing Frank's long-time advocacy of education & outreach and interest in rising talent at both the Observatory and the University of Texas at Austin, his colleagues Dan Jaffe and Neal Evans decided to celebrate his career by inviting a set of outstanding postdoctoral fellows from various subfields of astronomy to give review talks at UT Austin, creating a symposium that would emphasize "a look ahead at what will be the big issues and opportunities for progress in their fields over the next decade." Local UT postdocs served as panelists and moderators for extended discussion sessions after the talks, and the entire symposium brilliantly showcased both the current research and the research *vision* of the next generation of astronomers.

Bashfest 2003 was so successful that it inspired several UT astronomy professors, including Frank Bash, to pledge funds for a biennial series of similar conferences, this time organized by the UT postdocs themselves (with invaluable assistance from faculty and staff). The editors of this volume are all postdocs who served on the Scientific Organizing Committee and/or Local Organizing Committee for the first conference of the new series, "Frank N. Bash Symposium 2005: New Horizons in Astronomy," held October 16–18, 2005 on the campus of UT Austin.

Expanding on Bashfest 2003, the SOC of the 2005 Symposium chose to broaden local participation by inviting three UT postdocs to give review talks and by adding a contributed poster session open to both graduate students and postdocs. Happily, the poster session drew in many students and several people from outside institutions such as Cornell, UC Santa Cruz, and New Mexico State University. We hope that future symposia can grow in scope from this beginning. Participants enjoyed the rare opportunity to learn about research in very different subfields and to meet some of the most dynamic young researchers shaping those fields. Extended moderated discussions and informal break-time conversations fertilized new ideas and planted seeds for future collaboration. The LOC enhanced the symposium with a mix of social activities, including a fabulous banquet at the legendary Fonda San Miguel, two small-group graduate student Q&A lunches with speakers, and an unforgettable night on the town in "The Live Music Capitol of the World." The t-shirts were also a big hit!

Many people contributed to the success of Bash Symposium 2005. In addition to the scientific presenters featured in this volume, several others kept the meeting lively, including the opening and closing speakers (Associate Dean Peter Riley,

Astronomy Department Chair Don Winget, McDonald Observatory Director David Lambert, and of course Professor Frank Bash) and the session moderators (postdocs Judit Györgyey Ries, Michael Siegel, Martin Landriau, and Niv Drory). Behind the scenes and for many months before the conference, the organizing committees worked incredibly hard and efficiently to bring us a meeting full of memorable events inside the conference room and out. The conference planning was led primarily by postdocs (starred below) and one enthusiastic and valuable graduate student (double starred). The members were: *SOC* — *Seth Redfield (Chair), Frank Bash, Volker Bromm, *Niv Drory, *Sheila Kannappan, Pawan Kumar, *Martin Landriau; *LOC* — *Jacqueline Kessler-Silacci (Co-Chair), *Judit Györgyey Ries (Co-Chair), Monica Kidd, Mary Lindholm, *Douglas Mar, *Seth Redfield, and **Andrea Urban. Neal Evans and Tom Barnes served as helpful advisors to the SOC and LOC, respectively. Additional, much appreciated technical and administrative assistance came from Caroline Blakey, Rebecca Christian, Stephanie Crouch, and Lara Eakins. Jim Umbarger designed a fabulous web site and the graphic for the conference poster. Photos are courtesy of Martin Harris Photography.

The Bash Symposium could not have proceeded without the financial support of several institutions and individuals. Vital institutional funding came from the University of Texas Department of Astronomy, McDonald Observatory, and Dean Mary Ann Rankin and The College of Natural Sciences. The conference also relied on generous contributions from Professors Frank N. Bash, Rob Robinson, and Don Winget. Finally, we would like to extend a special thank you to the McDonald Observatory Board of Visitors for their support, and in particular to Richard E. Adams, Anne Marie Adkins, Wayne and Barbara Alexander, Paul Balmuth, Steve Bickerstaff and Charlotte Carter, Terry Bray, Michelle, G.W., and Sophia Brock, Frank Kell Cahoon, Clifton Caldwell, David F. Chappell, John L. Cotton, Jr., Harold Woody Davis, Marshall J. Doke, Jr., Catherine Bernell Estrada, Patricia Morrison Fleming, Walter Foxworth, Carol Whitcraft Fredericks, Jeffery L. Hart, John M. Heasley, Robert Jorrie, Jeffrey Kodosky, James A Kruger, Preston Edward Lindsey, Jr., Irl I. Nathan, Richard T. Schlosberg, Karen Goetting Skelton, F. Ford Smith, Jr., Sallie and Joe Tarride, Curtis Vaughan, Jr., Rom P. Welborn, John Wildenthal, Mark Richard Williams, Jim and Nancy Wood, Shirley C. Wozencraft, and Francis Wright. Support for SJK, SR, and JEK-S was provided by the NSF, Hubble, and Spitzer Postdoctoral Fellowship Programs, respectively, under NSF AST-0401547, STScI HST-HF-01190.01 (NASA contract NAS 5-26555), and NASA contract 1256316.

The Editors

Participants

J. J. ADAMS, Department of Astronomy, University of Texas at Austin, 1 University Station, C1400, Austin, TX, 78712-3451, USA ⟨ jjadams@astro.as.utexas.edu ⟩

A. M. ADKINS, Austin Astronomical Society, PO Box 12831, Austin, TX, 78711-2831, USA ⟨ anne@hadkins.com ⟩

S. AKIYAMA, Department of Astronomy, University of Texas at Austin, 1 University Station, C1400, Austin, TX, 78712-3451, USA ⟨ shizuka@astro.as.utexas.edu ⟩

C. ALLENDE PRIETO, McDonald Observatory and Department of Astronomy, University of Texas at Austin, 1 University Station, C1400, Austin, TX, 78712-3451, USA ⟨ callende@astro.as.utexas.edu ⟩

M. A. ALVAREZ, Department of Astronomy, University of Texas at Austin, 1 University Station, C1400, Austin, TX, 78712-3451, USA ⟨ marcelo@astro.as.utexas.edu ⟩

R. J. AVILA, Department of Astronomy, New Mexico State University, MSC 4500, Box 30001, Las Cruces, NM, 88003, USA ⟨ ravila@astronomy.nmsu.edu ⟩

B. J. BANICKI, Austin Astronomical Society, PO Box 12831, Austin, TX, 78711-2831, USA ⟨ bbanicki@hotmail.com ⟩

F. D. BARAZZA, Department of Astronomy, University of Texas at Austin, 1 University Station, C1400, Austin, TX, 78712-3451, USA ⟨ barazza@astro.as.utexas.edu ⟩

T. G. BARNES, McDonald Observatory, University of Texas at Austin, 1 University Station, C1400, Austin, TX, 78712-3451, USA ⟨ tgb@astro.as.utexas.edu ⟩

F. N. BASH, Department of Astronomy, University of Texas at Austin, 1 University Station, C1400, Austin, TX, 78712-3451, USA ⟨ fnb@astro.as.utexas.edu ⟩

A. E. BAUER, Department of Astronomy, University of Texas at Austin, 1 University Station, C1400, Austin, TX, 78712-3451, USA ⟨ amanda@astro.as.utexas.edu ⟩

J. L. BEAN, Department of Astronomy, University of Texas at Austin, 1 University Station, C1400, Austin, TX, 78712-3451, USA ⟨ bean@astro.as.utexas.edu ⟩

M. A. BITNER, Department of Astronomy, University of Texas at Austin, 1 University Station, C1400, Austin, TX, 78712-3451, USA ⟨ mbitner@astro.as.utexas.edu ⟩

J. BORNAK, Department of Astronomy, New Mexico State University, MSC 4500, Box 30001, Las Cruces, NM, 88003, USA ⟨ jbornak@nmsu.edu ⟩

V. BROMM, Department of Astronomy, University of Texas at Austin, 1 University Station, C1400, Austin, TX, 78712-3451, USA ⟨ vbromm@astro.as.utexas.edu ⟩

D. F. BROWN, McDonald Observatory Board of Visitors, 1 University Station, C1400, Austin, TX, 78712-3451, USA ⟨ david.brown@hughesluce.com ⟩

R. K. CAMPBELL, Department of Astronomy, New Mexico State University, MSC 4500, Box 30001, Las Cruces, NM, 88003, USA ⟨ cryan@nmsu.edu ⟩

L. S. CARTWRIGHT, Austin Astronomical Society, PO Box 12831, Austin, TX, 78711-2831, USA ⟨ lscar@sbcglobal.net ⟩

D. W. CHANDLER, Central Texas Astronomical Society, 528 Wildwood Trail, Lorena, TX, 76655, USA ⟨ chandler@vvm.com ⟩

J. M. CHAVEZ, Department of Astronomy, University of Texas at Austin, 1 University Station, C1400, Austin, TX, 78712-3451, USA ⟨ jchavez@astro.as.utexas.edu ⟩

C. H. CHEN, National Optical Astronomy Observatory, PO Box 26732, 950 N. Cherry Ave., Tuscon, AZ, 85726-6732, USA ⟨ cchen@noao.edu ⟩

J.-H. CHEN, Department of Astronomy, University of Texas at Austin, 1 University Station, C1400, Austin, TX, 78712-3451, USA ⟨ jhchen@astro.as.utexas.edu ⟩

L. A. CIEZA, Department of Astronomy, University of Texas at Austin, 1 University Station, C1400, Austin, TX, 78712-3451, USA ⟨ lcieza@astro.as.utexas.edu ⟩

V. P. DEBATTISTA, Department of Astronomy, University of Washington, Box 351580, Seattle, WA 98195-1580, USA ⟨ debattis@astro.washington.edu ⟩

H. L. DINERSTEIN, Department of Astronomy, University of Texas at Austin, 1 University Station, C1400, Austin, TX, 78712-3451, USA ⟨ harriet@astro.as.utexas.edu ⟩

N. DRORY, McDonald Observatory, University of Texas at Austin, 1 University Station, C1400, Austin, TX, 78712-3451, USA ⟨ drory@astro.as.utexas.edu ⟩

M. M. DUNHAM, Department of Astronomy, University of Texas at Austin, 1 University Station, C1400, Austin, TX, 78712-3451, USA ⟨ mdunham@astro.as.utexas.edu ⟩

N. J. EVANS, Department of Astronomy, University of Texas at Austin, 1 University Station, C1400, Austin, TX, 78712-3451, USA ⟨ nje@astro.as.utexas.edu ⟩

T. FEDER, Physics Today, American Center for Physics, One Physics Ellipse, College Park, MD, 20740-3842, USA ⟨ tfeder@wam.umd.edu ⟩

Y. FENNER, Institute for Theory and Computation, Harvard-Smithsonian Center for Astrophysics, 60 Garden Street, MS-51, Cambridge, MA, 02138, USA ⟨ yfenner@cfa.harvard.edu ⟩

E. R. FERNANDEZ, Department of Astronomy, University of Texas at Austin, 1 University Station, C1400, Austin, TX, 78712-3451, USA ⟨ beth@astro.as.utexas.edu ⟩

D. B. FISHER, Department of Astronomy, University of Texas at Austin, 1 University Station, C1400, Austin, TX, 78712-3451, USA ⟨ dbfisher@astro.as.utexas.edu ⟩

E. B. FORD, Department of Astronomy, Univeristy of California at Berkeley, 601 Campbell Hall, Berkeley, CA, 94720, USA ⟨ eford@berkeley.edu ⟩

A. D. FORESTELL, Department of Astronomy, University of Texas at Austin, 1 University Station, C1400, Austin, TX, 78712-3451, USA ⟨ amydove@astro.as.utexas.edu ⟩

E. J. GAWISER, Department of Astronomy, Yale University, PO Box 208101, New Haven, CT, 06520-8101, USA ⟨ gawiser@astro.yale.edu ⟩

L. HAO, Cornell University, 108 Space Science Building, Ithaca, NY, 14817, USA ⟨ haol@isc.astro.cornell.edu ⟩

M. K. HEMENWAY, Department of Astronomy, University of Texas at Austin, 1 University Station, C1400, Austin, TX, 78712-3451, USA ⟨ marykay@astro.as.utexas.edu ⟩

S. B. HOLMES, Department of Astronomy, University of Texas at Austin, 1 University Station, C1400, Austin, TX, 78712-3451, USA ⟨ sholmes@astro.as.utexas.edu ⟩

D. JAFFE, Department of Astronomy, University of Texas at Austin, 1 University Station, C1400, Austin, TX, 78712-3451, USA ⟨ dtj@astro.as.utexas.edu ⟩

E. J. JEFFERY, Department of Astronomy, University of Texas at Austin, 1 University Station, C1400, Austin, TX, 78712-3451, USA ⟨ ejeffery@astro.as.utexas.edu ⟩

J. S. KALIRAI, UCO/Lick Observatory and Department of Astronomy, University of California at Santa Cruz, 1156 High Street, Santa Cruz, CA, 95064, USA ⟨ jkalirai@ucolick.org ⟩

S. J. KANNAPPAN, McDonald Observatory, University of Texas at Austin, 1 University Station, C1400, Austin, TX, 78712-3451, USA ⟨ sheila@astro.as.utexas.edu ⟩

J. E. KESSLER-SILACCI, McDonald Observatory and Department of Astronomy, University of Texas at Austin, 1 University Station, C1400, Austin, TX, 78712-3451, USA ⟨ jes@astro.as.utexas.edu ⟩

M. KILIC, Department of Astronomy, University of Texas at Austin, 1 University Station, C1400, Austin, TX, 78712-3451, USA ⟨ kilic@astro.as.utexas.edu ⟩

A. B. KIM, Department of Astronomy, University of Texas at Austin, 1 University Station, C1400, Austin, TX, 78712-3451, USA ⟨ agnes@astro.as.utexas.edu ⟩

C. KNEZ, Department of Astronomy, University of Texas at Austin, 1 University Station, C1400, Austin, TX, 78712-3451, USA ⟨ claudia@astro.as.utexas.edu ⟩

E. KOMATSU, Department of Astronomy, University of Texas at Austin, 1 University Station, C1400, Austin, TX, 78712-3451, USA ⟨ komatsu@astro.as.utexas.edu ⟩

M. R. KRUMHOLZ, Department of Astrophysical Sciences, Princeton University, Peyton Hall, Ivy Lane, Princeton, NJ, 08544-1001, USA ⟨ krumholz@astro.princeton.edu ⟩

P. KUMAR, Department of Astronomy, University of Texas at Austin, 1 University Station, C1400, Austin, TX, 78712-3451, USA ⟨ pk@astro.as.utexas.edu ⟩

D. L. LAMBERT, McDonald Observatory, University of Texas at Austin, 1 University Station, C1400, Austin, TX, 78712-3451, USA ⟨ director@astro.as.utexas.edu ⟩

M. LANDRIAU, Department of Astronomy, University of Texas at Austin, 1 University Station, C1400, Austin, TX, 78712-3451, USA ⟨ landriau@astro.as.utexas.edu ⟩

D. F. LESTER, Department of Astronomy, University of Texas at Austin, 1 University Station, C1400, Austin, TX, 78712-3451, USA ⟨ dfl@astro.as.utexas.edu ⟩

P. L. LIM, Department of Astronomy, New Mexico State University, MSC 4500, Box 30001, Las Cruces, NM, 88003, USA ⟨ pllim0@nmsu.edu ⟩

M. E. LYON, Austin Astronomical Society, PO Box 12831, Austin, TX, 78711-2831, USA ⟨ sones27@yahoo.com ⟩

D. J. MAR, McDonald Observatory, University of Texas at Austin, 1 University Station, C1400, Austin, TX, 78712-3451, USA ⟨ djmar@astro.as.utexas.edu ⟩

H. MARION, Department of Astronomy, University of Texas at Austin, 1 University Station, C1400, Austin, TX, 78712-3451, USA ⟨ hman@astro.as.utexas.edu ⟩

S. B. MARKOFF, Astronomical Institute "Anton Pannekoek," University of Amsterdam, Kruislaan 403, 1098 SJ Amsterdam, The Netherlands ⟨ sera@science.uva.nl ⟩

B. E. MCARTHUR, Department of Astronomy, University of Texas at Austin, 1 University Station, C1400, Austin, TX, 78712-3451, USA ⟨ mca@barney.as.utexas.edu ⟩

N. M. MCCLURE-GRIFFITHS, CSIRO, Australia Telescope National Facility, PO Box 76, Epping, NSW, 1710, Australia ⟨ naomi.mcclure-griffiths@csiro.au ⟩

E. M. McMahon, Department of Astronomy, University of Texas at Austin, 1 University Station, C1400, Austin, TX, 78712-3451, USA ⟨ emcmahon@astro.as.utexas.edu ⟩

J. R. Middleton, Department of Engineering, University of Texas at Austin, 3208 Red River St Ste 200, Austin TX 78705-2650 ⟨ jmiddleton@ieee.org ⟩

J. M. Miller, Department of Astronomy, University of Michigan, 500 Church Street, Dennison 814, Ann Arbor, MI, 48109-1042, USA ⟨ jonmm@umich.edu ⟩

M. H. Montgomery, Department of Astronomy, University of Texas at Austin, 1 University Station, C1400, Austin, TX, 78712-3451, USA ⟨ mikemon@astro.as.utexas.edu ⟩

B. K. Moorthy, Department of Astronomy, New Mexico State University, MSC 4500, Box 30001, Las Cruces, NM, 88003, USA ⟨ bmoorthy@nmsu.edu ⟩

F. Mullally, Department of Astronomy, University of Texas at Austin, 1 University Station, C1400, Austin, TX, 78712-3451, USA ⟨ fergal@astro.as.utexas.edu ⟩

M. K. Nordhaus, Department of Astronomy, University of Texas at Austin, 1 University Station, C1400, Austin, TX, 78712-3451, USA ⟨ nordhaus@astro.as.utexas.edu ⟩

J. E. Pino, Institute for Fusion Studies, University of Texas at Austin, 1 University Station, C1600, Austin, Texas 78712-0264 ⟨ pino@mail.utexas.edu ⟩

E. Noyola, Department of Astronomy, University of Texas at Austin, 1 University Station, C1400, Austin, TX, 78712-3451, USA ⟨ eva@astro.as.utexas.edu ⟩

R. M. Quimby, Department of Astronomy, University of Texas at Austin, 1 University Station, C1400, Austin, TX, 78712-3451, USA ⟨ quimby@astro.as.utexas.edu ⟩

I. Ramirez, Department of Astronomy, University of Texas at Austin, 1 University Station, C1400, Austin, TX, 78712-3451, USA ⟨ ivan@astro.as.utexas.edu ⟩

M. A. Rankin, College of Natural Sciences, University of Texas at Austin, 1 University Station, G2500, Austin, TX, 78712-3451, USA ⟨ rankin@mail.utexas.edu ⟩

S. Redfield, McDonald Observatory and Department of Astronomy, University of Texas at Austin, 1 University Station, C1400, Austin, TX, 78712-3451, USA ⟨ sredfield@astro.as.utexas.edu ⟩

J. G. Ries, McDonald Observatory, University of Texas at Austin, 1 University Station, C1400, Austin, TX, 78712-3451, USA ⟨ moon@astro.as.utexas.edu ⟩

P. Riley, College of Natural Sciences, University of Texas at Austin, 1 University Station, G2500, Austin, TX, 78712-3451, USA ⟨ riley@mail.utexas.edu ⟩

I. U. ROEDERER, Department of Astronomy, University of Texas at Austin, 1 University Station, C1400, Austin, TX, 78712-3451, USA ⟨ iur@astro.as.utexas.edu ⟩

S. B. SALVIANDER, Department of Astronomy, University of Texas at Austin, 1 University Station, C1400, Austin, TX, 78712-3451, USA ⟨ triples@astro.as.utexas.edu ⟩

J. M. SCALO, Department of Astronomy, University of Texas at Austin, 1 University Station, C1400, Austin, TX, 78712-3451, USA ⟨ parrot@astro.as.utexas.edu ⟩

R. SHEN, Department of Astronomy, University of Texas at Austin, 1 University Station, C1400, Austin, TX, 78712-3451, USA ⟨ rfshen@astro.as.utexas.edu ⟩

J. SHEN, Department of Astronomy, University of Texas at Austin, 1 University Station, C1400, Austin, TX, 78712-3451, USA ⟨ shen@astro.as.utexas.edu ⟩

S. S. SHEPPARD, Department of Terrestrial Magnetism, Carnegie Institution of Washington, 5241 Broad Branch Rd. NW, Washington DC, 20015, USA ⟨ sheppard@dtm.ciw.edu ⟩

G. A. SHIELDS, Department of Astronomy, University of Texas at Austin, 1 University Station, C1400, Austin, TX, 78712-3451, USA ⟨ shields@astro.as.utexas.edu ⟩

M. H. SIEGEL, McDonald Observatory, University of Texas at Austin, 1 University Station, C1400, Austin, TX, 78712-3451, USA ⟨ siegel@astro.as.utexas.edu ⟩

P. D. STRYCKER, Department of Astronomy, New Mexico State University, MSC 4500, Box 30001, Las Cruces, NM, 88003, USA ⟨ strycker@nmsu.edu ⟩

S. E. THOMPSON, Department of Physics, Colorado College, 14 E. Cache La Poudre, Colorado Springs, CO, 80903, USA ⟨ sthompson@coloradocollege.edu ⟩

R. G. TULL, McDonald Observatory, University of Texas at Austin, 1 University Station, C1400, Austin, TX, 78712-3451, USA ⟨ rtull@austin.rr.com ⟩

A. URBAN, Department of Astronomy, University of Texas at Austin, 1 University Station, C1400, Austin, TX, 78712-3451, USA ⟨ aurban@astro.as.utexas.edu ⟩

P. A. VANDEN BOUT, North American ALMA Science Center, National Radio Astronomy Observatory, NRAO Headquarters, 520 Edgemont Road, Charlottesville, VA, 22903-2475, USA ⟨ pvandenb@nrao.edu ⟩

M. VIEL, Institute of Astronomy, Madingley Road, CB3 OHA, Cambridge, United Kingdom ⟨ viel@ast.cam.ac.uk ⟩

T. A. VON HIPPEL, Department of Astronomy, University of Texas at Austin, 1 University Station, C1400, Austin, TX, 78712-3451, USA ⟨ ted@astro.as.utexas.edu ⟩

C. O. VON HOLSTEIN-RATHLOU, Department of Astronomy, University of Texas at Austin, 1 University Station, C1400, Austin, TX, 78712-3451, USA ⟨christina_von_hr@yahoo.dk⟩

B. S. WALTER, Austin Astronomical Society, PO Box 12831, Austin, TX, 78711-2831, USA ⟨bwalter@activepower.com⟩

R. P. WELBORN, McDonald Observatory Board of Visitors, 1 University Station, C1400, Austin, TX, 78712-3451, USA ⟨rwelborn@austin.rr.com⟩

J. W. WELLHOUSE, Department of Astronomy, New Mexico State University, MSC 4500, Box 30001, Las Cruces, NM, 88003, USA ⟨jwell@nmsu.edu⟩

J. C. WHEELER, Department of Astronomy, University of Texas at Austin, 1 University Station, C1400, Austin, TX, 78712-3451, USA ⟨wheel@astro.as.utexas.edu⟩

B. J. WILLS, McDonald Observatory, University of Texas at Austin, 1 University Station, C1400, Austin, TX, 78712-3451, USA ⟨bev@astro.as.utexas.edu⟩

D. WINGET, Department of Astronomy, University of Texas at Austin, 1 University Station, C1400, Austin, TX, 78712-3451, USA ⟨dew@astro.as.utexas.edu⟩

J. L. WOOD, Department of Astronomy, University of Texas at Austin, 1 University Station, C1400, Austin, TX, 78712-3451, USA ⟨jlwood@astro.as.utexas.edu⟩

C. E. WU, Department of Astronomy, New Mexico State University, MSC 4500, Box 30001, Las Cruces, NM, 88003, USA ⟨catwu@nmsu.edu⟩

R. E. WYLLYS, School of Information, University of Texas at Austin, 1 University Station, D7000, Austin, TX, 78712-0390, USA ⟨wyllys@ischool.utexas.edu⟩

Part I

Frank N. Bash Symposium 2005: New Horizons in Astronomy
ASP Conference Series, Vol. 352, 2006
S. J. Kannappan, S. Redfield, J. E. Kessler-Silacci, M. Landriau, and N. Drory

Small Bodies in the Outer Solar System

Scott S. Sheppard[1]

Department of Terrestrial Magnetism, Carnegie Institution of Washington, Washington, DC, USA

Abstract.
The dynamical and physical characteristics of asteroids, comets, Kuiper Belt objects and satellites give us insight on the processes operating in the Solar System and allow us to probe the planet formation epoch. The recent advent of sensitive, wide-field CCD detectors are allowing us to complete the inventory of our Solar System and obtain detailed knowledge about the small bodies it contains. I will discuss the recent results with a focus on the new bodies being discovered beyond Neptune with a particular emphasis on the very distant orbit of (90377) Sedna and 2003 UB313, which is larger than Pluto.

1. Introduction

The idea of a small body population beyond Neptune can not really be attributed to any one individual but a slow realization that the Solar System did not end at Neptune or Pluto (Leonard 1930; Edgeworth 1949; Kuiper 1951). Oort (1950) realized that the long period comets must come from a spherical cloud at a distance greater than 50000 AU. Whipple (1950) first described comets as weak conglomerates of volatile and solid material (the dirty snowball). Everhart (1972) found that long period comets (Periods > 200 years) could become short period comets (Periods < 20 years) through interactions with Jupiter. Joss (1973) argued that this wasn't efficient enough to provide a sufficient source of all the observed short period comets. The most informative work on the origin of short-period comets was from Fernandez (1980). He determined that a Kuiper disk of objects with low inclinations just beyond Neptune would be a good source for short period comets which had mostly low inclinations. These are unlike the long period comets which have an isotropic distribution of inclinations and are most likely to come from the proposed Oort cloud.

Observational work with photographic plates was slow and very inefficient. An extensive photographic survey of the ecliptic by Kowal (1978) found only one outer Solar System object. This object is the first known Centaur, 2060 Chiron, which has an orbit completely within the giant planet region. Kowal's lone discovery was nevertheless important because it showed that some small bodies do have orbits entirely outside of Jupiter's orbit. Soon afterwards a number of stars having IR excesses was observed by IRAS (Aumann et al. 1984). It was quickly discovered that one of these IR excess stars, β Pic, has a large

---

[1]Hubble Fellow.

observable disk of dust that extended out to about 1000 AU in radius (Smith & Terrile 1984). Weissman (1984) suggested that the IR excess could be from dust produced by collisions in a distant small body population such as an Oort cloud or Kuiper Belt.

With the development of Charge Coupled Devices (CCDs) in the 1980's surveys could go much deeper and computers allowed for efficient image analysis. In 1992 the first Kuiper Belt object (KBO) other than Pluto was discovered, 1992 QB_1 (Jewitt & Luu 1993). In the mid to late 1990's moving object detection software on high performance computers and large field-of-view CCD detectors opened up the flood gates for discovering small bodies in the outer Solar System.

Over 1000 trans-Neptunian Objects (TNOs) have been detected since the discovery of 1992 QB_1. The Kuiper Belt (or Edgeworth-Kuiper Belt) is thought to be a relic from the original protoplanetary disk. It is estimated to contain about 80,000 objects with radii greater than 50 km (Trujillo, Jewitt, & Luu 2001). This "belt" has been collisionally processed and gravitationally perturbed throughout the age of the Solar System in ways that are not yet fully understood (Farinella & Davis 1996).

The Kuiper Belt is the largest known relatively stable ensemble of small bodies in the planetary region, outnumbering the main-belt asteroids and Jovian Trojans by a factor of ~300. The Kuiper Belt's vast distance from the Sun suggests the objects are chemically primitive, containing large amounts of volatiles. Thus we may be able to probe some aspects of the early history of the local solar nebula by studying the Kuiper Belt and related objects. The short-period comets and Centaurs are believed to originate from the Kuiper Belt (Fernandez 1980; Duncan, Quinn, & Tremaine 1988). Centaurs, which have semimajor axes and perihelia between Jupiter and Neptune, are currently in dynamically unstable orbits that will lead either to ejection from the Solar System, an impact with a planet or the Sun, or evolution into a short period comet (Dones et al. 1999; Tiscareno & Malhotra 2003). The long period comets likely come from the more distant and yet undetected Oort cloud.

2. Kuiper Belt Dynamical Structure

The Kuiper Belt has been found to be dynamically structured (see Figure 1). This structuring was a big surprise to most astronomers as it was assumed any Kuiper disk would be dynamically cold (i.e., most objects would have low eccentricities and inclinations). The mean velocity dispersion (~1.2 km s^{-1}) found for KBOs is larger than the escape speed of most KBOs. Thus the current Kuiper Belt is in a state of erosion (Davis & Farinella 1997). Theoretical and modeling work has largely been devoted to explaining the current dynamical structure seen in the Kuiper Belt. Most invoke the migration of Neptune (Malhotra 1995; Hahn & Malhotra 1999; Levison & Morbidelli 2003), the passage of giant planetesimals through the disk (Morbidelli & Valsecchi 1997), or a passing star (Ida, Larwood, & Burkert 2000; Levison, Morbidelli, & Dones 2004; Morbidelli & Levison 2004; Kenyon & Bromley 2004). Recent modeling suggests that some minor planets which formed between Jupiter and Neptune may have been scattered into the Kuiper Belt region causing the belt to look much more excited than if they were not present (Gomes 2003).

There are five known dynamical classes of KBOs (Figures 1 and 2). Classical KBOs have semi-major axis $40 < a < 50$ AU with moderate eccentricities ($e \sim 0.1$ to 0.2) and inclinations. Resonant KBOs are in mean motion resonances with Neptune and generally have higher eccentricities and inclinations than Classical KBOs (Chiang et al. 2003; Elliot et al. 2005). The 3:2 resonance currently has the most known members, but this may be a selection effect since this resonance is located at the inner regions of the Kuiper Belt. The KBOs in the 3:2 resonance are called Plutinos since Pluto occupies this resonance. The Neptune Trojans are in a 1 : 1 resonance with Neptune which means they have similar semi-major axes as Neptune but lead or trail the planet by about 60 degrees. Scattered disk objects have very large eccentricities with perihelia near the orbit of Neptune ($q \sim 30$ AU). The Scattered disk objects are believed to have been recently scattered out of the main Kuiper Belt through interactions with Neptune. Scattered disk objects are probably the progenitors of the Centaurs and short-period comets. A fifth class of KBO, the extended scattered disk (Figure 2), has only recently been recognized (Gladman et al. 2002; Morbidelli & Levison 2004) and to date only two members are known. These objects have large eccentricities but unlike the scattered disk objects the extended scattered disk objects have perihelia $q > 45$ AU which can not be directly caused by Neptune interactions alone. These objects may have obtained their orbits from a stellar passage near the Kuiper Belt (Morbidelli & Levison 2004). The bias corrected relative populations of classical : scattered : resonant are $1.0 : 0.8 : 0.1$ (Trujillo et al. 2001).

There is an edge to the classical population around 50 AU (Jewitt, Luu, & Trujillo 1998; Trujillo et al. 2001; Allen, Bernstein, & Malhotra 2002; Trujillo & Brown 2001). Only objects with very large eccentricities are known to have orbits which go beyond 50 AU. The only currently viable explanation known to date is that the disk was truncated by a passing star (Ida et al. 2000; Morbidelli & Levison 2004).

2.1. Sedna and the Extended Scattered Disk

How was the extended scattered disk and Sedna formed? Sedna's current large perihelion distance (~ 76 AU) along with its large eccentricity (~ 0.84) are hard to explain with what we currently know about the Solar System. If Sedna formed in its current location it must have initially been on a circular orbit otherwise accretion would not have been possible because the large relative velocities of colliding bodies would have been disruptive (Stern 2005). If Sedna obtained its large eccentricity through interactions with the currently known giant planets, somehow its perihelion must have been raised. Several theories on Sedna's history have been put forth (see Morbidelli & Levison 2004, for a review): 1) Sedna may have been scattered by an unseen planet in the outer Solar System. Neptune can only be invoked if an object has a perihelion distance of less than 50 AU (Gladman et al. 2002; Gomes 2003). 2) A single stellar encounter may have raised Sedna's perihelion (Brown, Trujillo, & Rabinowitz 2004; Morbidelli & Levison 2004). Galactic tides are too weak to raise the perihelion of an object that is so close to the Sun and only work for distant Oort cloud objects around 10,000 AU. The stellar encounter would have to have been very close, around 500 AU, in order for Sedna to be excited. This may hint that our Sun formed

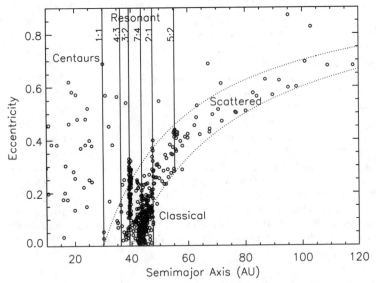

Figure 1.   Semimajor axis versus eccentricity of known trans-Neptunian objects which show four of the five distinct KBO populations as well as the Centaurs. Vertical solid lines show resonances with Neptune which includes the Neptune Trojans in the 1 : 1 resonance. Scattered disk objects reside between perihelia $30 < q < 40$ AU which are shown by dashed lines. Extended scattered disk objects have perihelia greater than 45 AU (see Figure 2). Classical objects are in the lower center portion of the figure. An edge around 50 AU can clearly be seen for low eccentricity objects.

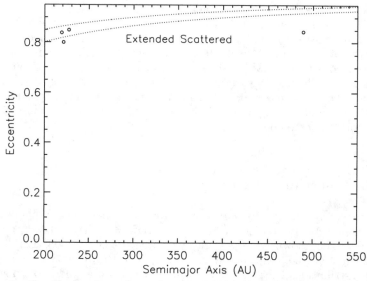

Figure 2.   Semimajor axis versus eccentricity of extended scattered disk objects which have perihelion greater than 45 AU (which fall below the dotted lines). Only two objects are known to be in the extended scattered disk; Sedna (seen in the upper right) and 2000 CR105. These objects with high semi-major axes and perihelion distances can not have obtained their current orbits with simple interactions with Neptune or any other known Solar System objects.

in a very dense stellar environment. In addition, this may have caused the edge we see today in the Kuiper Belt at 50 AU (Ida et al. 2000). A time constraint is required, that is, if the stellar encounter happened too soon Sedna would not have been formed but if it happened too late it would disrupt the Oort cloud. 3) A highly eccentric Neptune would have been able to produce Sedna. 4) If several massive planetary embryos were scattered into the outer Solar System during the planet formation epoch this may also produce Sedna's orbit as we see it today. 5) If the trans-Neptunian disk was thousands of times more massive it may have influenced Sedna's orbit. 6) Sedna may be a captured extrasolar planetesimal. To date, the stellar encounter scenario appears to work best.

2.2.   Neptune Trojans

As mentioned above, Trojans are objects which share a planet's semi-major axis but are ahead or behind the planet by about 60 degrees in orbital longitude, known as the leading (L4) and trailing (L5) Lagrangian regions. Recent dynamical studies show that Trojans are stable near the L4 and L5 points of Neptune for the age of the Solar System (Nesvorný & Dones 2002; Marzari, Tricarico, & Scholl 2003). The population mechanism and long-term stability of the Trojans is strongly linked to the physics and evolution of the Solar System. The Trojan's current dynamical properties constrain the formation, evolution and migration of the planets. It is still uncertain if the Trojans formed in their current locations or were captured at some point. In the present epoch there is no known efficient mechanism to permanently capture Trojans but Trojan capture could have occurred more easily towards the end of the the planet formation epoch. Trojan asteroids may have a very similar history as the irregular satellites of the planets (Jewitt et al. 2004). The effects of nebular gas drag (Peale 1993), collisions (Chiang & Lithwick 2005), planetary migration (Gomes 1998; Kortenkamp, Malhotra, & Michtchenko 2004), overlapping resonances (Marzari & Scholl 2000; Morbidelli et al. 2005), and the mass growth of the planets (Fleming & Hamilton 2000) all potentially influence the formation, capture efficiency and stability of these bodies.

3.   The Kuiper Belt Size Distribution

The Cumulative Luminosity Function (CLF) describes the sky-plane surface density of objects brighter than a given magnitude. For the Centaurs and KBOs the CLF is well known for $m_R < 26$ mags. (Jewitt et al. 1998; Trujillo et al. 2001; Allen et al. 2002; Millis et al. 2002; Trujillo & Brown 2003). To determine the size distribution of the KBOs we assume they follow a differential power-law radius distribution of the form $n(r)dr = \Gamma r^{-q} dr$, where Γ and q are constants, r is the radius of the KBO, and $n(r)dr$ is the number of KBOs with radii in the range r to $r + dr$. For the KBOs $q \sim 4$ which means the belt should be dominated by the smallest objects. A similar result is found for the Centaurs (Sheppard et al. 2000). The recently discovered scattered disk object, 2003 UB313, is slightly bigger than Pluto. This object has an orbit which is typical of the other known scattered disk objects. Discovery of such a large object as 2003 UB313 ($r \sim 1500$ km) was not unexpected (Trujillo & Brown 2003).

A recent deep survey using the Hubble Space Telescope reached a limiting magnitude of about 29th in the optical (Bernstein et al. 2004). Through extrapolation of the CLF to very small KBOs (radii ~ 10 km) the survey was expected to find hundreds of small KBOs. Only three new KBOs were detected in the survey. This strongly indicates that the Kuiper Belt does not follow the same power law size distribution found for larger objects ($r > 25$ km) at the smaller sizes. Thus, the Kuiper Belt may be very deficient in KBOs with small radii.

The inferred mass of the current Kuiper Belt is about 0.1 Earth masses (Trujillo et al. 2001). This is about one hundred times lower than would be expected from the solar nebula when augmenting the masses of the planets. In addition, the current population size structure, binary population and angular momentum of the Kuiper Belt suggests a more massive disk existed in the primordial Kuiper Belt (Stern 1996; Jewitt & Sheppard 2002; Astakhov, Lee, & Farrelly 2005). A detailed accretion model of an earlier denser Kuiper Belt by Kenyon & Luu (1999) showed that several Pluto sized objects could form within the age of the Solar System. The mass may have been lost through Neptune interactions (Hahn & Malhotra 1999) and collisional grinding (Stern 1996; Kenyon & Luu 1999). Infrared excesses found around some main sequence stars appears to corroborate dust production produced through collisions between small bodies in possible Kuiper Belt analogs (Weissman 1984; Aumann et al. 1984). This dust is unstable to radiation pressure from the host star and the Poynting Robertson effect and thus the dust must be constantly replenished through collisions.

3.1. Kuiper Belt Binaries

It appears that about 4% ± 2% of KBOs have companions with separations ≥0.15 " (Noll et al. 2002). To date most known KBO binaries have mass ratios near unity, though this may be an observational selection effect. These large nearly equal sized components probably did not form by simple direct collisions (Stern 2002). Formation of such binaries could occur through complex three-body interactions (Weidenschilling 2002; Funato et al. 2004; Astakhov et al. 2005) or when two bodies approach each other and energy is extracted either by dynamical friction from the surrounding sea of smaller KBOs or by a close third body (Goldreich, Lithwick, & Sari 2002). These processes mostly require that the density of KBOs was $\sim 10^2$ to 10^3 times greater than now.

4. KBO Physical Properties

4.1. Colors and Albedos

The surfaces of KBOs may have been altered over their lifetimes by collisions, cometary activity, and irradiation. The KBOs geometric albedos are still poorly sampled but the larger ones have been measured through detecting their thermal emission in the mid-infrared, submillimeter and millimeter wavelength regimes (Altenhoff, Bertoldi, & Menten 2004; Grundy, Noll, & Stephens 2005). It appears there is a large range of albedos with reflectivity being as dark as 4% and as bright as 60%. It is possible that the brighter surfaces are indicative of volatile ices while the darker surfaces may be more organic rich.

Colors of the KBOs have been found to be diverse, ranging from neutral to the reddest objects known in the Solar System ($V - R \sim 0.3$ to $V - R \sim 0.8$). The KBOs show signs of a possible correlation between colors and inclination at the ($\sim 3\sigma$) level (Trujillo & Brown 2002; Stern 2002; Tegler, Romanishin, & Consolmagno 2003; McBride et al. 2003; Peixinho et al. 2004; Fornasier et al. 2004; Barucci et al. 2005). This correlation has prompted the recent modeling work by Gomes (2003). In this scenario the current Kuiper Belt has two separate populations (Figure 3). The first are small bodies which formed near their current locations beyond Neptune and which have low to moderate inclinations. The second population are of bodies which formed in the giant planet region and were scattered into the Kuiper Belt. These objects would be on average larger, more neutral in color and have inclinations ranging from small to large. Objects which formed closer to the Sun could obtain larger sizes because the solar nebula was more dense there. Empirically, objects which formed closer to the Sun are more neutral in color, which is probably because they have less volatile material on their surfaces. This scenario is the result from the color observations mentioned above as well as that there appear to be two inclination distributions in the Kuiper Belt (Brown 2001) and that objects with higher inclinations appear larger (Levison & Stern 2001).

4.2. Spectra

To date the spectra of many KBOs have been mostly featureless with some showing 2 μm water ice absorptions (de Bergh et al. 2005). Very recent spectra on some of the largest and thus brightest KBOs have shown significant ice features.

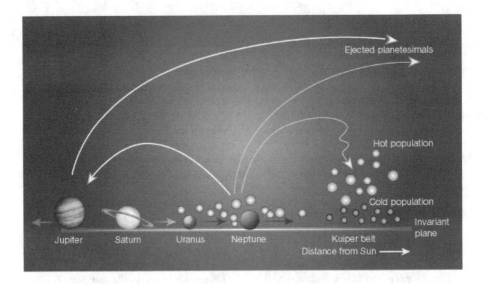

Figure 3.   There may be two populations in the Kuiper Belt. The "Cold" population formed near their present locations beyond Neptune and have low inclinations, are relatively small and red in color. The "Hot" population formed in the giant planet region and were scattered out to the Kuiper Belt. These objects range from small to large, are neutral to red in color and may have a wide range of inclinations (taken from Morbidelli & Levison 2003).

50000 Quaoar was found to have crystalline water ice on its surface (Jewitt & Luu 2004). This is surprising since crystalline water ice is not stable on the surface of these objects for long. Jewitt & Luu (2004) suggest that recent ice volcanism may have created the crystalline water ice. Spectra of 2003 UB313 make this object only the second, Pluto being the other, to have known methane ice on its surface (Brown, Trujillo, & Rabinowitz 2005). This may indicate that only the largest bodies can hold onto this volatile ice.

4.3.   Rotations and Phase Functions

Currently the most feasible way to determine KBO shapes and surface features is through their photometric light variations. The rotations and shapes of the KBOs may be a function of their size. Small KBOs (diameters $D < 100$ km) are thought to be collisionally produced (Farinella & Davis 1996). These objects retain no memory of the primordial angular momentum of their parent bodies. Instead, their spins are presumably set by the partitioning of kinetic energy delivered by the projectile responsible for break-up. Larger objects may be structurally damaged bodies held together by gravity (rubble piles). The spins of these objects should be much less influenced by recent impacts. A similar situation prevails in the main asteroid belt, where collisional modification of the rotations and shapes of the smaller objects is observationally well established (Catullo et al. 1984). The large objects in both the main-belt and the Kuiper Belt may provide a record of the primordial distribution of angular momenta imbued by the growth process. A key attribute of the Kuiper Belt is that the population is very large compared to the main asteroid belt, allowing access to a substantial sample of objects that are too large to have been influenced by recent collisions.

Time-resolved observations of KBOs show that ∼32% vary by ≥0.15 magnitudes, 18% by ≥0.40 magnitudes and 12% by ≥0.60 magnitudes (Sheppard & Jewitt 2002; Ortiz et al. 2003; Sheppard & Jewitt 2003). Statistically, the trans-Neptunian objects are less spherical than their main-belt asteroid counterparts, indicating a higher specific angular momentum perhaps resulting from the formation epoch (Sheppard & Jewitt 2002). Most KBO photometric variations with rotation can be explained by nonuniform surfaces but two types of objects stand out in rotation period and photometric range space. The first type of objects are 20000 Varuna and 2003 EL61 which have large amplitudes and short periods which are indicative of rotationally distorted, low density rubble piles (Jewitt & Sheppard 2002). The second type, of which only 2001 QG_{298} is a member of to date, shows an extremely large amplitude and slow rotation. Kuiper Belt object 2001 QG_{298} is the first known Kuiper Belt object, and only the third minor planet, with a radius >25 km to display a light curve with a range in excess of 1 magnitude and is best described as a contact binary with similar sized components (Sheppard & Jewitt 2004). By correcting for the effects of projection, it is estimated that the fraction of nearly equal sized component contact binaries in the Kuiper Belt is at least ∼10% to 20% with the true fraction of contact binaries probably much higher. Two objects, 1996 TO_{66} and 1995 SM_{55} may have variable amplitude light curves which may result from complex rotation, a satellite or cometary effects (Hainaut et al. 2000; Sheppard & Jewitt 2003).

Phase darkening coefficients for KBOs in the 0 to 2 degree phase angle range show steep linear slopes (0.16 magnitudes per degree) indicating backscatter from low albedo porous surface materials (Sheppard & Jewitt 2003). The similarity of the slopes of the phase functions of all small KBOs suggests comparative uniformity of the surface compositions, physical states, and albedos. The measured phase slopes are all distinct from that of Pluto, which has a much higher albedo surface due to frosts deposited from its tenuous atmosphere.

Future observations on smaller KBOs ($r < 50$ km) would be beneficial to determine if their rotation periods and amplitudes are similar to the larger objects observed to date. A transition between gravitational to mechanical structural domination should be observed for objects with radii between 50 and 100 km. The smaller objects ($r < 50$ km) should show a significantly different distribution of rotation periods and amplitudes than the larger objects ($r > 100$ km). KBOs with radii smaller than about 50 km are probably just collisional shards with shapes and rotations presumably set by the partitioning of kinetic energy delivered by the projectile responsible for break-up. Unlike the larger KBOs their rotation states are much more influenced from recent collisional events. These smaller KBOs would be much fainter than the larger objects and thus would require a number of nights on large class telescopes (6-10 meters) to obtain the signal-to-noise needed to detect their light curves.

Acknowledgments. Support for this work was provided by NASA through Hubble Fellowship grant # HF-01178.01-A awarded by the Space Telescope Science Institute, which is operated by the Association of Universities for Research in Astronomy, Inc., for NASA, under contract NAS 5-26555.

References

Allen, R. L., Bernstein, G. M., & Malhotra, R. 2002, AJ, 124, 2949
Altenhoff, W. J., Bertoldi, F., & Menten, K. M. 2004, A&A, 415, 771
Astakhov, S. A., Lee, E. A., & Farrelly, D. 2005, MNRAS, 360, 401
Aumann, H. H., et al. 1984, ApJ, 278, L23
Barucci, M. A., Belskaya, I. N., Fulchignoni, M., & Birlan, M. 2005, AJ, 130, 1291
Bernstein, G. M., Trilling, D. E., Allen, R. L., Brown, M. E., Holman, M., & Malhotra,
    R. 2004, AJ, 128, 1364
Brown, M. E. 2001, AJ, 121, 2804
Brown, M. E., Trujillo, C., & Rabinowitz, D. 2004, ApJ, 617, 645
Brown, M. E., Trujillo, C. A., & Rabinowitz, D. L. 2005, ApJ, 635, L97
Catullo, V., Zappala, V., Farinella, P., & Paolicchi, P. 1984, A&A, 138, 464
Chiang, E. I., & Lithwick, Y. 2005, ApJ, 628, 520
Chiang, E. I., et al. 2003, AJ, 126, 430
Davis, D. R., & Farinella, P. 1997, Icarus, 125, 50
de Bergh, C., Delsanti, A., Tozzi, G. P., Dotto, E., Doressoundiram, A., & Barucci,
    M. A. 2005, A&A, 437, 1115
Dones, L., Gladman, B., Melosh, H. J., Tonks, W. B., Levison, H. F., & Duncan, M.
    1999, Icarus, 142, 509
Duncan, M., Quinn, T., & Tremaine, S. 1988, ApJ, 328, L69
Edgeworth, K. E. 1949, MNRAS, 109, 600
Elliot, J. L., et al. 2005, AJ, 129, 1117
Everhart, E. 1972, Astrophys. Lett., 10, 131
Farinella, P., & Davis, D. R. 1996, Science, 273, 938
Fernandez, J. A. 1980, MNRAS, 192, 481
Fleming, H. J., & Hamilton, D. P. 2000, Icarus, 148, 479
Fornasier, S., et al. 2004, A&A, 421, 353
Funato, Y., Makino, J., Hut, P., Kokubo, E., & Kinoshita, D. 2004, Nature, 427, 518
Gladman, B., Holman, M., Grav, T., Kavelaars, J., Nicholson, P., Aksnes, K., & Petit,
    J.-M. 2002, Icarus, 157, 269
Goldreich, P., Lithwick, Y., & Sari, R. 2002, Nature, 420, 643
Gomes, R. S. 1998, AJ, 116, 2590
Gomes, R. S. 2003, Icarus, 161, 404
Grundy, W. M., Noll, K. S., & Stephens, D. C. 2005, Icarus, 176, 184
Hahn, J. M., & Malhotra, R. 1999, AJ, 117, 3041
Hainaut, O. R., et al. 2000, A&A, 356, 1076
Ida, S., Larwood, J., & Burkert, A. 2000, ApJ, 528, 351
Jewitt, D., & Luu, J. 1993, Nature, 362, 730
Jewitt, D., Luu, J., & Trujillo, C. 1998, AJ, 115, 2125
Jewitt, D. C., & Luu, J. 2004, Nature, 432, 731
Jewitt, D. C., Sheppard, S., & Porco, C. 2004, Cambridge Planetary Science, Vol. 1,
    Jupiter. The Planet, Satellites and Magnetosphere, ed. F. Bagenal, T. E. Dowl-
    ing, & W. B. McKinnon (Cambridge: Cambridge University Press), 263
Jewitt, D. C., & Sheppard, S. S. 2002, AJ, 123, 2110
Joss, P. C. 1973, A&A, 25, 271
Kenyon, S. J., & Bromley, B. C. 2004, Nature, 432, 598
Kenyon, S. J., & Luu, J. X. 1999, ApJ, 526, 465
Kortenkamp, S. J., Malhotra, R., & Michtchenko, T. 2004, Icarus, 167, 347
Kowal, C. T. 1978, Sciences, 18, 12
Kuiper, G. P. 1951, in Proceedings of a Topical Symposium Commemorating the 50th

Smooth

Anniversary of the Yerkes Observatory and Half a Century of Progress in Astrophysics, ed. J. A. Hynek (New York: McGraw-Hill), 357

Leonard, F. C. 1930, Leaflet of the Astronomical Society of the Pacific, 1, 121

Levison, H. F., & Morbidelli, A. 2003, Nature, 426, 419

Levison, H. F., Morbidelli, A., & Dones, L. 2004, AJ, 128, 2553

Levison, H. F., & Stern, S. A. 2001, AJ, 121, 1730

Malhotra, R. 1995, AJ, 110, 420

Marzari, F., & Scholl, H. 2000, Icarus, 146, 232

Marzari, F., Tricarico, P., & Scholl, H. 2003, A&A, 410, 725

McBride, N., Green, S. F., Davies, J. K., Tholen, D. J., Sheppard, S. S., Whiteley, R. J., & Hillier, J. K. 2003, Icarus, 161, 501

Millis, R. L., Buie, M. W., Wasserman, L. H., Elliot, J. L., Kern, S. D., & Wagner, R. M. 2002, AJ, 123, 2083

Morbidelli, A., & Levison, H. F. 2003, Nature, 422, 30

Morbidelli, A., & Levison, H. F. 2004, AJ, 128, 2564

Morbidelli, A., Levison, H. F., Tsiganis, K., & Gomes, R. 2005, Nature, 435, 462

Morbidelli, A., & Valsecchi, G. B. 1997, Icarus, 128, 464

Nesvorný, D., & Dones, L. 2002, Icarus, 160, 271

Noll, K. S., et al. 2002, AJ, 124, 3424

Oort, J. H. 1950, Bull. Astron. Inst. Netherlands, 11, 91

Ortiz, J. L., Gutiérrez, P. J., Casanova, V., & Sota, A. 2003, A&A, 407, 1149

Peale, S. J. 1993, Icarus, 106, 308

Peixinho, N., Boehnhardt, H., Belskaya, I., Doressoundiram, A., Barucci, M. A., & Delsanti, A. 2004, Icarus, 170, 153

Sheppard, S. S., & Jewitt, D. 2004, AJ, 127, 3023

Sheppard, S. S., & Jewitt, D. C. 2002, AJ, 124, 1757

Sheppard, S. S., & Jewitt, D. C. 2003, Earth Moon and Planets, 92, 207

Sheppard, S. S., Jewitt, D. C., Trujillo, C. A., Brown, M. J. I., & Ashley, M. C. B. 2000, AJ, 120, 2687

Smith, B. A., & Terrile, R. J. 1984, Science, 226, 1421

Stern, S. A. 1996, AJ, 112, 1203

Stern, S. A. 2002, AJ, 124, 2300

Stern, S. A. 2005, AJ, 129, 526

Tegler, S. C., Romanishin, W., & Consolmagno, S. J. 2003, ApJ, 599, L49

Tiscareno, M. S., & Malhotra, R. 2003, AJ, 126, 3122

Trujillo, C. A., & Brown, M. E. 2001, ApJ, 554, L95

Trujillo, C. A., & Brown, M. E. 2002, ApJ, 566, L125

Trujillo, C. A., & Brown, M. E. 2003, Earth Moon and Planets, 92, 99

Trujillo, C. A., Jewitt, D. C., & Luu, J. X. 2001, AJ, 122, 457

Weidenschilling, S. J. 2002, Icarus, 160, 212

Weissman, P. R. 1984, Science, 224, 987

Whipple, F. L. 1950, ApJ, 111, 375

Ted von Hippel, Jason Kalirai, and Eric Ford discuss planets around white dwarf stars.

Frank N. Bash Symposium 2005: New Horizons in Astronomy
ASP Conference Series, Vol. 352, 2006
S. J. Kannappan, S. Redfield, J. E. Kessler-Silacci, M. Landriau, and N. Drory

What Do Multiple Planet Systems Teach Us about Planet Formation?

Eric B. Ford[1]

*Department of Astronomy, University of California at Berkeley,
Berkeley, CA, USA*

Abstract. For centuries, our knowledge of planetary systems and ideas about planet formation were based on a single example, our solar system. During the last thirteen years, the discovery of $\simeq 170$ planetary systems has ushered in a new era for astronomy. I review the surprising properties of extrasolar planetary systems and discuss how they are reshaping theories of planet formation. I focus on how multiple planet systems constrain the mechanisms proposed to explain the large eccentricities typical of extrasolar planets. I suggest that strong planet-planet scattering is common and most planetary systems underwent a phase of large eccentricities. I propose that a planetary system's final eccentricities may be strongly influenced by how much mass remains in a planetesimal disk after the last strong planet-planet scattering event.

1. Introduction

For centuries, theories of planet formation had been designed to explain our own Solar System, but the first few discoveries of extrasolar planetary systems were wildly different than our own. These discoveries led to the realization that planet formation theory must be generalized to explain a much wider range of planetary systems. For example, traditional theories predicted that giant planets would form at several AU and beyond, where temperatures are cold enough for ices to initiate the growth of grains and planetesimals (Lissauer 1993, 1995). Now, we know of over 70 giant planets inside 1 AU and 40 inside 0.1 AU (http://www.obspm.fr/planets). Theorists have proposed numerous possible mechanisms to explain the existence of these planets. Typically, they assume that the giant planet formed beyond a few AU, but then migrated inwards through a protoplanetary or planetesimal disk to their currently observed locations (e.g., Goldreich & Tremaine 1980; Lin et al. 1996; Ward 1997; Murray et al. 1998; Cionco & Brunini 2002) and stopped before being accreted on the star (e.g., Trilling et al. 1998; Ford & Rasio 2006). Similarly, it had long been assumed that planets formed in circular orbits due to strong eccentricity damping in the protoplanetary disk and remained on nearly circular orbits (i.e., eccentricity ≤ 0.1; Lissauer 1993, 1995). However, over half of the extrasolar planets beyond 0.1 AU have eccentricities ≥ 0.3, and one is as large as $\simeq 0.95$. Theorists have suggested numerous mechanisms to excite the orbital eccentricity of giant planets (e.g., Rasio & Ford 1996; Weidenschilling & Marzari 1996; Lin & Ida 1997; Holman et al. 1997; Murray et al. 1998; Ford, Havlickova & Rasio 2001;

---

[1] Miller Research Fellow.

Kley 2000; Kley et al. 2004; Chiang & Murray 2002; Lee & Peale 2002; Marzari & Weidenschilling 2002; Ford, Rasio & Yu 2003; Adams & Laughlin 2003; Veras & Armitage 2004; Namouni 2005). In recent years, improved observations of a few multiple planet systems have allowed theorists to determine their current orbital configuration and use that to place strong constraints on the formation of a few planetary systems (Lee & Peale 2002; Ford, Lystad & Rasio 2005).

We review some of the mechanisms proposed to explain orbital migration in disks in §2 and eccentricity excitation in §3. In §4, we review the current knowledge of three particularly well-studied multiple planet systems. We conclude with a discussion of the implications of these multiple planet systems for theories of orbital migration in §5.

2.   Orbital Migration

2.1.   Interactions with a Gaseous Disk

Well before the discovery of extrasolar planets, analytic studies of a planet in a gaseous protoplanetary disk indicated that torques could lead to rapid orbital evolution (Goldreich & Tremaine 1979, 1980). Initially, it was not clear if the net torque would lead to inward or outward migration, but subsequent investigations indicated that the net torque typically leads to an inward migration for a single planet in a quiet disk (Ward 1997). Recently, numerous researchers have conducted detailed hydrodynamic models to better understand the details of the torques occurring at various locations in the disk. While early work focused on torques exerted at Linblad resonances, it is now clear that one must also consider torques occurring at corotation resonances and accretion onto the planet, even once the planet has cleared a gap in the disk (Artymowics & Lubow 1996; Bate et al. 2003; D'Angelo et al. 2003). Unfortunately, these complications demand that simulations include physics spanning a large range of physical scales, and this remains a computational challenge. While multiple groups have found qualitatively similar results, the details remain a matter of active research (e.g., Bryden et al. 1999, Kley 1999). Further complicating matters, recent work has suggested that turbulent fluctuations in the disk may be critical for understanding migration (Rice & Armitage 2003, Laughlin et al. 2004).

Shortly after the discovery of giant planets in very short orbital periods, it was realized that the planets likely formed at several AU, but migrated to their current small orbital periods. Torques from a gaseous disk are widely believed to be responsible, as the torques appear more than adequate to cause such large scale migrations. Indeed, the main challenge to such theories is to explain why the migration process is halted before the planet is accreted onto the star. Naively, one would expect the rate of migration to increase with decreasing orbital period and the planets to accrete onto the star. Several halting mechanisms have been proposed (e.g., Trilling et al. 1998), but it is not yet clear to what extent each of these mechanisms is significant. Many migration scenarios require some degree of fine tuning (e.g., disk mass or lifetime) in order to halt the migration at orbital periods of only 1.5–4 d.

2.2.   Interactions with a Planetesimal Disk

A disk of small solid bodies (e.g., protoplanets, planetesimals, pebbles) can remain long after the gaseous protoplanetary nebula disperses (Goldreich, Lithwick, Sari 2004). If this disk is sufficiently massive, then a giant planet could migrate through the disk by scattering planetesimals (Murray et al. 1998; Cionco & Brunini 2002; Del Popolo & Eks 2002). Migration all the way to a few stellar radii requires that the mass of planetesimal in the disk be large compared to the observed disk masses of protoplanetary disks in Taurus and Ophiuchus (Beckwith & Sargent 1996). Still, typical disk masses are expected to result in a smaller amount of migration. For a single giant planet, the planetesimals that can be scattered at a given time come from a relatively small range of semi-major axes near mean-motion resonances, and the density of planetesimals must exceed a significant threshold to power an extended period of migration. When there is more than one planet, the dynamics can become significantly more complex and the feeding zones significantly enlarged. For example, in our own solar system, Saturn, Uranus, and Neptune are inefficient at ejecting planetesimals, but efficiently scatter them inwards, enabling Jupiter to eject them from the Solar System (Fernandez & Ip 1984, Malhotra 1995).

3. Eccentricity Excitation

3.1. Mutual Planetary Perturbations

Mutual gravitation perturbations in multiple planet systems can lead to significant orbital evolution.

Secular Planetary Perturbations Secular perturbation theory approximates each planet as a ring of mass smeared out over the planet's orbit. In the secular approximation, the semi-major axes remain constant, but the eccentricities, inclinations, and orientations of the orbits evolve with time (Murray & Dermott 1999). If the orbital planes are highly inclined ($\geq 40°$), then even a system with initially circular orbits can undergo large eccentricity oscillations (the "Kozai effect"; Kozai 1962; Holman et al. 1997; Ford, Kozinsky & Rasio 2000). While this effect is almost certainly important for some planets orbiting stars that have a wide stellar binary companion, dissipation in the protoplanetary disk makes it very unlikely for giant planets to form with large relative inclinations (Lissauer 1993). In the low-inclinations and low-eccentricity regime, the eccentricity and inclination oscillations decouple to lowest order, and angular momentum is exchanged between the various planets on long timescales (Murray & Dermot 1999). The low-inclination, high-eccentricity regime can be studied by the octupole approximation (Ford, Kozinsky, Rasio 2000; Lee & Peale 2003) or by a numerical averaging procedure (Michtechenko & Malhotra 2004). In both approximations, the inclinations remain small, and conservation of angular momentum requires that secular perturbations can only transfer angular momentum from one orbit to another. Therefore, secular planetary perturbations can only excite significant eccentricities, if there is already at least one eccentric planet in the system.

Strong Planet-Planet Scattering If planet formation commonly results in planetary systems with multiple planets, then it should be expected that the initial configurations will not be dynamically stable for time spans orders of magnitude longer than the timescale for planet formation (e.g., Levison, Lissauer & Duncan

1998). When protoplanetary cores form, they do not know how much gas they will eventually accrete, so planets will accrete too much mass to remain stable for the lifetime of their star. Additionally, giant planets must form while there is still significant gas in the protoplanetary disk, so they are likely subject to significant eccentricity damping which prevents eccentricity growth. Once the protoplanetary disk disperses, the eccentricity damping is removed and mutual gravitational perturbations can start exciting eccentricities that will eventually lead to close encounters.

In multiple planet systems which are dynamically unstable, close encounters and strong planet-planet scattering can produce large eccentricities (Rasio & Ford 1996; Weidenschilling & Marzari 1996). For systems of two giant planets initially on nearly circular orbits, dynamical instabilities are typically resolved by two planets colliding and producing a more massive giant planet in another low-eccentricity orbit or by one planet being ejected from the system, leaving behind the other planet in an eccentric orbit. For comparable mass planets, this typically results in large eccentricities (Ford, Havlickova & Rasio 2001), but this same mechanism naturally produces lower eccentricities when the planet mass ratio differs from unity (Ford, Rasio & Yu 2003). While the distribution of eccentricities depends on the planet mass ratio distribution, the two planet scattering model predicts a maximum eccentricity of $\simeq 0.8$, independent of the mass ratio distribution. This compares favorably with the observed distribution of extrasolar planet eccentricities, since only one of the $\simeq 170$ known extrasolar planets has an eccentricity greater than 0.8 (and the exceptional planet is in a known binary). The fraction of systems which result in ejections and eccentric planets depends on the orbital distance and effective radius for collisions (Ford, Havlickova & Rasio 2001), as well as the ratio of planet masses (Ford, Rasio & Yu 2003). While ejections dominate for giant planets at several AU, collisions are more frequent for comparable planets inside ~ 1 AU. Therefore, strong planet-planet scattering can easily produce the large eccentricities of giant planets at large separations, but by itself would predict that low-eccentricity orbits would be more frequent at small separations.

Simulations of planet-planet scattering often begin with closely spaced giant planets (e.g., Rasio & Ford 1996; Ford, Havlickova & Rasio 2001). This is necessary for dynamical instabilities to occur in systems with only two planets initially on circular orbits. While such systems facilitate the systematic study of the relevant physics, real planetary systems likely have more than two massive bodies. In planetary systems with multiple planets, dynamical stabilities are common even for systems with large initial separations (Chambers, Wetherill & Boss 1996; Marzari & Weidenschilling 2002). Additionally, such systems can persist uneventfully for $\sim 10^{6-8}$ yr, before chaos leads to close encounters and strong planet-planet scattering.

Dynamical Relaxation   If protoplanetary disks form many planets nearly simultaneously, then planet-planet scattering may lead to a phase of dynamical relaxation. Several researchers have numerically investigated the dynamics of planetary systems with 10–100 planets (Lin & Ida 1997; Papaloizou & Terquem 2001, 2002; Adams & Laughlin 2003; Barnes & Quinn 2004). Initially, such systems are highly chaotic and close encounters are common. The close encounters lead to planets colliding (creating more massive planet) and/or planets being ejected from the system, depending on the orbital periods and planet radii. Either process results in the number of planets in the system being reduced and the typical separations between planets increasing. The system gradually evolves from a rapidly unstable state to states which will endure longer before the next collision or ejection. Such systems typically evolve to a state with 1–3 eccentric giant planets which will persist for the lifetime of the star. In systems with at least two remaining planets, the ratio of semi-major axes of the innermost planets is typically large, but shows considerable variation across different systems, $\langle a_2/a_1 \rangle = 25 \pm 24$ and 11 ± 7.8 for two different mass distributions (Table 4 of Adams & Laughlin 2003). These distributions of final eccentricities in such systems display a breadth comparable to the observed distribution of eccentricities of extrasolar planets, but underproduce planets with small eccentricities. Although dynamical relation does not predict a strict upper limit for the eccentricities generated (as does planet-planet scattering with two planets initially on circular orbits), extreme eccentricities are unlikely ($p(e > 0.8) \leq 0.1$; see Fig. 7 of Adams & Laughlin 2003), since the final eccentricities are the result of a succession of ejections and/or collisions.

Since the initial evolution is strongly chaotic, the results of such simulations are relatively insensitive to the exact choice of initial conditions, but bounded by conservations of energy and angular momentum. This partially explains the similar results of several groups using different initial conditions. However, nearly all such simulations have considered purely gravitational forces. In fact, planetary systems may evolve via dynamical relaxation while the disk still has a significant amount of mass in gas or planetesimals. Either a gas or planetesimal disk is likely to provide a significant amount of dissipation which could significantly alter the evolution of the system. While some work has investigated the effects of a dissipative gaseous disk that drives convergent migration between two planets and lead to close encounters (Adams & Laughlin 2003; Moorhead & Adams 2005), much more work remains to be done to explore the wide range of parameter space which exists for systems with multiple planets and a dissipative disk.

4.   Three Multiple-Planet Systems

First, we review recent research on the history of three well-studied multiple planet systems orbiting three solar type stars: the Sun, GJ 876, and Upsilon Andromedae (v And). Several other multiple planet systems have been discovered by radial velocity searches, but either the planets interact too weakly to provide dynamical constraints on planet formation or the published observations are not yet sufficient to precisely constrain their dynamics. Even though high precision measurements are also available for the planets orbiting pulsar PSR 1257+12, we do not include this system, since it's formation may have been very different than planet formation around solar type stars.

4.1.   The Solar System

Despite centuries of study and *in situ* measurements by space probes, the formation of giant planets in our solar system remains a matter of significant debate. In particular, it is not certain whether giant planets form via the gradual accretion of a rocky core or via direct gravitational collapse. According to the gravitational instability model, giant planets are formed by gravitational instabilities in the protoplanetary disk, much like binary stars (Boss 1995, 1996). These simulations are very computationally challenging, so they are not able to include all the relevant physics. Whether or not giant planets form depends on the simplifying assumptions used for the simulation. While some numerical simulations form massive giant planets in a few orbital times, these typically start from disks that are violently unstable. Further, these typical integrations are run for such a short period of time that they can not start from plausible initial conditions. Recent simulations have considered disks that start from a stable state and gradually approach instability via cooling (Pickett et al. 2003; Mejia et al. 2005). These simulations form rings and can temporarily fragment, if the cooling time is sufficiently rapid, but they have not resulted in forming stable giant planets. In principle, the main advantage of the gravitational instability model is that it might be able to form giant planets rapidly, even at large orbital separations. Another potential advantage is that the giant planets would typically be formed in eccentric orbits. Thus, the significant eccentricities of extrasolar planets could be explained without invoking any additional mechanisms for eccentricity excitation.

According to the competing model of core accretion, collisions between rocky planetesimals result in the gradual growth of a rocky core (Lissauer 1993). Once the core becomes sufficiently massive, it accretes a large quantity of gas from the protoplanetary disk (Pollack et al. 1996). Several details of this model remain active areas of research (e.g., "Why do collisions between planetesimals result in accretion rather than shattering?" and "How do small planetesimals avoid rapid orbital decay in the protoplanetary disk?"). Still, there is little doubt that core accretion must explain the formation of the terrestrial planets, asteroids, and other small bodies in the solar system. However, there is active debate whether core accretion could have formed the cores of Uranus and Neptune before the gas disk dissipated. This has led some researchers to propose that Uranus and Neptune, and perhaps all four giant planets, may have formed via gravitational instability. Other researchers have proposed refinements to the core accretion model that could allow for the more rapid formation of Uranus and Neptune. Here we summarize two recent attempts to explain the formation of Uranus and Neptune within the core accretion framework.

Two similar scenarios for forming Uranus and Neptune via core accretion both suggest that they initially formed at much smaller orbital distances, where the timescales relevant for planet formation are shorter. In one version, Thommes, Duncan & Levison (1999) proposed that Uranus and Neptune formed much closer to the Sun than their current orbital separations, perhaps even between Jupiter and Saturn. As the disk began to dissipate, planet-planet scattering excited large eccentricities and caused their orbits to extend well beyond Saturn. Then dynamical friction in the protoplanetary disk would have circularized their orbits at orbital separations comparable to those we see today. In a slightly refined version, Uranus and Neptune again would have initially formed closer to the Sun than their current orbital separations (but still beyond Saturn). This closely packed system could survive for an extended period of time if the eccentricities of all four giant planets were significantly smaller than they are today. Planetesimal scattering would have caused Jupiter to migrate slightly

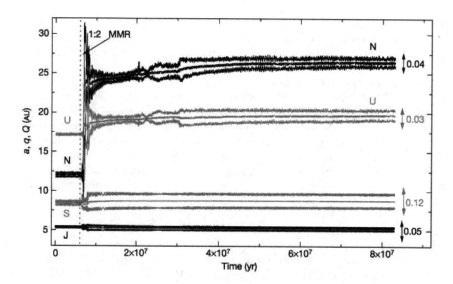

Figure 1. Orbital evolution of a hypothetical planetary system similar to the Solar System. The lines show the semimajor axis (middle lines), periastron distance (q; lower lines), and apastron distance (Q; upper lines) for each planet. This n-body simulation started with the giant planets closer together than the Solar System giant planets are today. The planets migrated due to scattering planetesimals from a 35 $M_\oplus$ disk extending out to 30 AU. The vertical dotted line marks the epoch where Jupiter and Saturn crossed their 1:2 mean motion resonance. After this point, large eccentricities were excited and the planets underwent close encounters and strong planet-planet scattering. For example, the orbits of planets U and N cross. Continued planetesimal scattering damps the eccentricities to near the present values for the solar system giant planets. The values at the right indicate the maximum eccentricities of each planet over the last 2 Myr. Reprinted by permission from Macmillan Publishers Ltd: *Nature* (Tsiganis et al. 2005), copyright 2005.

inwards, while Saturn, Uranus, and Neptune would have migrated outwards. The eccentricities would have remained small until Saturn crossed the 2:1 mean motion resonance with Jupiter. This divergent resonant crossing would not result in resonance capture, but would excite significant eccentricities that would propagate throughout the system. Uranus and Neptune would be scattered outwards, but could have circularized near their current orbits due to dynamical friction with a planetesimal disk (Fig. 1; Tsiganis et al. 2005). This scenario is particularly appealing, since n-body simulations show that it can also reproduce several other observed properties of the solar system (Morbidelli et al. 2005; Gomes et al. 2005; Strom et al. 2005).

Another possibility is that a collisional cascade maintained a significant fraction of the disk mass in small rocky bodies, even after protoplanets had formed (Goldreich, Lithwick & Sari 2004). In this scenario, several Uranus and Neptune-mass protoplanets could have formed near the current location of Uranus and Neptune, since dynamical friction damped the random velocities and gravitational focusing allowed them to accrete more rapidly than conventionally assumed in the core accretion model. Eventually, the mass in the small bodies must have decreased to the point where dynamical friction was no longer sufficient to prevent the protoplanets from exciting each other's eccentricities. Then the protoplanets could have close encounters and scatter each other. In the solar system, several massive proto-planets would have been scattered from near Neptune inward to Uranus, then on to Saturn and Jupiter, before being ejected from the Solar System. Once Uranus and Neptune were the only remaining massive bodies remaining, both planets would be expected to have had large eccentricities from scattering nearly comparable mass protoplanets inwards (Chambers 2001). Therefore, some mechanism for eccentricity damping would be necessary to explain their current low eccentricity orbits. The circularization could be caused by dynamical friction and planetesimal scattering in what remains of the planetesimal disk.

4.2.   GJ 876

Three planets have been discovered around the M4 dwarf, GJ 876 (Marcy et al. 2001; Rivera et al. 2005). The most recently discovered planet (d) has a minimum mass of $\simeq 6$ $M_\oplus$ and orbits at 0.02 AU, but is not essential for our subsequent discussion of the orbital evolution of the outer two planets. The two more massive planets (b & c) have minimum masses of 1.9 and 0.6 M_{Jup} and orbit at 0.21 and 0.13 AU, respectively. The middle planet (c) has an eccentricity $\simeq 0.2$, but the outer planet's eccentricity is much smaller (≤ 0.03). These planets are particularly interesting, since they are near a 2:1 mean motion resonance, and mutual planetary perturbations have already been observed (Laughlin et al. 2005).

Since mean motion resonances occupy only a small fraction of the available phase space, one might naively assume that it is unlikely for two planets to form in a mean motion resonance. However, if significant planetary migration and multiple planet systems are both common, then planets could form away from mean motion resonances and differential migration could cause the planets to approach a mean motion resonance. If the migration is both smooth and convergent, then as planets approach mean-motion resonances, they can be efficiently captured into a low-order mean-motion resonance. Thus, the pair of planets in GJ 876 suggests that significant migration is likely to have occurred in that planetary system. If the migration were to continue after resonant capture, then both planets would migrate together, leading to significant eccentricity evolution (Peale 1986). Indeed, hydrodynamic simulations of two planets embedded in a gaseous disk confirm this behavior (Bryden et al. 2000; Kley 2000; Snellgrove, Papaloizou, & Nelson 2001; Papaloizou 2003; Kley et al. 2005). Therefore, eccentricity excitation via resonance capture is a natural explanation for the observed eccentricities for those extrasolar planetary systems which participate in low-order mean-motion resonances. This possibility has been studied intensively in the context of GJ 876 (Lee & Peale 2002; Snellgrove, Papaloizou & Nelson 2001; Kley et al. 2005), as well as extrasolar planetary systems more generally (Lee 2004; Nelson & Papaloizou 2002).

Lee & Peale (2002) studied the evolution of GJ 876b & c, assuming initially well-separated circular orbits and a smooth convergent migration leading to capture in the 2:1 mean motion resonance. This naturally leads to eccentricity excitation and can easily generate the observed eccentricity of planet c and a small eccentricity for planet b. In fact, the eccentricity excitation due to resonance capture is so efficient that this places significant constraints on the migration history. In one scenario, the migration would have led to capture in the 2:1 mean motion resonance, but the migration must have halted very shortly afterwards. Lee & Peale (2002) estimate that the semi-major axis of the outer planet could only decrease by 7% after resonance capture, requiring the protoplanetary nebula to dissipate at nearly the same time as the capture into resonance. Since this scenario would require an unlikely fine-tuning of parameters, they develop an alternative model which includes eccentricity damping due to interactions with the disk of the form $\dot{e}/e = -K\dot{a}/a$, where e is the eccentricity, a is the semimajor axis, the dots represent time derivatives, and K is a numerical constant. Significant eccentricity damping could slow the excitation of eccentricities and allow the planets to migrate by more than 7% after the resonance capture, somewhat reducing the level of fine-tuning needed. If $K \sim 100$, then the eccentricities start to grow after resonance capture, but saturate at near the currently observed eccentricities, eliminating the need for the migration to halt shortly after resonance capture (see Fig. 2, left). More detailed hydrodynamic simulations confirm this finding (Papaloizou 2003; Kley et al. 2004; Kley et al. 2005). Kley et al. 2005 used the revised orbital fits from Laughlin et al. (2004) and found that $K \simeq 40$–170 was needed for the eccentricity excitation to saturate near the current values, depending on the inclination of the system (but assuming coplanar orbits with an inclination relative to the plane of the sky greater $35°$, as suggested by radial velocity constraints).

4.3. *υ* Andromedae

The system of three giant planets orbiting $υ$ And (Butler et al. 1999) also offers clues to the history of orbital migration. Like GJ876, one planet (b) has a short orbital period (4.6 d) and is not essential for understanding the dynamics of the outer two planets. The outer two planets (c & d) have orbital periods of 241 d and 1301 d and eccentricities of 0.26 and 0.28, respectively (Ford, Lystad & Rasio 2005). Soon after their discovery, it was realized that mutual planetary perturbations could cause significant secular evolution of the eccentricities and longitudes of periastron for the outer two planets (Stepinski, Malhotra & Black 2000; Chiang, Tabachnik & Tremaine 2001; Lissauer & Rivera 2001).

Two models were proposed to explain the eccentricities and longitudes of pericenter for planets c & d. Chiang & Murray (2002) proposed that a protoplanetary disk beyond planet d could *adiabatically* torque planet d. If the longitudes of periastron were initially circulating, then this torque would drive the system towards solutions where the longitudes of periastron librate about an aligned configuration. Once the system was in the librating regime, the torque would damp the libration amplitude. Thus, this model would predict that the the pericenters of the outer two planets would currently be librating with small amplitude about an aligned configuration and that the secular evolution would cause only small variations in the eccentricities. Malhotra (2002) proposed an alternative model in which the outer planet was perturbed *impulsively*, as would be expected if it had a close encounter with another (undetected) planet. In this scenario, the two planets could be either librating or circulating, depending on

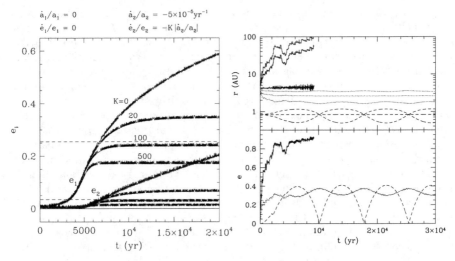

Figure 2.   Left: Model for the eccentricity evolution of the outer two planets in GJ 876 due to smooth convergent migration. The solid curves show how the eccentricities are excited following capture into the 2:1 mean motion resonance for different assumptions about the rate of eccentricity damping. The horizontal dashed lines show the approximate observed eccentricities for the planets. Unless there was strong eccentricity damping, continued migration after capture into the 2:1 mean motion resonance would rapidly cause the eccentricities to exceed their observed values. In this model, an outer disk is assumed to torque only the outer planet. More sophisticated models give similar results (e.g., Kley et al. 2005). Note that the inner planet is referred to as 1 in the figure and c in the text, and the outer planet is referred to as 2 in the figure and b in the text. Reproduced by the kind permission of the AAS (Lee & Peale 2002). Right: Dynamical evolution of a hypothetical planetary system similar to v And. The top panel shows the semimajor axes (middle lines) and periastron and apastron distance (lower and upper lines) for planets similar to the middle (C, dashed line) and outer (D, dotted line) planets around v And, as well as a hypothetical fourth planet (E, solid line). The innermost planet, B, is not shown, as it plays a negligible role. The lower panel shows the eccentricity evolution for the same numerical integration. After a brief period of dynamical instability, planet E is ejected, leaving the other two in a configuration that is very similar to that presently observed for v And c and d. Note that the timescale to completely eject the outer planet from the system (after $\simeq 9{,}000$ yr in this particular simulation) is much longer than the timescale of the initial strong scattering ($\simeq 100$ yr). After this initial brief phase of strong interaction, the perturbations on the outer planet are too weak to affect significantly the coupled secular evolution of v And C and D. Thus, the "initial" eccentricity of v And C for the secular evolution is determined by its value at the end of the strong interaction phase, rather than that at the time of the final ejection. Reprinted by permission from Macmillan Publishers Ltd: *Nature* (Ford, Lystad & Rasio 2005), copyright 2005.

the relative phases at the time of the impulsive perturbation. If the system were librating, then this model would generally predict that the libration amplitude would be large and that there would be significant eccentricity oscillations.

The best-fit orbital solution to the early observations suggested that the pericenter directions of the outer two planets were very nearly aligned ($\leq 10°$; Butler et al. 1999), favoring the model for adiabatic perturbations from a disk (Chiang & Murray 2002). However, subsequent observations show that the pericenters are less well aligned than previously thought ($\Delta\omega = 37.6° \pm 4.8°$; Ford, Lystad & Rasio 2005), favoring an impulsive perturbation due to planet scattering.

The planetary system around υ And has an even more remarkable property. This system lies very close to the boundary between librating and circulating solutions. As a result, the eccentricity of the middle planet undergoes very large oscillations with e ranging from from 0.34 to very nearly zero (see Fig. 2, right, after 10^4 yr). Stepinski, Malhotra & Black (2000) recognized that this was possible for *some* orbital solutions consistent with the radial velocity observations. Ford, Lystad & Rasio (2005) used a rigorous Bayesian statistical analysis to demonstrate that the eccentricity of the middle planet periodically returns to nearly zero for *all allowed* orbital solutions (see Fig. 2 of Ford, Lystad & Rasio 2005). This provides a strong constraint on the timescale for eccentricity excitation in υ And ($\simeq 100$ yr). For a planet-disk interaction to excite an eccentricity of $\simeq 0.3$ would require a very massive disk (≥ 40 M_{Jup}) exerting a very strong torque only to abruptly stop after $\simeq 100$ yr. Thus, this peculiar orbital configuration would be extremely unlikely, unless both planets were initially on circular orbits and the outer planet were perturbed impulsively by strong planet-planet scattering (Malhotra 2002).

5. Implications for Planet Formation

5.1. Orbital Migration

Regardless of how the giant planets formed, the large number of Kuiper belt objects in mean-motion resonances with Neptune provides strong evidence for significant outward migration of Neptune via planetesimal scattering (Hahn & Malhotra 1999). Numerical simulations have shown that the necessary migration is naturally explained via planetesimal scattering for reasonable disk masses. Only Jupiter is efficient at ejecting planetesimals from the Solar System, but together Neptune, Uranus, and Saturn can scatter planetesimals from near Neptune's orbit inwards to Jupiter. Therefore, Jupiter migrated inwards (slightly due to its large mass), while Saturn, Uranus, and Neptune migrated outwards (Fernandez & Ip 1984; Malhotra 1995).

Initially, theoretical difficulties for forming giant planets at orbital separations of $\simeq 0.05$ AU helped rekindle models of planet migration. The detection of pairs of planets in 2:1 mean motion resonances (e.g., GJ 876 b & c) suggests that smooth convergent migration likely occurred in these systems. Additionally, the fact that migration models can simultaneously match the observed eccentricities for both planets b & c provides further evidence for migration in this system.

5.2. Eccentricity Excitation via Orbital Migration

It is natural to ask if the large torques presumed responsible for orbital migration could also be responsible for exciting orbital eccentricities.

Migration in Planetesimal Disk   Analytical arguments suggest that the planetesimals typically provide a source of dynamical friction (Goldreich et al. 2004). Simulations of a single-planet scattering planetesimals in the Opik approximation also show that eccentricities are usually damped (Murray et al. 1998), although eccentricity excitation may be possible for sufficiently massive planets (≥ 10 M_{Jup}). In our own solar system, it is also believed that scattering of planetesimals may have damped the eccentricities of the outer planets after violent events. Finally, direct simulations of our solar system also demonstrate that planetesimal scattering typically damps eccentricities (Hahn & Malhotra 1999; Thommes, Duncan & Levison 1999, 2002; Tsiganis et al. 2005).

Migration in Gaseous Disk   While the dissipative nature of a gaseous disk naturally leads to eccentricity damping (Artymowicz 1993), a few researchers have suggested that excitation may also be possible. Artymowicz (1992) found that a sufficiently massive giant planet (≥ 10 M_{Jup}) can open a wide gap, leading to torques which excite eccentricities. More recently, Goldreich & Sari (2003) have suggested that a gas disk could excite eccentricities even for less massive planets via a finite amplitude instability. This claim is controversial, as 3-d numerical simulations have not been able to reproduce this behavior (e.g., Papalouizou et al. 2001; Ogilvie & Lubow 2003). Given the large dynamic ranges involved and the complexity of the simulations, one might question the accuracy of current simulations. For example, three dimensional simulations have suggested that the gaps induced by giant planets might not be as well cleared as assumed in many two dimensional disk models (Bate et al. 2003; D'Angelo et al. 2003). We believe that further theoretical and numerical work is needed to better understand planet-disk interactions. In the mean time, we look to the observations for guidance on the question of eccentricity damping or excitation.

Empirical Evidence   In the GJ876 system, the observed eccentricities are not consistent with eccentricity excitation via interactions with the disk. The current observed eccentricities could be readily explained if interactions with a gas disk led to strong eccentricity damping $K = \dot{e}a/e\dot{a} \gg 1$ (Lee & Peale 2002; Kley et al. 2005). This is in sharp contrast to current hydrodynamic simulations of migration that suggest $K \simeq 1$ and theories that predict $K < 0$ (e.g., Goldreich & Sari 2003; Ogilvie & Lubow 2003). While other planetary systems are not yet as well constrained or studied as GJ 876, the moderate eccentricities of other extrasolar planetary systems near the 2:1 mean motion resonance suggest that GJ 876 is not unique.

The υ And system also provides a constraint on eccentricity excitation during migration. If the outer two planets migrated to their current locations (0.8 and 2.5 AU), then they must have been in nearly circular orbits at the time of the impulsive perturbation in order for the middle planet's eccentricity to periodically return to nearly zero. While this does not demonstrate a need for rapid eccentricity damping as in GJ 876, this is inconsistent with models which predict significant eccentricity excitation. Since dynamical analyses severely limit the possibility of eccentricity excitation in both the GJ 876 and v And systems, we conclude that orbital migration does not typically excite eccentricities, at least for a planet-star mass ratio less than ~0.003–0.006 (those of the most massive planet in v And and GJ 876).

5.3. Origin of Eccentricities

Empirical constraints that suggest that interactions with a gaseous disk do not typically excite eccentricities (§5.1). For GJ 876 (and other planetary systems near mean motion resonances) continued migration after resonance capture could excite the eccentricities of the outer two planets. However, this mechanism is insufficient for explaining the eccentricities of extrasolar planets in general, since the majority of observed multiple planet systems are not near a low-order mean-motion resonance. The dramatic eccentricity oscillations of v And c provide an upper limit on the timescale for eccentricity excitation in v And (~100 yr) and strong evidence for planet-planet scattering in this system (Ford, Lystad & Rasio 2005). Planet-planet scattering in either few-planet systems (Ford, Rasio & Yu 2003) or many-planet systems (Adams & Laughlin 2003) could produce an eccentricity distribution quite similar to that observed for extrasolar planets.

A complete theory of planet formation must explain both the eccentric orbits prevalent among extrasolar planets and the nearly circular orbits in the Solar System. Despite significant uncertainties about giant planet formation, all three mechanisms for forming the Solar System's giant planets (see §4.1) agree that the giant planets in the Solar System went through a phase of large eccentricities. If Uranus and Neptune formed closer to the Sun, then close encounters are necessary to scatter them outwards to their current orbital distances. During this phase, their eccentricities can exceed ≃0.5 (Tsiganis et al. 2005). Alternatively, if Uranus and Neptune were able to form near their current locations due to eccentricity damping from a disk of small bodies, then several other ice giants should have formed contemporaneously in the region between Uranus and Neptune. The scattering necessary to to remove these extra ice giants would have excited sizable eccentricities in Uranus and Neptune (Goldreich, Lithwick & Sari 2004). Finally, the gravitational instability model predicts that most giant planets form with significant eccentricities. Therefore, it seems most likely that even the giant planets in our Solar System were once eccentric.

Perhaps the question, "What mechanism excites the eccentricity of extrasolar planets?" should be replaced with "What mechanism damps the eccentricities of giant planets?" Unless giant planets form via gravitational instability, interactions with a gas disk are not an option, since the eccentricities would have been excited after the gas was cleared. Both dynamical friction within a planetesimal disk and planetesimal scattering could damp eccentricities in both the Solar System and other planetary systems. Dynamical friction alone would not clear the small bodies, so either accretion or ejection would be required to satisfy observational constraints (Goldreich, Lithwick & Sari 2004). Planetesimal scattering provides a natural mechanism to simultaneously damp eccentricities and remove small bodies from planetary systems.

Perhaps, *the key parameter that determines whether a planetary system will have eccentric or nearly circular orbits is the amount of mass in planetesimals at the time of the last strong planet-planet scattering event.* The chaotic evolution of multiple planet systems naturally provides a large dispersion in the time until dynamical instability results in close encounters (Chambers, Wetherill & Boss 1996; Ford, Havlickova & Rasio 2001; Marzari & Weidenschilling 2002). Unfortunately, this could significantly complicate the interpretation of the observed eccentricity distribution for extrasolar planets. On a positive note, this model might naturally explain both the eccentric orbits of extrasolar planets and the circular orbits in the Solar System. Future numerical investigations will be necessary to test this model further.

Acknowledgments.   E.B.F. thanks E.I. Chiang, G. Laughlin, M.H. Lee, H. Levison, G.W. Marcy, A. Morbidelli, J.C.B. Papaloizou, S. Peale, F.A. Rasio, and J. Wright for useful discussions. E.B.F. acknowledges the support of the Miller Institute for Basic Research.

References

Adams, F. C., & Laughlin, G. 2003, Icarus, 163, 290
Artymowicz 1992, PASP, 104, 769
Artymowicz 1993, ApJ, 419, 116
Artymowicz, P. & Lubow, S. H. 1996, ApJ, 476, L77
Barnes, R. & Quinn, T. 2004, ApJ, 611, 494
Bate, M. R., Lubow, S. H., Ogilvie, G. I., & Miller, K. A. 2003, MNRAS, 341, 213
Beckwith, S. V. W. & Sargent, A. I. 1996, Nature, 383, 189
Boss, A. P. 1995, Science, 267, 360
Boss, A. P. 1996, L&PS, 27, 139
Bryden, G., Rózyczka, M., Lin, D. N. C., & Bodenheimer, P. 2000, ApJ, 540, 1091
Bryden, G., et al. 1999, ApJ, 514, 334
Butler, R. P., et al. 1999, ApJ, 526, 916
Chambers, J. E. 2001, Icarus, 152, 205
Chambers, J. E., Wetherill, G. W., & Boss, A. P. 1996, Icarus, 119, 261
Chiang, E. I. & Murray, N. 2002, ApJ, 576, 473
Chiang, E. I., Tabachnik, S, & Tremaine, S. 2001, AJ, 122, 1607
Cionco, R. G. & Brunini, A. 2002, MNRAS, 334, 77
D'Angelo, G., Kley, W, & Henning, T. 2003, ApJ, 586, 548
Del Popolo, A. & Eks, I.,K. Y. 2002, MNRAS, 332, 485
Fernandez, J. A. & Ip, W.-J. 1984, Icarus, 58, 109
Ford, E. B., Havlickova, M., & Rasio, F. A. 2001, Icarus, 150, 303
Ford, E. B., Kozinsky, B., & Rasio, F. A. 2000, ApJ, 535, 385
Ford, E. B., Lystad, V., & Rasio, F. A. 2005, Nature, 434, 873
Ford, E. B. & Rasio, F. A. 2006, ApJL, 638, 45
Ford, E. B., Rasio, F. A., & Yu, K. 2003, in ASP Conf. Ser. 294: Scientific Frontiers
    in Research on Extrasolar Planets, ed. D. Deming & S. Seager (San Francisco:
    ASP), 181
Goldreich, P., Lithwick, Y., & Sari, R. 2004, ApJ, 614, 497
Goldreich, P. & Sari, R. 2003, ApJ, 585, 1024
Goldreich, P. & Tremaine, S. 1979, ApJ, 233, 857
Goldreich, P. & Tremaine, S. 1980, ApJ, 241, 425
Gomes, R., Levison, H. F., Tsiganis, K., & Morbidelli, A. 2005, Nature, 435, 466
Hahn, J. M. & Malhotra, R. 1999, AJ, 117, 3041
Holman, M., Touma, J., & Tremaine, S. 1997, Nature, 386, 254
Kley, W. 1999, MNRAS, 303, 696
Kley, W. 2000, MNRAS, 313, L47
Kley, W., Lee, M. H., Murray, N., & Peale, S. J. 2005, A&A, 437, 727
Kley, W., Peitz, J., & Bryden, G. 2004, A&A, 414, 735

Kozai, Y. 1962, AJ, 67, 591
Laughlin, G., Butler, R. P., Fischer, D. A., Marcy, G. W., Vogt, S. S., & Wolf, A. S. 2005, ApJ, 622,1182
Laughlin, G., Steinacker, A., & Adams, F. C. 2004, ApJ, 608, 489
Lee, M. H. 2004, ApJ, 611, 517
Lee, M. H. & Peale, S. J. 2002, ApJ, 567, 596
Lee, M. H. & Peale, S. J. 2003, ApJ, 592, 1201
Levison, H. F., Lissuaer, J. J., & Duncan, M. J. 1998, AJ, 116, 1998
Lin, D. N. C., Bodenheimer, P., & Richardson, D. C. 1996, Nature, 380, 606
Lin, D. N. C. & Ida, S. 1997, ApJ, 447, 781
Lissauer, J. J. 1993, ARA&A, 31, 129
Lissauer, J. J. 1995, Icarus 114, 217
Lissauer, J. J. & Rivera, E. J. 2001, ApJ, 554, 1141
Malhotra, R. 1995, AJ, 110, 420
Malhotra, R. 2002, ApJ, 575, 33
Marcy, G., et al. 2001, ApJ, 555, 418
Marzari, F. & Weidenschilling, S. J. 2002, Icarus, 156, 570
Mejia, A. C., Durisen, R. J., Pickett, M. K., & Cai, K. 2005, ApJ, 619, 1098
Michtechenko, R. A. & Malhotra, R. 2004, Icarus, 168, 237
Moorhead, A. V. Adams, F. C. 2005, Icarus, 178, 517
Morbidelli, A., Levison, H. F., Tsiganis, K., & Gomes, R. 2005, Nature, 435, 462
Murray, C. D. & Dermott, S. F. 1999, Solar System Dynamics (New York: Cambridge University Press)
Murray, N. Hansen, B., Holman, M., & Tremaine, S. 1998, Science, 279, 69
Namouni, F. 2005, AJ, 130, 280
Nelson, R. P. & Papaloizou, J. C. B. 2002, MNRAS, 333, L26
Ogilvie, G. I. & Lubow, S. H. 2003, ApJ, 587, 398
Papaloizou, J. C. B. 2003, CeMDA, 87, 53
Papaloizou, J. C. B. & Terquem, C. 2001, MNRAS, 325, 221
Papaloizou, J. C. B. & Terquem, C. 2002, MNRAS, 332, L39
Papalouizou, J. C. B., Nelson, R. P. & Masset, F. 2001, A&A, 366, 263
Peale, S. J. 1986, in Satellites, ed. J. A. Burns & M. S. Matthews (Tucson: Univ. Arizona Press), 159
Pickett, B. K., Majia, A. C., Durisen, R. H., Cassen, P. M., Berry, D. K., & Link, R. P. 2003, ApJ, 590, 1060
Pollack, J. B., Hubickyj, O., Bodenheimer, P., Lissauer, J. J., Podolak, M. & Greenzweig, Y. 1996, Icarus, 124, 62
Rasio, F. A. & Ford, E. B. 1996, Science, 274, 954
Rice, W. K. M. & Armitage, P. J. 2003, ApJ, 598, 55
Rivera, E. J. et al. 2005, ApJ, 634, 625
Snellgrove, M. D., Papaloizou, J. C. B., & Nelson, R. P. 2001, A&A, 374, 1092
Stepinski, T F., Malhotra, R., & Black, D. C. 2000, ApJ, 545, 1044
Strom, R. G., Malhotra, R., Ito, T., Yoshida, F., & Kring, D. A. 2005, Science, 572, 1847
Thommes, E. W., Duncan, M. J., & Levison, H. F. 1999, Nature, 402, 635
Thommes, E. W., Duncan, M. J., & Levison, H. F. 2002, AJ, 123, 2862
Trilling, D. E., Benz, W., Guillot, T., Lunine, J. I., Hubbard, W. B., & Burrows, A. 1998, ApJ, 500, 428
Tsiganis, K., Gomes, R., Morbidelli, A., & Levison, H. F. 2005, Nature, 435, 459
Veras, D. & Armitage, P. J. 2004, Icarus, 172, 349
Ward, W. R. 1997, Icarus, 126, 261
Weidenschilling, S. J. & Marzari, F. 1996, Nature, 384, 619

Eric Ford listens to Mark Krumholz describe recent work on star formation during a panel discussion.

Frank N. Bash Symposium 2005: New Horizons in Astronomy
ASP Conference Series, Vol. 352, 2006
S. J. Kannappan, S. Redfield, J. E. Kessler-Silacci, M. Landriau, and N. Drory

Massive Star Formation: A Tale of Two Theories

Mark R. Krumholz[1]

Department of Astrophysical Sciences, Princeton University, Princeton, NJ, USA

Abstract. The physical mechanism that allows massive stars to form is a major unsolved problem in astrophysics. Stars with masses $\gtrsim 20\ M_\odot$ reach the main sequence while still embedded in their natal clouds, and the immense radiation output they generate once fusion begins can exert a force stronger than gravity on the dust and gas around them. They also produce huge Lyman continuum luminosities, which can ionize and potentially unbind their parent clouds. This makes massive star formation a more daunting problem than the formation of low mass stars. In this review I present the current state of the field, and discuss the two primary approaches to massive star formation. One holds that the most massive stars form by direct collisions between lower mass stars and their disks. The other approach is to see if the radiation barrier can be overcome by improved treatment of the radiation-hydrodynamic accretion process. I discuss the theoretical background to each model, the observational predictions that can be used to test them, and the substantial parts of the problem that neither theory has fully addressed.

1. Introduction

Observations indicate that stellar initial mass function (IMF) is an unbroken power law out to masses of about 150 $M_\odot$ (Elmegreen 2000; Weidner & Kroupa 2004; Figer 2005; Oey & Clarke 2005), and there is no evidence for variation of either the limit or the index of the mass function with metallicity or other properties of the star-forming environment (Massey 1998). Why the mass limit for stars is so high, what physics sets it, and why the mass spectrum seems to be universal are major unsolved problems in astrophysics. Their solution requires a model for how massive stars form, which at present is lacking due to both observational and theoretical challenges.

Massive stars form in the densest regions within molecular clouds. We detect these massive star-forming clumps as infrared dark clouds (e.g., Rathborne et al. 2005) or as millimeter sources (e.g., Plume et al. 1997; Shirley et al. 2003). The clumps have extremely high column densities and velocity dispersions ($\Sigma \sim 1$ g cm^{-2}, $\sigma \sim 4$ km s^{-1} on scales of $\lesssim 1$ pc), and appear to be approximately virialized. However, the structures within the clumps that are the progenitors of single massive stars or small-multiple systems are only now becoming accessible to observations (Reid & Wilson 2005; Beuther et al. 2005). Observations

---

[1]Hubble Fellow

continue to improve, but are hampered by large distances, heavy obscuration, and confusion due to high densities.

On the theoretical side the problem is perhaps even more difficult. Stars with masses $\gtrsim 20$ $M_\odot$ have short Kelvin-Helmholtz times that enable them to reach the main sequence while still accreting from their natal clouds (Shu et al. 1987). The resulting nuclear burning produces a huge luminosity and a correspondingly large radiation pressure force on dust grains suspended in the gas surrounding the star. Early spherically symmetric calculations found that the radiation force becomes stronger than gravity, and sufficient to halt further accretion, once a star reaches a mass of roughly $20 - 40$ $M_\odot$ (Kahn 1974; Wolfire & Cassinelli 1987) for typical Galactic metallicities. More recent work has loosened this constraint by considering the effect of an accretion disk. Disks concentrate the incoming gas into a smaller solid angle, while shadowing most of it from direct exposure to starlight (Nakano 1989; Nakano et al. 1995; Jijina & Adams 1996). Cylindrically symmetric numerical simulations with disks find that they allow accretion to continue up to just over 40 $M_\odot$ before radiation pressure reverses the inflow (Yorke & Sonnhalter 2002).

Ionization from a massive star presents a second problem to be overcome. The escape speed in a massive star-forming core is considerably smaller than the sound speed of 10 km s^{-1} in ionized gas, so if a star is able to ionize its parent core into an HII region, the core will be unbound and accretion will stop (Larson & Starrfield 1971; Yorke & Kruegel 1977). Only if the ionization is quenched near the stellar surface, where the escape speed is larger than the sound speed, can accretion continue. Early work on the problem of massive star formation found that ionization was the dominant mechanism in setting an upper mass limit on stars, although the later realization that dust will absorb much of the ionizing radiation shifted theoretical attention more towards the effects of radiation pressure.

Today, there are two dominant models of massive star formation. In § 2., I present the competitive accretion model, in which stars are born small and grow by accretion of unbound gas and by collisions. In § 3., I discuss the turbulent radiation-hydrodynamic model, which suggests that massive stars form from massive, turbulent cores, and that neither radiation pressure nor ionization prevents accretion onto a massive star. Finally, I discuss the missing pieces of the picture that neither model is yet able to supply in § 4., and summarize the state of the field and future prospects in § 5.

2.   Competitive Accretion

2.1.   The Model

The competitive accretion model for massive star formation begins with the premise that all stars are born small, with an initial mass ranging from as much as ~ 0.5 $M_\odot$ (Bonnell et al. 2004) to as little as ~ 3 Jupiter masses (Bate & Bonnell 2005), depending on the particular variant of the theory. These "seeds" are born in a dense molecular clump, and they immediately begin accreting gas to which they were not initially bound. Stars near the center of the clump are immersed in the highest density, lowest velocity dispersion gas, and accrete most rapidly (Bonnell et al. 2001a,b). The clump is globally unstable to collapse, and it contracts to stellar densities of $\sim 10^6 - 10^8$ pc^{-3}. At this point low mass stars begin to merge, either through direct collisions (Bonnell et al. 1998), or because gas drag and continuing accretion of low angular momentum gas causes binary systems to inspiral (Bonnell & Bate 2005). For example, a simulation of a 1000 $M_\odot$ clump with a radius of 0.5 pc by Bonnell et al. (2003) produces a nearly 30 $M_\odot$ binary system whose members approach within $\sim$20 AU of one another. In the simulation gravity is softened on scales of 160 AU, so it is unclear how the system would really evolve. However, Bonnell & Bate (2005) argue that it would likely merge, leading to the formation of a 20–30 $M_\odot$ star.

One particularly appealing feature of the merger scenario is that it provides a natural explanation for the observation that O and B stars form solely (or almost solely) in rich clusters (Lada & Lada 2003) that are strongly mass segregated (Hillenbrand & Hartmann 1998). Since the rates of competitive accretion and mergers are highest in cluster centers, and both processes can only occur in clusters, this model qualitatively reproduces the observations automatically. The model also naturally produces a high proportion of close, massive binaries, since for every binary that merges there are several more that come close (Bonnell & Bate 2005; Pinsonneault & Stanek 2006).

Most work on competitive accretion to date uses no physics beyond hydrodynamics and gravity. Dale et al. (2005) make a preliminary effort to include ionization effects, but they focus more on the scale of clusters than on individual stars, so their simulations do not have the resolution to study how photoionization might affect accretion onto a single star. No competitive accretion model to date includes either magnetic fields or radiation pressure. The latter omission is particularly important, since it means there is no evidence that competitive accretion by itself resolves the radiation pressure problem – only mergers do that. Indeed, simulations of Bondi-Hoyle accretion with radiation find that radiation pressure halts accretion onto stars with masses $\gtrsim 8\ M_\odot$ (Edgar & Clarke 2004) – although these results appear questionable in light of the more realistic simulations we discuss in § 3. If Edgar & Clarke's results do hold, though, so that radiation pressure limits Bondi-Hoyle accretion (but not accretion from a core) onto a massive binary, then the only way for massive stars to grow in a competitive accretion model is by direct collisions, rather than drag-induced binary mergers. This requires stellar densities of 10^8 pc^{-3}, $\sim$3 orders of magnitude larger than any observed to date in the Galactic plane.

2.2. Observational Evidence

There are several potential direct observational signatures of the competitive accretion scenario. Bally & Zinnecker (2005) suggest two approaches: collisions would produce both infrared flares lasting years to centuries and explosive, poorly-collimated outflows. At present there is no data set available where one could search for the flares. For the outflows, there is one known example that roughly fits the description that Bally & Zinnecker propose (the OMC-1 outflow), but there has been no detailed modeling of how the outflow from a collision would actually appear, and, as we discuss below, it is unclear how common such poorly collimated outflows are. A third direct test is to search for embedded clusters with densities of $\sim 10^8$ pc^{-3}, which are a required component of the competitive accretion picture. Such objects should be short-lived and therefore rare, but their high column densities would give them a distinct spectral shape that might be observable with Spitzer, and should be easily observable by JWST or SOFIA (S. Chakrabarti & C. F. McKee, 2006, in preparation).

One can also use more indirect tests to look for evidence of mergers, and here the competitive accretion picture runs into considerable difficulty. If massive stars form via collisions, the collision process should truncate their accretion disks or disrupt them entirely. The collision itself may give rise to a fat torus, but is unlikely to produce a thin disk (Bally & Zinnecker 2005). Thus, the collisional formation model predicts that massive stars should not have disks hundreds of AU in size, as are observed for low mass stars. However, there are now at least two known examples of massive stars with such large disks (Jiang et al. 2005; Patel et al. 2005). Since thin disks are probably required to create well-collimated MHD outflows, the collision scenario also predicts that massive protostars should lack well-collimated outflows. However, interferometric observations of young massive stars reveal that outflows for stars as massive as early B usually are well-collimated (Beuther & Shepherd 2005, and references therein). Position-velocity diagrams (Beuther et al. 2004) and near-IR images (Davis et al. 2004) of outflows, as well as correlations between outflow momentum and luminosity of the driving star (Richer et al. 2000), also point to a common driving mechanism for low mass and high mass protostellar outflows, inconsistent with the competitive accretion / collision scenario.

2.3.   Theoretical Difficulties

The apparent conflict between competitive accretion models and observations has led to theoretical reconsideration of the problem. For competitive accretion to be effective a small "seed" protostar in a molecular clump must be able to accrete its own mass or more within a dynamical time of its parent clump. The process by which the protostar gathers gas from the clump is Bondi-Hoyle accretion in a turbulent medium, a process for which Krumholz et al. (2005a, 2006b) give a general theory supported by simulations. Using this result, together with an analysis of the possibility that protostars might gain mass by capturing other cores in their parent clump, Krumholz et al. (2005c) determine what properties a star-forming molecular clump must have for competitive accretion within it to be effective. They show that competitive accretion only works in clumps with $\alpha_{\mathrm{vir}}^2 M \lesssim 10~M_\odot$, where M is the clump mass and α_{vir} is the clump virial parameter (Bertoldi & McKee 1992; Fiege & Pudritz 2000), roughly its ratio of kinetic energy to gravitational potential energy. For observed star-forming clumps, $\alpha_{\mathrm{vir}} \sim 1$ and $M \sim 1000~M_\odot$, so competitive accretion will not operate. It occurs in simulations only because the regions simulated have smaller virial parameters and masses than observed regions. In some cases the virial parameters are too small to begin with (Bonnell et al. 2001a,b), and in others the virial parameters start near unity, but decay of turbulence quickly reduces them to smaller values (Bonnell et al. 2004; Bate et al. 2002a,b, 2003). The results of Krumholz et al. (2005c) strongly suggest that competitive accretion plus mergers cannot be the mechanism by which massive stars form.

3.   Turbulent Radiation Hydrodynamic Models

Given the difficulties with the competitive accretion scenario, one must ask whether it might be possible to form massive stars in roughly the same way as low mass stars, via collapse from a coherent core and disk accretion. Such a scenario must overcome three serious challenges: one requires a plausible model for the origin and structure of massive cores, a method to allow accretion to occur despite radiation pressure feedback, and an explanation for why ionization does not destroy the protostellar core before the massive star is fully assembled.

3.1.   Massive Cores

Theoretical arguments predict that fragmentation in a turbulent medium produces a spectrum of bound fragment masses that resembles the stellar IMF (Padoan & Nordlund 2002). If these arguments are correct, then massive cores are simply the tail of the distribution of core masses. Simulations of fragmentation in a turbulent medium do roughly concur with analytic models (Li et al. 2004), and observations also support the idea that cores have a mass distribution that parallels the stellar IMF, and that cores with masses $\gg M_\odot$ exist (e.g., Motte et al. 1998; Testi & Sargent 1998; Johnstone et al. 2001; Reid & Wilson 2005; Beuther et al. 2005). Thus, massive cores may naturally arise from turbulent fragmentation.

Massive cores, however, must be structured somewhat differently than the cores that give rise to low mass stars. The thermal Jeans mass in star forming regions is ~ 1 $M_\odot$, so massive cores cannot be supported primarily by thermal pressure. Instead, they must be turbulent. McKee & Tan (2003) present a self-similar model of massive, turbulent cores that are in rough pressure balance with the high pressure environments they form. This gives them surface densities ~ 1 g cm^{-2} and pressures $P/k \sim 10^8$ K cm^{-3}, much larger than the mean column density and pressure in giant molecular clouds. These high pressures cause the cores to be extremely compact, with radii $\lesssim 0.1$ pc, and the correspondingly high density produces accretion rates of $\sim 10^{-3}$ $M_\odot$ yr^{-1} onto embedded stars, allowing massive stars to form in $\sim 10^5$ yr.

One important question for models of massive cores is whether they will produce one or a few massive stars, or fragment to produce numerous low mass stars instead. Dobbs et al. (2005) simulate centrally condensed turbulent cores with structures that follow the McKee & Tan (2003) model, and find that they form many low mass stars rather than a single massive star. However, their simulations do not include radiation. Krumholz et al. (2006a) perform similar simulations including radiative transfer, and find that the combination of high accretion luminosity and high optical depth that occur in high-density cores produce rapid heating that inhibits fragmentation. Of course the massive core models used by both Dobbs et al. and Krumholz et al. are highly idealized, so the question of to what extent real massive cores fragment remains open.

3.2.   Accretion with Radiation Pressure

The Flashlight Effect   Once a massive protostar reaches ~ 15 $M_\odot$, the pressure exerted by its radiation field will begin to have a significant effect on the accretion flow. Two dimensional radiation-hydrodynamic simulations by Yorke &

Sonnhalter (2002) find that, once the radiation field becomes significant, it begins to reverse inflow along the poles. Accretion continues through an accretion disk in the equatorial plane, and the disk serves to collimate the radiation field and beam it preferentially in the polar direction. This collimation is called the flashlight effect. However, in Yorke & Sonnhalter's simulations this is not enough to allow very massive stars to form. As the protostellar mass and luminosity increase, inflow stops over a wider and wider range of angles about the pole. Eventually, no more material is able to fall onto the disk, and soon thereafter the radiation field disperses the disk itself. Yorke & Sonnhalter find a maximum final mass of the star of $\approx 20\ M_\odot$ in simulations with gray radiation, and ≈ 40 $M_\odot$ in simulations with a multi-frequency treatment of the radiation field. The difference in outcome is likely due to enhancement of the flashlight effect by the more realistic multi-frequency radiation model.

More recent three-dimensional radiation-hydrodynamic simulations, however, demonstrate a qualitatively new effect that allows accretion to higher masses than two-dimensional work suggests. Krumholz et al. (2006a) find that at masses $\lesssim 17\ M_\odot$, the radiation field is too weak to reverse the inflow, and massive cores evolve much as Yorke & Sonnhalter (2002) find. At larger masses, the radiation field begins to inhibit accretion along the poles, driving bubbles into the accreting gas. However, the three-dimensional simulations show that bubbles grow asymmetrically due to an instability that is suppressed in Yorke & Sonnhalter's two-dimensional, single quadrant (i.e. assuming symmetry about the xy plane) simulations. Figure 1a shows this effect. Since the gas is extremely optically thick to stellar radiation, the bubbles are able to collimate the radiation field and beam it preferentially in the polar direction, as shown in Figure 2a. At the time shown in the Figure, the flux of radiation in polar direction at the edge of the bubble is larger than the flux in the equatorial plane by more than an order of magnitude. The strong flux in the polar direction deflects gas that reaches the bubble walls to the side. As the velocity field in Figure 1a shows, it then travels along the bubble wall and falls onto the disk, where it is shielded from the effects of radiation by the disk's high optical depth. The gas then accretes onto the star.

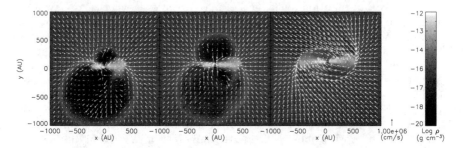

Figure 1.   The plot shows a simulation of the collapse of a 100 $M_\odot$ core by Krumholz et al. (2006a). Each panel is a slice in the XZ plane at a different time, showing the density (grayscale) and velocity (arrows). The times of the three slices are 1.5×10^4 (*left*), 1.65×10^4 (*center*), and 2.0×10^4 yrs (*right*), and the stellar masses at those times are 21.3 $M_\odot$, 22.4 $M_\odot$, and 25.7 $M_\odot$.

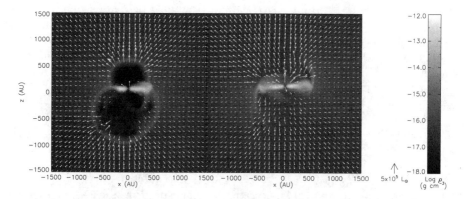

Figure 2. The plot shows a simulation by Krumholz et al. (2006a). The panels show the same times as the center and right panels of Figure 1, but the arrows show radiation flux rather than velocity. The flux vectors are multiplied by $4\pi r^2$, where r is the distance from the central star. For clarity, flux vectors from inside the optically thin bubble interior in panel (a) have been omitted.

Eventually the instability becomes violent enough for the bubbles to start collapsing (Figure 1b), leaving behind remnant bubble walls (Figure 1c). These dense walls serve to collimate the radiation and shield the gas from it, as shown in Figure 2b, allowing it to reach the accretion disk and then the star. The collapse is in essence a radiation Rayleigh-Taylor instability, caused by the inability of radiation, a light fluid, to hold up the heavy gas. Simulations to date have reached masses of $\approx 34\ M_\odot$ onto the star, with $5 - 10\ M_\odot$ in the disk. Thus far there are no sign of accretion being reversed, and the simulations are continuing as of this writing. Note that these calculations use gray radiative transfer, a case for which Yorke & Sonnhalter (2002) found a limit of $\approx 20\ M_\odot$; Yorke & Sonnhalter's results suggest collimation of the radiation field would be even more effective, and accretion correspondingly easier, with multi-frequency radiation.

Protostellar Outflows   The simulations discussed above all ignore the presence of protostellar outflows. However, Krumholz et al. (2005b) point out that outflows can also have a strong effect on the radiation field. Outflows from massive stars are launched from close to the star, where radiation heats the gas to the point where all the dust sublimes. As a result, outflows are dust-free and very optically thin when they are lunched. Because outflows leave the vicinity of the star at high speeds ($\gtrsim 500$ km s^{-1}), there is no time for dust within the outflow cavity to re-form before the gas is well outside the collapsing core. Because the core around it is very optically thick, the outflow cavity collimates radiation and carries it away very efficiently. It effectively becomes a pressure-release valve for the radiation. Monte Carlo radiative transfer calculations show that, for outflow cavities similar to those observed from massive protostellar outflows, the presence of a cavity can reduce the radiation pressure force on infalling gas by an order of magnitude. This can shift the inflow from a regime where radiation pressure is stronger than gravity to one where it is weaker. Krumholz et al.

(2005a) show that, even for a 50 $M_\odot$ star embedded in a 50 $M_\odot$ envelope, an outflow cavity would make radiation pressure weaker than gravity over a quarter of the solid angle onto the star.

It is unclear how radiation collimation by outflows will interact with the radiation bubbles and Rayleigh-Taylor instability that occur in simulations where an outflow is not present. However, the overall conclusion one may draw from both effects is that, in an optically thick core, it is very easy to collimate radiation. Collimation reduces radiation pressure over much of the available solid angle, and allows accretion to continue to higher masses than naive estimates suggest. Radiation pressure is not a significant barrier to accretion.

3.3. Ionization

The third puzzle to solve in an accretion mechanism for massive star formation is ionization. For spherically symmetric accretion, Walmsley (1995) shows that accretion above a critical rate

$$\dot{M}_{\rm crit} = \sqrt{\frac{4\pi GMSm_H^2}{\alpha^{(2)}}} \approx 2 \times 10^{-5} \left(\frac{M}{10\ M_\odot}\right)^{1/2} \left(\frac{S}{10^{49}\ {\rm s}^{-1}}\right)^{1/2} M_\odot\ {\rm yr}^{-1},\ (1)$$

where M is the stellar mass, S is the ionizing luminosity (in photons s^{-1}), and $\alpha^{(2)}$ is the recombination coefficient to excited levels of hydrogen, will trap all ionizing photons near the stellar surface. Since estimated accretion rates for massive stars are much higher than this, if accretion is spherically symmetric then the star will be unable to ionize its parent core.

Observations support the idea that HII regions can be confined near their source stars for long periods, and that they are therefore not able to stop accretion. The argument comes from statistics: ultracompact HII regions (roughly those $\lesssim 0.1$ pc in size) have dynamical times $\sim 10^3$ yr, but a census of the number of HII regions versus their size implies that ultracompact HII regions must survive for times closer to $\sim 10^5$ yr (Wood & Churchwell 1989; Kurtz et al. 1994). An extended phase during which HII regions are confined by accretion, and which lasts for a time comparable to the star formation time ($\sim 10^5$ yr), is consistent with the data, while the idea that HII regions expand dynamically and halt accretion is not. In addition, observations of inflow signatures in ionized gas in some systems (Sollins et al. 2005) provide direct evidence that accretion can continue even after the formation of an HII region.

The exact mechanism by which HII regions are confined is still uncertain. Keto (2002, 2003) presents spherically symmetric hydrodynamic accretion models with ionizing radiation. In the models, when the ionizing luminosity is low, accretion traps all ionizing photons at the stellar surface and there is no HII region. As the stellar mass and ionizing luminosity increase, the HII region is able to lift off the stellar surface, but it remains trapped in a region where the thermal pressure of the ionized gas is insufficient to escape. Accretion continues through the ionization front for a long time, but eventually the ionizing luminosity rises high enough for the HII region to expand outward and reach the point where it halts further accretion. While this model seems to be consistent with the observational data, there are two possible problems. First, it is spherically symmetric, while massive stars form primarily in very turbulent regions. Second, the long trapped HII region phase requires that the ionizing luminosity and the accretion rate rise together. Ionizing luminosity rises sharply with stellar mass, and in Keto's models (which assume Bondi accretion) so does the accretion rate. However, if the accretion rate were a weaker function of mass, as is expected for more realistic core models (e.g., McKee & Tan 2003), then it is unclear the confinement would work.

Xie et al. (1996) offer another possibility: HII regions may be confined by turbulent pressure, which far exceeds thermal pressure in the dense regions where massive stars form. However, Xie et al.'s model is purely analytic, and it is unclear if turbulent confinement of ionization can work in reality. Recent simulations by Dale et al. (2005) of collapsing regions suggest that it will not, because ionizing radiation will escape through low-density regions of the turbulent flow. In Dale et al.'s simulations the turbulent pressure is far lower than it should be due to turbulent decay (§ 2.3.), but the results suggest at a minimum that turbulent confinement of HII regions needs further study. Tan & McKee (2003) offer the alternative model that HII regions may be confined by the dense outflows of massive stars. In this picture, the ionized region is confined to the dense walls of an outflow cavity that has been largely evacuated of gas by magnetic fields. They find that the ionization stays confined near the star as long as the star is type B or later. However, it is unclear what this model predicts will happen for more massive stars.

Regardless of which model for trapping is correct, it is important to note that *some* mechanism for confining HII regions for many dynamical times seems to be required by the observational data. All the proposed explanations thus far involve accretion, winds, or turbulence in some form. There is no plausible solution for the long lifetimes of HII regions in the competitive accretion picture, which lacks these elements and generally predicts gas (as opposed to collisional) accretion rates onto massive stars that are too low to confine ionization (Dale et al. 2005; Dobbs et al. 2005). This is another serious argument against competitive accretion.

4. Missing Pieces

Thus far, I have argued that the turbulent radiation hydrodynamic model provides a good solution to the problem of massive star formation. However, there are several elements of the problem for which neither that model nor the competitive accretion model have made much progress.

4.1. Magnetic Fields

None of the simulations of massive star formation performed to date have included magnetic fields, and analytic models have included them only in the most cursory fashion. It is unclear how serious an omission this is. The dynamical importance of the magnetic field is determined by the mass to magnetic flux ratio, M/Φ. For a given flux there is a maximum mass M_Φ than flux can support against gravitational collapse, so for a cloud of mass M it is natural to define the magnetic support parameter $\lambda = M/M_\Phi$. Values of $\lambda > 1$ are termed supercritical, and correspond to configurations where magnetic support cannot prevent the cloud from collapsing dynamically; in the subcritical regime, $\lambda < 1$, magnetic support can prevent collapse. If typical massive star forming cores are subcritical or critical, then omission of magnetic fields is a serious error.

In principle one can determine λ directly from observations. In practice, however, magnetic fields are extremely difficult to detect even in low-mass star forming regions, which are generally closer and suffer much less from confusion and extinction than massive regions. Crutcher (2006) reviews the observations of magnetic fields in massive star-forming regions that exist, and concludes that $\lambda \approx 1$ is typical, indicating the magnetic effects are significant.

However, this conclusion is plagued by large systematic uncertainties. First, to determine λ from observations, one must assume a geometry for the cloud. Crutcher's conclusion assumes that cores are two-dimensional disks. However, observations show that cores are roughly triaxial (Jones et al. 2001), with ratios of long to short axis of $\sim 2 : 1$. This would give $\lambda > 1$ for the vast majority of observed regions. A second difficulty stems from uncertainty in where within a core one is measuring a magnetic field. The most common and reliable way to detect magnetic fields in molecular gas is via Zeeman splitting in OH or CN. However, both of these molecules are biased tracers of the mass, due to excitation threshold effects and freeze-out onto dust grains (Tafalla et al. 2002). It is unclear what systematic biases observing the field in these biased tracers might produce. Methods such as the Chandrasekhar-Fermi effect, which are based on polarization of dust grains, are not affected by freeze-out, but are affected by uncertainty as to where along a line of sight a polarized signal is arising. It is not clear whether these effects will systematically increase or decrease λ. A third bias is that in many cases observations do not detect a magnetic field at all, and at least some non-detections remain unpublished. If such regions are not properly included in statistical analyses, this can artificially raise estimates of λ (Bourke et al. 2001). In summary, observations are quite ambiguous as to whether magnetic fields are dynamically significant in regions of massive star formation. Ideally they should be included in models, but limitations of algorithms have prevented their inclusion thus far.

4.2.   Masers

Decades ago observations established that massive star forming regions are often host to water, methanol, OH, and SiO masers. Although they were originally thought to arise from shocks at the edges of HII regions, high resolution observations show that they are generally offset from HII regions, and are more closely associated with infrared sources (Hofner & Churchwell 1996). Masers are particularly useful because they provide spatial resolutions of milliarcseconds, far higher than any other technique possible for deeply embedded sources. The resolution in space and time is sufficiently high that multi-epoch observations can often detect proper motions of individual maser spots.

The primary difficulty with maser observations is that they are difficult to interpret. Maser spots often show linear or arc-like arrangements, which early observers interpreted as tracing edge-on disks (Norris et al. 1993). This would have been interesting, because at the time no disks around massive stars were known. Even today, it would provide us with a powerful tool for tracing the dynamics of massive accretion disks on very small scales. However, more recent work that has combined maser data with observations in other wavelengths provides little support to the disk hypothesis. In a few cases, such as Orion BN/KL (Greenhill et al. 2004), linear arrangements of maser spots are perpendicular to molecular outflows, as one would expect were masers tracing a disk. More often, however, maser spots are parallel to the direction of outflows (De Buizer 2003; De Buizer & Minier 2005; De Buizer et al. 2006), suggesting that they trace outflows rather than disks. The primary lesson is that masers cannot be used as diagnostics of the massive star formation process without more complete models of how and where maser emission arises.

While difficult, making such models is likely to yield new insights into the star formation process, particularly when applied to some of the more unusual arrangements of masers that appear consistent with neither a disk nor an outflow. For example, Figure 3 reproduces Figure 1a of Torrelles et al. (2001). The dots show the positions of water maser spots observed in three epochs in the Cepheus A region, a site of massive star formation. At each epoch the spots fit a circle roughly 62 AU in radius around the same center to an accuracy of 0.1%. The change in radius with time implies that the circle is expanding at 9 km s^{-1}. The best explanation for this geometry is that we are seeing a limb-brightened, expanding spherical shell, and there is no obvious way that either a disk or an ordinary bipolar outflow could explain the data. One intriguing possibility is that the masers could be tracing the wall of a radiation bubble, as seen in the simulations of Krumholz et al. (2006a). The bubbles are quite spherical when they are at such small radii, the expansion velocities are roughly consistent with what is seen in the simulations, and the densities and temperatures in the bubble walls are roughly what would be needed to produce maser emission.

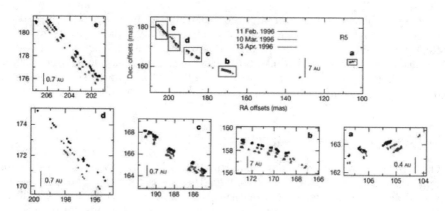

Figure 3.   Dots show spots of water maser emission observed by Torrelles et al. (2001) in the Cepheus A star-forming region. Figure appears by the kind permission of the Nature Publishing Group.

4.3. The Stellar Mass Limit

A final observational result that neither model has been able to incorporate or explain thus far is the existence of an upper mass limit to the stellar IMF. Statistical arguments applied to the Galaxy as a whole (Elmegreen 2000; Oey & Clarke 2005) and direct star counts in individual massive clusters (Weidner

& Kroupa 2004; Figer 2005) both show that the IMF cannot continue to have a Salpeter slope out to arbitrarily high masses. Instead, there must be a fairly sharp turn-down at around 150 $M_\odot$. It is difficult to see how such a cutoff could occur in the competitive accretion model. Collisions between point particles should be a scale-free process, producing a featureless power-law distribution of masses. It is possible that the "microphysics" of the collision process could provide a break in the power-law – for example collisions between stars with a total mass above 150 $M_\odot$ might lead directly to intermediate mass black holes rather than to stars that we could observe – but there is at present no evidence to support this hypothesis.

The upper mass limit is not much easier to understand in the context of the turbulent radiation hydrodynamic model. One might think that the increasing strength of radiation pressure feedback with mass could produce a cutoff, but this explanation faces two serious objections. First, at masses $\gtrsim 100\ M_\odot$, stellar luminosity is almost directly proportional to mass, since at such masses stars are supported primarily by internal radiation pressure. Thus, the ratio of radiation pressure force to gravitational attraction does not change significantly for stars larger than $\sim 100\ M_\odot$. Why, then, should be there be a change in the IMF at 150 $M_\odot$? A second problem with an explanation based on radiation pressure is the absence of evidence for a variation in the stellar IMF with metallicity. Since the strength of the radiation pressure force is directly proportional to the metallicity, if radiation pressure set the stellar mass limit then one would expect the high mass end of the IMF to change with metallicity, which should be observable as a change in IMF with Galactocentric radius. Such a change has not been observed (Massey 1998).

One possible way out would be if the mass limit is unrelated to the formation process, and is instead set by stellar stability. Humphreys & Davidson (1979, 1994) investigate the structure of very massive stars, and find that they are often pulsationally unstable. This instability can cause rapid mass loss, which might set a stellar upper mass limit. However, whether such a limit really exists, and if so whether it coincides with the observed mass limit, is at present unknown.

5.   Conclusions and Prospects

Our knowledge of the physical mechanism of massive star formation is still quite limited, as evidenced by the fact that for the last decade there have been two very different models for it that observations could not definitively distinguish. However, theoretical and observational work over the last year or two have advanced the field to the point where we can begin to decide between the models. Observations of disks and outflows from young massive stars point to accretion from a core rather than collision as the mechanism by which massive stars form, and theoretical work strongly suggests that competitive accretion does not operate in observed star-forming clouds, consonant with observations favoring accretion from cores. Moreover, the problem of how to make massive stars despite radiation feedback, one of the original motivations for the competitive accretion and collision model, seems to be receding. Recent simulations and analytic work show that both radiation pressure force and ionization are much less effective at inhibiting accretion than had previously been assumed.

In the next decade, observations should be able to settle definitively the issue by searching for more direct indicators of collision, such as very high column density embedded clusters and infrared flares from collisions. They also promise to give us a window into the massive star formation process on much smaller scales, where the effects of radiation pressure and ionization should be more obvious. Masers have started to provide data with high resolution in space and time, but interpreting maser data still requires much theoretical work. The next generation of millimeter interferometers, such as ALMA, will enable us to resolve disks around massive stars, and possibly to see dense shells of material shaped by protostellar radiation and outflows on $\lesssim 1000$ AU scales. These observations should be much easier to interpret.

On the theoretical side, progress will depend primarily on improving computational models, and should focus on four problems with the current generation of simulations. First, no simulation of massive star formation to date has included outflows. This is a major omission, since we know that outflows are present, and that they can have profound effects on the formation process. Outflows may also be responsible for driving turbulence in star-forming clumps, and should therefore be included in simulations of cluster formation to avoid the problem of unphysically small virial parameters identified by Krumholz et al. (2005c). Improving the computations from hydrodynamics to magnetohydrodynamics is a second potential advance. The major difficulty here is knowing what initial conditions to use, since the strength and geometry of the magnetic field in regions of massive star formation is so poorly known. Third, simulations could be improved by starting from larger scales. Both competitive accretion and turbulent radiation hydrodynamic simulations of massive star formation start with extremely unrealistic initial conditions. A better approach would be to simulate a cluster-forming clump ~ 4000 $M_\odot$ in size, typical of the Plume et al. (1997) sample, follow the formation of a massive core, and then simulate the subsequent collapse of the core at high resolution using adaptive mesh refinement or adaptive smoothed particle hydrodynamics.

A final area ripe for improvement is radiative transfer. Thus far simulations have either used multi-frequency radiation in two dimensions (Yorke & Sonnhalter 2002), or gray radiation in three dimensions (Krumholz et al. 2006a). Since the simulations show that both multi-frequency and three-dimensional effects are important, it is critical to do three-dimensional multi-frequency radiative transfer simulations. A natural outgrowth of this is modeling ionization, since in principle one can treat Lyman continuum photons as just another frequency group and then add a chemistry update step to recompute the ionization fraction after a radiation update. A final improvement to the radiation would be to use an approximation better than flux-limited diffusion, which may produce errors inside low optical-depth radiation bubbles. The major obstacle here is computational cost. Three-dimensional gray flux-limited diffusion calculations require months of supercomputer time on present computers, and improvements to the radiation physics without significant advances in processor or algorithmic speed would make the problem impossible to run.

Perhaps the best opportunities for progress now come not purely from theory or from observation, but from work that makes detailed comparisons of the two. Hopefully in the future more observers and theorists will collaborate to post-process simulations so that they can make definite comparisons to observations, and use the results of those comparisons to refine theoretical models. In the next decade, work of this sort should be able to provide us with at least the basic outline of how massive stars form.

Acknowledgments. I thank S. C. Chakrabarti, C. F. McKee, and J. C. Tan for helpful discussions. Support for this work was provided by NASA through Hubble Fellowship grant #HSF-HF-01186 awarded by the Space Telescope Science Institute, which is operated by the Association of Universities for Research in Astronomy, Inc., for NASA, under contract NAS 5-26555.

References

Bally, J. & Zinnecker, H. 2005, AJ, 129, 2281
Bate, M. R. & Bonnell, I. A. 2005, MNRAS, 356, 1201
Bate, M. R., Bonnell, I. A., & Bromm, V. 2002a, MNRAS, 332, L65
—. 2002b, MNRAS, 336, 705
—. 2003, MNRAS, 339, 577
Bertoldi, F. & McKee, C. F. 1992, ApJ, 395, 140
Beuther, H., Schilke, P., & Gueth, F. 2004, ApJ, 608, 330
Beuther, H. & Shepherd, D. 2005, in Cores to Clusters, ed. M. S. N. Kumar, M. Tafalla,
        & P. Caselli (Berlin: Springer), 105
Beuther, H., Sridharan, T. K., & Saito, M. 2005, ApJ, 634, L185
Bonnell, I. A. & Bate, M. R. 2005, MNRAS, 362, 915
Bonnell, I. A., Bate, M. R., Clarke, C. J., & Pringle, J. E. 2001a, MNRAS, 323, 785
Bonnell, I. A., Bate, M. R., & Vine, S. G. 2003, MNRAS, 343, 413
Bonnell, I. A., Bate, M. R., & Zinnecker, H. 1998, MNRAS, 298, 93
Bonnell, I. A., Clarke, C. J., Bate, M. R., & Pringle, J. E. 2001b, MNRAS, 324, 573
Bonnell, I. A., Vine, S. G., & Bate, M. R. 2004, MNRAS, 349, 735
Bourke, T. L., Myers, P. C., Robinson, G., & Hyland, A. R. 2001, ApJ, 554, 916
Chakrabarti, S. C., & McKee, C. F. 2006, in preparation
Crutcher, R. M. 2006, in IAU 227: Massive Star Birth: A Crossroads of Astrophysics, ed.
        R. Cesaroni, E. Churchwell, M. Felli, & C. M. Walmsley (Cambridge University
        Press), in press
Dale, J. E., Bonnell, I. A., Clarke, C. J., & Bate, M. R. 2005, MNRAS, 358, 291
Davis, C. J., Varricatt, W. P., Todd, S. P., & Ramsay Howat, S. 2004, A&A, 425, 981
De Buizer, J. M. 2003, MNRAS, 341, 277
De Buizer, J. M. & Minier, V. 2005, ApJ, 628, L151
De Buizer, J. M., Radomski, J. T., Telesco, C. M., & Piña, R. K. 2006, in IAU 227:
        Massive Star Birth: A Crossroads of Astrophysics, ed. R. Cesaroni, E. Church-
        well, M. Felli, & C. M. Walmsley (Cambridge University Press), in press (astro-
        ph/0506156)
Dobbs, C. L., Bonnell, I. A., & Clark, P. C. 2005, MNRAS, 360, 2
Edgar, R. & Clarke, C. 2004, MNRAS, 349, 678
Elmegreen, B. G. 2000, ApJ, 539, 342
Fiege, J. D. & Pudritz, R. E. 2000, MNRAS, 311, 85
Figer, D. F. 2005, Nat, 434, 192
Greenhill, L. J., Reid, M. J., Chandler, C. J., Diamond, P. J., & Elitzur, M. 2004, in IAU
        Symposium 221: Star Formation at High Angular Resolution, ed. M. Burton, R.
        Jayawardhana & T. Bourke (San Fransisco: ASP), 155
Hillenbrand, L. A. & Hartmann, L. W. 1998, ApJ, 492, 540
Hofner, P. & Churchwell, E. 1996, A&AS, 120, 283
Humphreys, R. M. & Davidson, K. 1979, ApJ, 232, 409
—. 1994, PASP, 106, 1025
Jiang, Z., Tamura, M., Fukagawa, M., Hough, J., Lucas, P., Suto, H., Ishii, M., & Yang,
        J. 2005, Nat, 437, 112

Jijina, J. & Adams, F. C. 1996, ApJ, 462, 874
Johnstone, D., Fich, M., Mitchell, G. F., & Moriarty-Schieven, G. 2001, ApJ, 559, 307
Jones, C. E., Basu, S., & Dubinski, J. 2001, ApJ, 551, 387
Kahn, F. D. 1974, A&A, 37, 149
Keto, E. 2002, ApJ, 580, 980
—. 2003, ApJ, 599, 1196
Krumholz, M. R., Klein, R. I., & McKee, C. F. 2006a, in IAU 227: Massive Star Birth: A Crossroads of Astrophysics, ed. R. Cesaroni, E. Churchwell, M. Felli, & C. M. Walmsley (Cambridge University Press), in press (astro-ph/0510432)
Krumholz, M. R., McKee, C. F., & Klein, R. I. 2005a, ApJ, 618, 757
—. 2005b, ApJ, 618, L33
—. 2005c, Nat, 438, 332
—. 2006b, ApJ, 638, 369
Kurtz, S., Churchwell, E., & Wood, D. O. S. 1994, ApJS, 91, 659
Lada, C. J. & Lada, E. A. 2003, ARA&A, 41, 57
Larson, R. B. & Starrfield, S. 1971, A&A, 13, 190
Li, P. S., Norman, M. L., Mac Low, M., & Heitsch, F. 2004, ApJ, 605, 800
Massey, P. 1998, in ASP Conf. Ser. 142: The Stellar Initial Mass Function (38th Herstmonceux Conference), ed. G. Gilmore & D. Howell, 17
McKee, C. F. & Tan, J. C. 2003, ApJ, 585, 850
Motte, F., Andre, P., & Neri, R. 1998, A&A, 336, 150
Nakano, T. 1989, ApJ, 345, 464
Nakano, T., Hasegawa, T., & Norman, C. 1995, ApJ, 450, 183
Norris, R. P., Whiteoak, J. B., Caswell, J. L., Wieringa, M. H., & Gough, R. G. 1993, ApJ, 412, 222
Oey, M. S. & Clarke, C. J. 2005, ApJ, 620, L43
Padoan, P. & Nordlund, Å. 2002, ApJ, 576, 870
Patel, N. A., et al. 2005, Nat, 437, 109
Pinsonneault, M. H. & Stanek, K. Z. 2006, ApJ, submitted (astro-ph/0511193)
Plume, R., Jaffe, D. T., Evans, N. J., Martin-Pintado, J., & Gomez-Gonzalez, J. 1997, ApJ, 476, 730
Rathborne, J. M., Jackson, J. M., Chambers, E. T., Simon, R., Shipman, R., & Frieswijk, W. 2005, ApJ, 630, L181
Reid, M. A. & Wilson, C. D. 2005, ApJ, 625, 891
Richer, J. S., Shepherd, D. S., Cabrit, S., Bachiller, R., & Churchwell, E. 2000, in Protostars and Planets IV, ed. V. Mannings, A. P. Boss, & S. S. Russell (Tucson: University of Arizona Press), 867
Shirley, Y. L., Evans, N. J., Young, K. E., Knez, C., & Jaffe, D. 2003, ApJS, 149, 375
Shu, F. H., Adams, F. C., & Lizano, S. 1987, ARA&A, 25, 23
Sollins, P. K., Zhang, Q., Keto, E., & Ho, P. T. P. 2005, ApJ, 624, L49
Tafalla, M., Myers, P. C., Caselli, P., Walmsley, C., & Comito, C. 2002, ApJ, 569, 815
Tan, J. C. & McKee, C. F. 2003, in IAU Symposium 221: Star Formation at High Angular Resolution, ed. M. Burton, R. Jayawardhana & T. Bourke (San Fransisco: ASP), 274
Testi, L. & Sargent, A. I. 1998, ApJ, 508, L91
Torrelles, J. M., et al. 2001, Nat, 411, 277
Walmsley, M. 1995, in Rev. Mexicana Astron. Astrofis. Conf., 1, 137
Weidner, C. & Kroupa, P. 2004, MNRAS, 348, 187
Wolfire, M. G. & Cassinelli, J. P. 1987, ApJ, 319, 850
Wood, D. O. S. & Churchwell, E. 1989, ApJS, 69, 831
Xie, T., Mundy, L. G., Vogel, S. N., & Hofner, P. 1996, ApJ, 473, L131
Yorke, H. W. & Kruegel, E. 1977, A&A, 54, 183
Yorke, H. W. & Sonnhalter, C. 2002, ApJ, 569, 846

Jackie Kessler-Silacci and Shardha Jogee share a laugh.

Frank N. Bash Symposium 2005: New Horizons in Astronomy
ASP Conference Series, Vol. 352, 2006
S. J. Kannappan, S. Redfield, J. E. Kessler-Silacci, M. Landriau, and N. Drory

Probing Chemistry During Star and Planet Formation

Jacqueline E. Kessler-Silacci[1]

Department of Astronomy, University of Texas, Austin, TX, USA

Abstract. The overlap between the fields of chemistry and astronomy is continually expanding. Molecular-line emission has long been used as a tracer of the density and temperature of star-forming environments. More recently, observations tracing the evolution of material in young stellar objects (YSOs) have begun to provide valuable insights into the development of complex molecules and ultimately, the origin of life. Observations of the composition of ice and dust in disks additionally suggest that chemistry plays a major role in determining the type of planetary system, if any, that can be formed around a given star. Thus, chemical models have now been constructed to interpret observations of solid and gaseous molecules in each stage of star formation, from molecular clouds to circumstellar disks. In this summary, I will outline the current understanding of the chemistry in these objects and the promise of future instrumentation and computation.

1. Introduction

The formation and ubiquity of life on Earth is a long-standing question. The presence of numerous organic molecules in meteoritic samples and solar system bodies, has led to the theory that life, or prebiotic matter, was brought to Earth by comets or meteorites. Observations of a wealth of fairly complex organic molecules in the interstellar medium (ISM), in particular in star-forming regions, indicates the possibility that complex prebiotic molecules may be abundant and can be formed in various stages of the star-formation process.

An amazing variety of organics are found in carbonaceous chondrite meteorites, including a wide variety of aliphatic and aromatic hydrocarbons, carboxylic acids, amino acids, purines, pyrimidines, and sugars (see summary in Ehrenfreund & Charnley 2000). The presence of such complex prebiotic molecules has led some to suggest that the life on Earth was seeded by meteorites (Chyba & Sagan 1992), but it is not clear whether the meteoritic organics were produced in the ISM or later, in the solar nebula. The isotopic composition of presolar graphite grains, diamonds, SiC grains, and GEMS (Glasses with Embedded Metal and Sulfide) isolated from meteorites show evidence of interstellar origins (Ott 1993; Bradley et al. 1999). Furthermore, the large deuterium abundances in chondritic material are quite similar to material formed in the ISM (Irvine 1998). Additionally, racemic (even handed) mixtures of amino acids are found in most meteorites, in contrast to the primarily "left-handed" amino acids found on Earth. The structural diversity of chondrites further indicates

---

[1]Spitzer Fellow.

that they represent a highly processed sample of interstellar matter. Detailed comparisons of meteoritic organics with those in the ISM and near young stellar objects (YSOs) are required to discern the formation mechanism of individual prebiotic molecules.

Organic material may have also been brought to the early Earth via comets. Comparisons of the D/H fractionation in comets with ocean water indicate that comets may have delivered $\leq 10\%$ of the Earth's water (Morbidelli et al. 2000). Simple prebiotic molecules (and possibly amino acids) may have also been deposited. Comets are composed of organic "CHON" particles, silicate dust, and volatile ices (see summary by Despois et al. 2002). The volatiles are mostly water (70%), carbon monoxide (CO, 2–20%), and ($\sim$5%) carbon dioxide (CO_2) and methanol (CH_3OH). Less than 1% of cometary volatiles are attributed to other molecules, including organics, nitriles, and alkynes (Bockelée-Morvan et al. 2000). The origin of cometary organics is difficult to determine. Some molecules observed in cometary comae are secondary products of degradations of larger species or gas-phase reactions in the coma (e.g., CS, CN, HCN), while others (e.g., H_2O) may simply be evaporated directly from the nucleus (Biver et al. 2002; Charnley et al. 2002). The organics (HCOOH, $HCOOCH_3$, HC_3N, CH_3CN) observed in cometary comae are similar to those in hot cores (see discussion in Charnley et al. 2002), and models of gas-phase chemistry in comet comae (Rodgers & Charnley 2001) cannot produce the large abundances of organics observed. Thus, at least some of the material observed in comet comae may be directly evaporated from ice in the nuclei, which are remnants from earlier phases of star formation. Cometary organic material is thus linked to the chemistry in the ISM and star-forming regions.

Observations indicate that a large degree of chemical complexity can indeed be reached in the early phases of star formation. More than 135 molecules have currently been detected in the gas-phase interstellar and circumstellar medium (see Ehrenfreund & Charnley 2000, for summary). These molecules vary in complexity from 2 to 13 atoms, including aliphatic and aromatic carbon chains and rings. In diffuse clouds, ($T \approx 100$ K, $n = 100$–200 cm^{-3}), simple molecules (HCO^+, CO, OH, C_2, HCN, CN, CS, H_2CO, C_2H, C_2H_2) are abundant and $\sim$1% of the mass is in solid dust grains, including carbonaceous and silicon-based grains and polycyclic aromatic hydrocarbons (PAH). In cold, dense clouds ($T = 10$–30 K, $n = 10^{4-8}$ cm^{-3}), cold gas chemistry leads to the additional production of larger C-chains and grain-surface reactions lead to production of CO_2 and CH_3OH. Protostellar sources greatly influence chemistry in these molecular clouds, via thermal, and likely ultraviolet (UV) processing of interstellar ice and gas. Thus, a complete understanding of the formation of organic, and possibly prebiotic, molecules in our solar system requires astronomical observations of such objects, as well as theoretical models and laboratory experiments to aid in the interpretation of these observations. This review will discuss theories for the production of organics involving grain-mantle and gas-phase reactions (§2), observational evidence for chemical complexity in interstellar ices (§3) and the current state of observations and models of gas-phase chemistry in pre- and protostellar cores (§4) and protoplanetary disks (§5).

2.   The Formation of Organic Molecules

At the low temperatures of dense molecular clouds, gaseous species (atoms and molecules) rapidly accrete onto grain surfaces at roughly the collision rate (Tielens & Allamandola 1987) forming ice mantles that are ~ 1 μm thick. These ices contain C,N,O,H atoms, H_2 and various radicals, including OH, CH_2OH, and NH_2. Laboratory experiments indicate that C,N,O atoms and H_2 can quickly

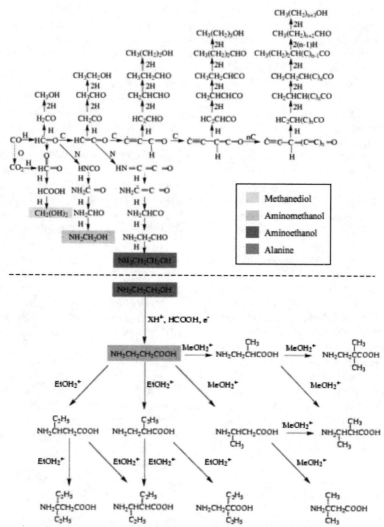

Figure 1.   Interstellar organic chemistry, adapted from Charnley et al. (2002) and Widicus Weaver (2005). Potential grain surface H,C,N,O-atom addition reactions forming a range of alcohols including, aminoethanol, aminomethanol and methanediol are shown above the dashed line. Below the dashed line, subsequent gas-phase production of prebiotic molecules from grain mantle evaporates (in this case aminoethanol) are shown.

hop across grain surfaces and H and D atoms tunnel through barriers, scanning the entire grain surface in $\sim 10^{-7}$ s. Thus, hydrogenation is efficient on these cold grains and models suggest that some degree of chemical complexity can be achieved through simple, single-atom addition reactions (Figure 1, review by Charnley et al. 2002). The evaporation of ices containing ammonia, methanol, ethanol (and higher alcohols) can then lead to ethers and N-bearing organics, as seen in star-forming regions, through alkyl cation transfer reactions. Thus, dust grains are vital in prebiotic chemistry, as they serve as chemical catalysts.

The effectiveness of this catalysis depends on several factors, including: 1) condensation rates, 2) atomic and molecular grain-surface mobility, 3) photolysis and ion-irradiation rates, and 4) desorption rates. Thus, laboratory experiments are required to place further constraints on grain-mantle chemistry. Addition reactions, in particular, are highly dependent on mobility. As H atoms are the most mobile (mobility: $H \gg O,C > NH_2 > CO > CH_3 > OH \gg CH_2OH$, Tielens & Hagen 1982), H-atom addition is common and molecules produced on grain surfaces are likely to be fully hydrogenated. Carbon and oxygen addition reactions can be expected to be frequent as well, leading to long carbon chains and molecules with $C=O$ functional groups (e.g., carboxylic acids, alcohols, ethers). Because NH_2 is the most mobile molecule, nitrogenated organics should also be abundant. Furthermore, icy molecules should be enhanced in deuterium due to longer residence of D atoms on grain surfaces (Lipshtat, Biham, & Herbst 2004).

Depending on the local environment, the chemistry of ices may be further enriched by energetic processing, such as ultraviolet irradiation and cosmic rays. Laboratory UV photolysis of "astronomical" methanol containing ices produces a variety of complex molecules, including alcohols, nitriles and isonitriles, hexamethylenetetramine (HMT), polyoxymethylene (POM), amides and ketones, on grain surfaces (Allamandola et al. 1999; Bernstein et al. 1995). Of particular interest is the recent production of glycine and other amino acids via laboratory UV photolysis of interstellar ice analogs (Bernstein et al. 2002; Muñoz Caro et al. 2002). Laboratory experiments have successfully used irradiation to produce radicals, complex molecules and organic refractory material (Wu et al. 2002; Moore & Hudson 1998; Greenberg et al. 2000), and such studies are continuing to improve our understanding of the formation mechanisms in more detail (e.g., Watanabe et al. 2004; Moore, Hudson, & Ferrante 2003). The thermal modification and sublimation of grain-mantle ices also largely effects the resulting gas-phase chemistry and there is much laboratory work being done now to better characterize heating and desorption mechanisms of isolated and layered ice species (e.g., Öberg et al. 2005; Viti et al. 2004; Collings et al. 2003).

3. Observations of Interstellar and Circumstellar Ices

Our current inventory of grain-mantle species and understanding of the processing of these ices was established through a combined effort of laboratory simulations and infrared (IR) observations of ice compositions in molecular clouds and near YSOs. For astronomically relevant materials, the most important solid state bands lie between 2 and 50 μm. Studies of interstellar ices were thus initially hindered by atmospheric absorption lines in the same regions. The Infrared Observatory (*ISO*, Kessler et al. 1996), launched in 1996, greatly advanced the study of interstellar ices through 2.5–25 μm observations of a large suite of ices in the ISM and near massive protostars. The study of ices in regions with high extinction and near low-mass protostars became increasingly feasible in recent years with the advent the Spitzer Space Telescope (*Spitzer*, Werner et al. 2004). Additionally, CO, H_2O and NH_4^+ ices have been detected in a single edge-on disk (CRBR 2422-3423, Pontoppidan et al. 2005) for the first time with *Spitzer*, but such studies require specific conditions (inclination, clear line-of-sight, etc.) that may limit the number of disks that can be studied in this manner. The major components of ices in molecular clouds, young high-mass and low-mass stellar objects and comets are shown in Table 1.

Table 1.   Abundances of ices relative to H_2O along lines of sight toward the background stars Elias 16 and CK 2, the low-mass stars Elias 29 and HH 46, the high-mass stars NGC 7538 IRS 9 and W33A and comets (Gibb et al. 2000b; Boogert et al. 2000; Knez et al. 2006).

Species	Elias 29	CK 2	N7538	W33A	HH 46	Elias 16	Comets
H_2O	100	100	100	100	100	100	100
CO	26	36–57	16	8	25	30	5–30
CO_2	18–24	33	22	13	18	32	3
CH_3OH	<2.3	<2.1	5	18	<3	7.0	0.3–5
CH_4	<3	<3	2	1.5	...	4	1
H_2CO	<0.02	...	4	6	...	...	0.2–1
HCOOH	<0.01	1.9	3	7	...	<8.7	0.05
OCS	<0.001	...	...	0.2	<0.2	...	0.5
NH_3	≤ 8	≤ 8	13	15	≤ 9	17	0.1–1.8
OCN^-	<2.3	...	1	2	<2	≤ 0.7	0.01–0.4

Observations of ices in ambient regions of molecular clouds allow us to obtain an inventory of the ice composition prior to star-formation. Ice features are observed in absorption against the stellar emission of stars located behind the cloud. Results from *ISO* (Gerakines et al. 1999) and *Spitzer* (Bergin et al. 2005b; Knez et al. 2006) indicate that interstellar ice contains large abundances of CO_2, and that the CO_2 absorption shows little evidence of energetic processing. The total ice abundances, indicated by water ice, appear to increase with increasing visual extinction (A_V), but are also dependent on the local temperature and density of the line-of-sight being probed. Perhaps most surprising is the presence of absorption features near 5.8 and 6.8 μm, which may be due to formic acid (HCOOH) and ammonium ions (NH_4^+), as they indicate the importance of acid-base chemistry even in remote areas of molecular clouds.

Inventories of ices near embedded high-mass YSOs were also obtained by *ISO* (see reviews by Boogert & Ehrenfreund 2004; Gibb et al. 2004), and have been extended to low-mass and sub-stellar YSOs with *Spitzer* (Noriega-Crespo et al. 2004; Boogert et al. 2004), as shown in Figure 2. Complementary 2–5 μm ground-based observations probe the H_2O and CO absorption features toward both high- and low-mass protostars. These observations identify the major organic species as H_2O, CO, CO_2 and CH_3OH, while minor species (such as CH_4, OCS, H_2CO, HCOOH, and OCN^-) comprise only ~1% of the water ice

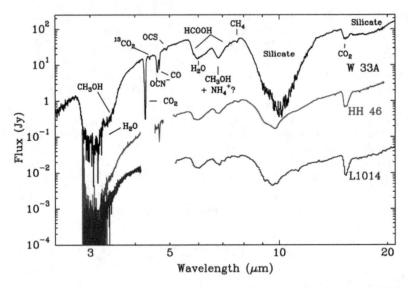

Figure 2.   IR spectra of ice absorption features toward a high-mass YSO (W33A, Gibb et al. 2000b), low-mass YSO (HH 46, Boogert et al. 2004) and sub-stellar mass YSO (L1014, Boogert et al., in preparation). The spectrum of W33A was observed with *ISO* and the spectra of HH 46 and L1014 are a combination of *Spitzer* (5–25 μm) and ground-based (2–5 μm) observations.

(Whittet et al. 1996; Dartois et al. 1999; Gibb et al. 2000b). The abundances in low-mass protostars are in general similar to, within a factor of 2, those of high-mass cores (e.g., for CO_2). However, some molecules (e.g., CH_3OH, NH_3, OCN^-) show significant variations, differing by more than a factor of 10 in some cases (Boogert et al. 2004; van Broekhuizen et al. 2005). *ISO* identified systematic trends in band profiles with increasing temperature, such as changes in band profiles of CO_2 (2 and 15 μm) and NH_4^+ (6.8 μm) and variations in gas/solid ratios (see, e.g., Gerakines et al. 1999; Keane et al. 2001). *Spitzer* and ground-based observations find similar trends for low-mass YSOs (Pontoppidan et al. 2003; Boogert et al. 2004, Pontoppidan et al., in preparation). Thermal processing experiments in laboratories can tie these trends to the modifications of the structure of the grain-mantle ices and interactions between polar and apolar grain ices (Ehrenfreund et al. 1998). New laboratory databases of pure, mixed and layered CO-CO_2 ices (e.g., van Broekhuizen et al. 2006) will play a vital role in the interpretation of these observations.

There is a strong similarity between the abundances of interstellar ices and those in comets, with some dark clouds showing enhanced abundances of NH_3, H_2S, CH_4, and $HCOOH$ (Turner 1996; Boogert et al. 1998; Knez et al. 2006). Observations of gas in hot-cores (Yanti & Snyder 1997), indirect probes of the ices in warm star forming regions, tend to show larger abundances of more complex, saturated molecules, such as $(CH_3)_2CO$, CH_3COOH, C_2H_5CN, and CH_3OH, suggesting the efficiency of gas-grain interactions and grain-surface reactions (Keane & Tielens 2003). Studies of ices are thus complementary to observations of the gas-phase composition near YSOs and protoplanetary disks.

4. Chemical Evolution in YSOs

Pre-protostellar cores are regions of extremely dense gas and dust within molecular clouds that are potential sites for star formation. Several clumps or globules in molecular clouds that were previously identified as regions of high extinction in optical surveys, can now be characterized as unstable or collapsing cores through interpretation of near-IR photometry and molecular line observations (see discussion in Kandori et al. 2005). The characteristic densities and time scales for freeze-out in these cores can be constrained from physical models of dust continuum data and observations of CO isotopes (Shirley, Evans, & Rawlings 2002; Young et al. 2003; Jørgensen et al. 2002). The increased sensitivity of *Spitzer* in the IR has enabled more complete spectral energy distributions (SEDs) to be obtained (e.g., Young et al. 2004; Jørgensen et al. 2005b), improving physical models of younger and lower-mass cores.

Chemical models of cores are also becoming more robust. Millimeter, submillimeter, and infrared emission lines from a given molecule trace different layers of the core via both chemistry and excitation (see Figure 3). In a (very) simplified model, the chemical composition of these cores can be described in terms of 3 layers of increasing density or depth into the core (Bergin, Maret, & van der Tak 2005a; Jørgensen et al. 2002). The outer, less extincted, region of the core can be penetrated by the ambient radiation field, creating a photodissociation dominated layer, similar to classical photon-dominated regions (PDRs). Additionally, the time scale for freeze-out is long in the outer regions of the core, and molecules like CN, CS and CO (and its isotopes) are good probes of this region. In the center of the core ($A_V \geq 10$ mag), the high densities and low temperatures lead to heavy freeze-out of most molecules (CO in particular). These regions are characterized by enhanced abundances of deuterated species (e.g., N_2D^+, DCO^+, D_2CO), as reaction with CO is the primary destruction mechanism for HD and deuteration reactions proceed more efficiently at low temperature. Models combining the chemistry and dynamical collapse of these cores are consistent with the basic 3 layer models described above and indicate that chemistry in this early phase is dominated by freeze-out onto grain surfaces (Lee, Bergin, & Evans 2004; Lee, Evans, & Bergin 2005).

The chemistry becomes even more complex in cores containing protostars. The protostar heats the surrounding core and leads to evaporation of ices in the inner regions. The gas-phase chemistry in the inner part of the core is thus highly dependent on the composition of the grain-mantle ice. Such hot, dense cores ($T > 100$ K, $n > 10^6$ cm^{-3}) were previously observed around massive stars (Gibb et al. 2000a,b; Sutton et al. 1993) and could be identified by several emission lines from complex organics (CH_3CN, CH_3OCH_3, $HCOOCH_3$, etc). The large deuterium-to-hydrogen ratios and saturation of these molecules indicate that they are second-generation products of gas-phase reactions with evaporated molecules (Charnley, Tielens, & Millar 1992, §2). These hot cores also show large abundances of other deuterated species and often multiply deuterated molecules (e.g., D_2CO and CH_2DO), reflecting their icy origins (Jacq et al.

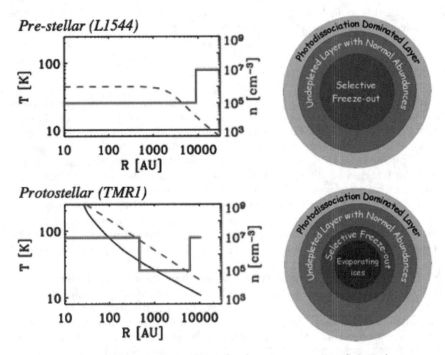

Figure 3.   Evolution of pre-stellar (top) and protostellar (bottom) cores, adapted from Bergin et al. (2005a) and Jørgensen et al. (2002). The left panel shows the temperature (thin black lines) and density (dashed grey lines) as a function of radius for the typical low-mass pre-stellar (L1544) and protostellar (TMR1) objects. The thick grey lines indicate the derived abundance structures. The right panel shows the predicted chemical layers of the pre-stellar and protostellar cores.

1993; Loinard et al. 2001; Charnley, Tielens, & Rodgers 1997). Observations of isomers (e.g., acetic acid, methyl formate and glycol aldehyde) can trace formation mechanisms, and suggest that both grain-mantle formation and gas-phase methyl cation transfer are occurring (Hollis et al. 2000; Hollis et al. 2004).

More recently, organics and other complex molecules have been observed in the gas-phase toward a few low-mass cores, labeled "hot corinos" (IRAS 16293-2422, Cazaux et al. 2003; NGC 1333 IRAS4A, Bottinelli et al. 2004). The origin of organics in low-mass protostellar cores is even more uncertain than in hot cores because the transit time for gas in the low-mass hot core region is too short ($\sim$200 yr, Schöier et al. 2002) for "standard" gas-phase hot core chemistry (10^4–10^5 yr; Charnley et al. 1992). Additionally, high-resolution submillimeter interferometric observations of a few low-mass cores indicate complex physical environments (even in the "pre-protostellar" stage), including the presence of shocks associated with accretion (L1157; Velusamy, Langer, & Goldsmith 2002) and outflows (IRAS 16293-2422; Chandler et al. 2005). These observations, and new models of IRAS 16293-2422 (Jørgensen et al. 2005b), which suggest that only a small fraction of the core is actually hot ($T > 100$ K), indicate that shocks may be the dominant mode of releasing molecular ices into the gas phase in low-mass corinos, rather than thermal desorption of ambient material as in hot cores. Furthermore, in at least one low-mass protostellar core, some of the molecules detected may reside in an accretion disk (NGC1333-IRS2A; Jørgensen et al. 2005a). The interpretation that low-mass cores are merely scaled-down versions of hot cores is thus very uncertain.

The physical complexity of actively collapsing protostellar cores, containing disks and/or outflows, makes chemical modeling extremely difficult. Outflows likely play an important role in processing and releasing ices through grain sputtering. Irradiation by stellar UV (200–300 AU) and X-rays (>1000 AU) also play a role: increasing abundances of ions, radicals, and hydrides and reducing abundances of CO_2, H_2O (Stäuber et al. 2004, 2005). Several different "chemical clocks" can be used to trace the influence of these processes on the chemistry (e.g., Fuente et al. 2005). For example, the ratio of $SiO/C^{34}S$ is useful in distinguishing between sources with powerful outflows (traced by sputtering of silicates to form SiO) and sources with warm material (traced by $C^{34}S$). Observations of deuterated molecules (DCO^+/HCO^+, D_2CO/DCO^+) are good probes of the ice-evaporation zone. Finally, CN and HCN are useful as photochemical tracers and CN/N_2H^+ and HCN/N_2H^+ were found to be good indicators of envelope dissipation, rising by factors of ~6 for sources with thinner outer envelopes.

Observations of star-forming cores have greatly improved in the last few years, enabling detailed study the physical and chemical structure of these objects. Further progress requires a good understanding of the physics of the gas-grain interaction, including grain-molecule/atom binding energies, desorption processes and yields, grain-surface chemistry, and gas-phase branching ratios/rates at low temperatures. Additionally, models of hot cores need to incorporate more realistic physical structures. The temperature and power-law density profiles most often used are reasonable down to ~100 AU from the star, but on smaller spatial scales departures from spherical symmetry, outflows and disks become important. Furthermore, the combination of chemical networks with dynamical models of core collapse and disk formation are necessary for the study of the chemical evolution of these systems (Lee et al. 2005). The advent of new IR telescopes (*Herschel* and SOFIA) will enable observations of several gas-phase water lines and UV/X-ray enhanced hydrides, with improvements of orders of magnitude in sensitivity, spatial and spectral resolution. The Atacama Large Millimeter Array (ALMA), will provide imaging sensitivity down to spatial scales of tens of AU, enabling the measurement of chemical gradients in the inner envelope directly, as well as the relationship with circumstellar disk and outflow material, and size and geometry of the region affected by UV/X-rays.

5. Chemical Evolution in Protoplanetary Disks

A significant fraction of 1–5 Myr-old low-mass T Tauri stars and intermediate-mass Herbig Ae stars have been shown to possess remnant accretion disks with masses of $\sim 10^{-3}$–$10^{-1} M_\odot$ and sizes of hundreds of AU, comparable to those inferred for the primitive solar nebula (Beckwith & Sargent 1996). It is out of this material that planets, comets and asteroids are formed, and the physical and chemical properties of these disks will largely effect those of the resulting

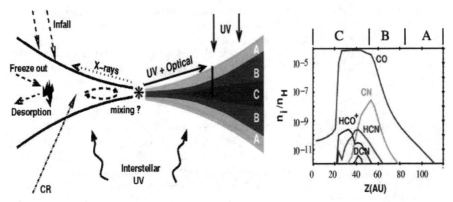

Figure 4.   Protoplanetary disk chemistry, adapted from van Zadelhoff et al. (2003). The left panel shows the relevant dynamical processes and heating/ionization sources. As discussed in the text, the disk can be described as 3 layers, as A) PDR-like layer, B) molecular layer, and C) desorption layer. The corresponding abundances of several molecules as a function of disk height (Z) are shown in the right panel.

system. With this in mind, there has been a great effort to model disk chemistry (Willacy & Langer 2000; Aikawa et al. 2002), thermodynamics (Chiang et al. 2001; D'Alessio et al. 2005) and radiative transfer (Hogerheijde & van der Tak 2000; de Gregorio-Monsalvo, Gómez, & D'Alessio 2004). Millimeter and submillimeter wave interferometers have imaged the gas and dust surrounding over a dozen T Tauri and Herbig Ae stars at $0.''3$–$1.''0$ resolution (e.g., Mannings & Sargent 1997; Duvert et al. 2000; Qi et al. 2004), leading to dramatic improvements in the quantitative understanding of their physical properties.

Protoplanetary disks are complex physical environments (see Figure 4). Circumstellar accretion disks radially transport material inward and angular momentum outward. Radiation from the central star and the liberation of gravitational energy produce strong density and temperature gradients in the disk, which directly effect the chemical composition. Models of disk SEDs (e.g., Chiang & Goldreich 1997) and scattered light images of disks (Burrows et al. 1996) indicate that the outer regions of optically thick disks are flared (disk height increases as a function of radius). Thus dust on the disk surface can absorb radiation from the star (down to the optically thick $\tau \approx 1$ limit) and re-emit in the infrared, forming a warmer optically thin surface layer on top of a cooler optically thick midplane. In the outer, colder regions, abundances of the (C,N,O,S)-bearing gas in the disk midplane should be depleted as the gas is frozen out into an icy mantle on the surfaces of dust grains. There is a large change in chemical composition as matter accretes toward the star; molecular gas abundances increase as the increased temperature causes ice mantle evaporation ($R < 10$ AU), complex molecules form due to dust destruction ($R < 0.3$ AU), and free atoms are formed via complete molecular dissociation ($R < 0.05$ AU).

Due to the differing conditions as a function of radius, chemical models have been developed separately for the inner and outer disk. These models consider variations in disk chemistry as a function of scale height in addition to radial variations. Recent 1+1D models of the outer regions of static flaring disks (Jonkheid et al. 2004; Kamp & Dullemond 2004), show that gas temperatures in the surface layer can become much hotter than the dust, in excess of 100 K out to 100 AU (Chiang et al. 2001). Thus, the chemistry in this layer (van Zadelhoff et al. 2003) should be similar to that of a PDR. Molecules other than CO and H_2, which are self-shielding, are readily destroyed in this region and the ionization fraction, as traced by C^+, can be quite high ($\sim 10^{-4}$). In contrast, gas in the cold disk midplane (at least in the outer disk) should be heavily depleted (particularly CO) due to freeze-out onto grain surfaces and more abundant in cold gas-tracers, such as deuterated molecules (e.g., H_2D^+). Between these two layers lies a warm layer in which most molecules (CO, CN, HCN, HCO^+) remain in the gas phase and photodissociation rates are low. The location of this warm molecular layer is highly dependent on the temperature structure of the disk and the extent of the UV field.

The rotational transitions of several simple gas-phase molecules have been observed with millimeter and submillimeter single-dish and interferometric observatories. Emission from CO isotopes, due to their ubiquity and ease of detection, has long been used to constrain disk masses, sizes and velocities. HCO^+, which is produced from hydrogenation of CO, is also abundant and a good tracer of ionization fraction. CN and HCN are good tracers of photon processes and have been detected in a small sample of disks. Emission lines from deuterated molecules (DCO^+, HDO, H_2D^+) have to date been detected in few disks (van Dishoeck, Thi, & van Zadelhoff 2003; Kessler, Blake, & Qi 2003; Ceccarelli & Dominik 2005), with abundances ranging between those of molecular clouds and hot cores. Low abundances of complex molecules, such as H_2CO, CH_3OH, have also been observed in a few disks (Thi, van Zadelhoff, & van Dishoeck 2004). Due to limited spatial resolution, observations in the millimeter and submillimeter are only sensitive to >50 AU, and so probe only the outer radii of these disks. Additionally, thermal information derived from studying multiple transitions of the same molecule indicate that the lines observed originate in the warm (20–40 K) layer and trace moderate density gas ($n = 10^6$–10^8 cm^{-3}). Correspondingly, masses calculated from disk-averaged CO abundances are "depleted" by a factor of 5–100, when compared with gas masses derived using the dust mass from dust continuum observations and assuming standard gas-to-dust ratios of ~ 100 (Dutrey et al. 1996; Dartois, Dutrey, & Guilloteau 2003; Qi et al. 2003). Complementary tracers are H_2, to probe the slightly deeper into the disk, and C and C^+, to probe the surface photodissociation layer.

Infrared observations of molecular ro-vibrational transitions have the potential to probe the inner disk (<50 AU), but have heretofore been limited to observations of CO and H_2 in more massive disks due to sensitivity constraints (e.g., Bitner et al. and Chen in this volume). Regardless of the lack of observational constraints, chemical models of the inner disk are developing rapidly. Chemical models of inner disks have existed since the 1950's and were based on solar system observations, with differences in planetary composition defining changes in chemistry as a function of disk radius (see summary in Encrenaz 2001). These models were dominated by thermal equilibrium not kinetics. Since then, other important processes have been included: ice evaporation (4–20 AU), dust evaporation (<1 AU), accretion shocks, stellar UV/X-rays, radial/vertical mixing, and ionization by radionuclides (^{26}Al, ^{60}Fe). The most recent (1+1D,

2D) models include mass transport (Ilgner et al. 2004; Gail 2004) or photochemistry (Markwick et al. 2002; Nomura & Inutsuka 2004), but typically not both. However, mass transport models show that vertical and radial mixing can have large effects on chemical abundances and should be included. These studies are currently limited by a lack of observational constraints.

The observations and models discussed above apply primarily to young, optically thick disks. As disks evolve, and gas is accreted or dispersed and dust settles toward the disk midplane, gas-phase chemistry will be dominated by photochemistry and ice chemistry will be dominated by processes linked to equilibration of planetesimals (e.g., aqueous alteration). The chemistry in these older disks will be largely dependent on the method of planet formation (core accretion or gravitational instability, see Ford, this volume). If planets form by core accretion, the increased temperatures and pressures of accretion flows in the planet forming regions of these disks ($\sim$0.5–30 AU), where densities are highest, will transform and ultimately vaporize previously accreted ices, and the resulting material may later be recondensed in planetary atmospheres. Material in the outer disk may remain unaffected by these processes and later form cometary bodies with ice compositions similar to the ISM. In the near future, improvements in sensitivity and spectral coverage of millimeter arrays, such as CARMA and the SMA, will enable detections of more diverse chemical species, tracing the evolution of organic material in disks. Later, observations of protoplanetary disks with ALMA will resolve disks on planet-formation scales, providing valuable information (density, thermal history, and composition) about initial conditions of solar systems like our own.

6.   Summary

Interstellar dust, gas, and ice are the building blocks of stars, planets, and comets. Modification of this material during the star formation process is believed to result in the chemical complexity of grain-surface ices, and desorption of these ices in turn enhances the complexity of gas-phase chemistry. Interstellar materials are precursors to those in circumstellar disks in which planets and comets are formed. The material in the outer radii of these disks is likely unprocessed and similar to its interstellar origin, while material near the star has been significantly modified by star formation. Thus observations tracing the evolution of material in YSOs are essential for addressing such issues as the formation of comets and the origin of life.

Complex organics have been observed in hot cores, comets, and meteorites, but it is still not clear whether the organic molecules detected in comets are remnants of earlier phases of star-formation preserved in the ice, or chemically young molecules formed in the solar nebula. Only very simple organics (CH_3OH, H_2O) have been detected in protoplanetary disks so far, due to weak rotational transitions, but infrared observations probing warmer gas should be more fruitful. Observations indicate that chemistry can be used as tracer of evolution in cores, with chemical clocks discerning between the dense cold material, hot core, and outflow activity. Tracing chemistry in disks has been limited by both spatial and spectral resolution in the (sub)millimeter and by sensitivity in the infrared. Observations of ice composition in disks is a new field and recent IR observations of ices in a few edge-on disks are promising.

Chemical models of cores and disks are becoming more sophisticated. Models are dependent on knowledge of the density and temperature structure, deduced by photometry and imaging of dust, spectroscopic observations of the UV radiation field, X-ray detections, and observations of mass tracers (CO, H_2, C^+). Progress also requires an improved understanding of the physics of gas-grain interactions via laboratory determinations of binding energies, photodesorption yields, grain-surface chemistry, and gas-phase branching ratios/rates at low temperatures. Although some fully integrated chemical and dynamical models exist for embedded cores, they have not yet been developed for disks.

Observationally, the future holds great promise with the development of ALMA in the southern hemisphere and CARMA and SMA in the north. These (sub)mm observatories offer increased wavelength coverage, collecting area, and spatial resolution and will enable us to resolve disks and cores on planetary scales. In the infrared, we are just starting to see the benefits of improved spectral resolution (with spectrographs such as TEXES) for studying gas-phase emission lines and sensitivity (*Spitzer*) for studying gas, ice and dust features. In the near future, SOFIA and *Herschel* will be sensitive to emission from water and hydrides, and the torsional modes of organic molecules.

Acknowledgments. Support for JEK-S was provided by the Spitzer Space Telescope under awards #1256316 and #1224608. The author also thanks Edwin Bergin, Ewine van Dishoeck, Neal Evans, Jes Jørgensen, Claudia Knez, and Susanna Widicus Weaver for help in preparing this submission.

References

Aikawa, Y., van Zadelhoff, G. J., van Dishoeck, E., & Herbst, E. 2002, A&A, 386, 622
Allamandola, L., Bernstein, M., Sandford, S., & Walker, R. 1999, Space Science Reviews, 90, 219
Beckwith, S. V. W. & Sargent, A. I. 1996, Nature, 383, 139
Bergin, E. A., Maret, S., & van der Tak, F. 2005a, in The Dusty and Molecular Universe: A Prelude to Herschel and ALMA, ed. A. Wilson (Noordwijk: ESA Publications Devision), 185
Bergin, E. A., Melnick, G. J., Gerakines, P. A., Neufeld, D. A., & Whittet, D. C. B. 2005b, ApJ, 627, L33
Bernstein, M., Sandford, S., Allamandola, L., & Chang, S. 1995, ApJ, 454, 327
Bernstein, M. P., Dworkin, J. P., Sandford, S. A., Cooper, G. W., & Allamandola, L. J. 2002, Nature, 416, 401
Biver, N., et al. 2002, Earth Moon and Planets, 90, 323
Bockelée-Morvan, D., et al. 2000, A&A, 353, 1101
Boogert, A. C. A. & Ehrenfreund, P. 2004, in ASP Conf. Ser. 309: Astrophysics of Dust, ed. A. Witt, G. Clayton, & B. Draine (Estes Park: ASP), 547
Boogert, A. C. A., Helmich, F. P., van Dishoeck, E. F., Schutte, W. A., Tielens, A. G. G. M., & Whittet, D. C. B. 1998, A&A, 336, 352
Boogert, A. C. A., et al. 2004, ApJS, 154, 359
Boogert, A. C. A., et al. 2000, A&A, 360, 683
Bottinelli, S., et al. 2004, ApJ, 615, 354
Bradley, J. P., et al. 1999, Science, 285, 1716
Burrows, C. J., et al. 1996, ApJ, 473, 437
Cazaux, S., et al. 2003, ApJ, 593, L51
Ceccarelli, C. & Dominik, C. 2005, A&A, 440, 583

Chandler, C. J., Brogan, C. L., Shirley, Y. L., & Loinard, L. 2005, ApJ, 632, 371

Charnley, S., Tielens, A., & Millar, T. 1992, ApJ, 399, 71

Charnley, S., Tielens, A., & Rodgers, S. 1997, ApJ, 482, 203

Charnley, S. B., Rodgers, S. D., Butner, H. M., & Ehrenfreund, P. 2002, Earth Moon and Planets, 90, 349

Chiang, E. I. & Goldreich, P. 1997, ApJ, 490, 368

Chiang, E. I., et al. 2001, ApJ, 547, 1077

Chyba, C. F. & Sagan, C. 1992, Nature, 355, 125

Collings, M. P., Dever, J. W., Fraser, H. J., McCoustra, M. R. S., & Williams, D. A. 2003, ApJ, 583, 1058

D'Alessio, P., Merín, B., Calvet, N., Hartmann, L., & Montesinos, B. 2005, Revista Mexicana de Astronomia y Astrofisica, 41, 61

Dartois, E., Demyk, K., d'Hendecourt, L., & Ehrenfreund, P. 1999, A&A, 351, 1066

Dartois, E., Dutrey, A., & Guilloteau, S. 2003, A&A, 399, 773

de Gregorio-Monsalvo, I., Gómez, J. F., & D'Alessio, P. 2004, Ap&SS, 292, 445

Despois, D., Crovisier, J., Bockelée-Morvan, D., & Biver, N. 2002, in ESA SP-518: Exo-Astrobiology, ed. H. Lacoste (Noordwijk: ESA Publications Devision), 123

Dutrey, A., et al. 1996, A&A, 309, 493

Duvert, G., Guilloteau, S., Ménard, F., Simon, M., & Dutrey, A. 2000, A&A, 355, 165

Ehrenfreund, P. & Charnley, S. B. 2000, ARA&A, 38, 427

Ehrenfreund, P., Dartois, E., Demyk, K., & D'Hendecourt, L. 1998, A&A, 339, L17

Encrenaz, T. 2001, in Lect. Notes Phys. 577, Solar and Extra-Solar Planetary Systems, ed. I. P. Williams, & N. Thomas (Berlin: Springer), 76

Fuente, A., Rizzo, J. R., Caselli, P., Bachiller, R., & Henkel, C. 2005, A&A, 433, 535

Gail, H.-P. 2004, A&A, 413, 571

Gerakines, et al. 1999, ApJ, 522, 357

Gibb, E., Nummelin, A., Irvine, W. M., Whittet, D. C. B., & Bergman, P. 2000a, ApJ, 545, 309

Gibb, E., et al. 2000b, ApJ, 536, 347

Gibb, E. L., Whittet, D. C. B., Boogert, A. C. A., & Tielens, A. G. G. M. 2004, ApJS, 151, 35

Greenberg, J. M., et al. 2000, ApJ, 531, L71

Hogerheijde, M. R. & van der Tak, F. F. S. 2000, A&A, 362, 697

Hollis, J., Lovas, F., & Jewell, P. 2000, ApJ, 540, L107

Hollis, J. M., Jewell, P., Lovas, F., Remijan, A., & Møllendal, H. 2004, ApJ, 610, L21

Ilgner, M., Henning, T., Markwick, A. J., & Millar, T. J. 2004, A&A, 415, 643

Irvine, W. M. 1998, Origins of Life and Evolution of the Biosphere, 28, 365

Jacq, T., Walmsley, C. M., Mauersberger, R., Anderson, T., Herbst, E., & de Lucia, F. C. 1993, A&A, 271, 276

Jonkheid, B., Faas, F. G. A., van Zadelhoff, G.-J., & van Dishoeck, E. F. 2004, A&A, 428, 511

Jørgensen, J. K., Bourke, T. L., Myers, P. C., Schöier, F. L., van Dishoeck, E. F., & Wilner, D. J. 2005a, ApJ, 632, 973

Jørgensen, J. K., et al. 2005b, ApJ, 631, L77

Jørgensen, J. K., Schöier, F. L., & van Dishoeck, E. F. 2002, A&A, 389, 908

Kamp, I. & Dullemond, C. P. 2004, ApJ, 615, 991

Kandori, R., et al. 2005, AJ, 130, 2166

Keane, J. V., Boogert, A. C. A., Tielens, A. G. G. M., Ehrenfreund, P., & Schutte, W. A. 2001, A&A, 375, L43

Keane, J. V. & Tielens, A. G. G. M. 2003, in SFChem 2002: Chemistry as a Diagnostic of Star Formation, ed. C. Curry & M. Fich (Ottawa: NRC Press), 229

Kessler, J. E., Blake, G. A., & Qi, C. 2003, in SFChem 2002: Chemistry as a Diagnostic of Star Formation, ed. C. Curry & M. Fich (Ottawa: NRC Press), 188

Kessler, M. F., et al. 1996, A&A, 315, L27

Knez, C., et al. 2005, ApJ, 635, L145

Lee, J.-E., Bergin, E. A., & Evans, N. J. 2004, ApJ, 617, 360
Lee, J.-E., Evans, N. J., & Bergin, E. A. 2005, ApJ, 631, 351
Lipshtat, A., Biham, O., & Herbst, E. 2004, MNRAS, 348, 1055
Loinard, L., Castets, A., Ceccarelli, C., Caux, E., & Tielens, A. G. G. M. 2001, ApJ, 552, L163
Mannings, V. & Sargent, A. I. 1997, ApJ, 490, 792
Markwick, A. J., Ilgner, M., Millar, T. J., & Henning, T. 2002, A&A, 385, 632
Moore, M. H. & Hudson, R. L. 1998, Icarus, 135, 518
Moore, M. H., Hudson, R. L., & Ferrante, R. F. 2003, Earth Moon and Planets, 92, 291
Morbidelli, A., et al. 2000, Meteoritics and Planetary Science, 35, 1309
Muñoz Caro, G. M., et al. 2002, Nature, 416, 403
Nomura, H. & Inutsuka, S. 2004, in ASP Conf. Ser. 321: Extrasolar Planets: Today and Tomorrow, ed. J.-P. Beaulieu, A. des Etangs, & C. Terquem (Paris: ASP), 335
Noriega-Crespo, A., et al. 2004, ApJS, 154, 352
Öberg, K. I., van Broekhuizen, F., Fraser, H. J., Bisschop, S. E., van Dishoeck, E. F., & Schlemmer, S. 2005, ApJ, 621, L33
Ott, U. 1993, Nature, 364, 25
Pontoppidan, K. M., et al. 2005, ApJ, 622, 463
Pontoppidan, K. M., et al. 2003, A&A, 408, 981
Qi, C., et al. 2004, ApJ, 616, L11
Qi, C., Kessler, J. E., Koerner, D. W., Sargent, A. I., & Blake, G. 2003, ApJ, 597, 986
Rodgers, S. D. & Charnley, S. B. 2001, MNRAS, 320, L61
Schöier, F. L., Jørgensen, J. K., van Dishoeck, E. F., & Blake, G. 2002, A&A, 390, 1001
Shirley, Y. L., Evans, N. J., & Rawlings, J. M. C. 2002, ApJ, 575, 337
Stäuber, P., Doty, S. D., van Dishoeck, E. F., & Benz, A. O. 2005, A&A, 440, 949
Stäuber, P., Doty, S. D., van Dishoeck, E. F., Jørgensen, J. K., & Benz, A. O. 2004, A&A, 425, 577
Sutton, E. C., Peng, R. S., Danchi, W. C., Jaminet, P. A., Sandell, G., & Russell, A. P. G. 1993, Bulletin of the American Astronomical Society, 25, 899
Thi, W.-F., van Zadelhoff, G.-J., & van Dishoeck, E. F. 2004, A&A, 425, 955
Tielens, A. & Allamandola, L. 1987, in Interstellar Processes, ed. D. Hollenbeck & H. Thronson (Dordrecht: D. Reidel Publishing Co.), 397
Tielens, A. G. G. M. & Hagen, W. 1982, A&A, 114, 245
Turner, B. 1996, ApJ, 468, 694
van Broekhuizen, F. A., Pontoppidan, K. M., Fraser, H. J., & van Dishoeck, E. F. 2005, A&A, 441, 249
van Broekhuizen, F. A., Groot, I. M. N., Fraser, H. J., van Dishoeck, E. F., & Schlemmer, S. 2006, A&A, in press (astro-ph/0511815)
van Dishoeck, E. F., Thi, W.-F., & van Zadelhoff, G.-J. 2003, A&A, 400, L1
van Zadelhoff, G.-J., Aikawa, Y., Hogerheijde, M. R., & van Dishoeck, E. F. 2003, A&A, 397, 789
Velusamy, T., Langer, W. D., & Goldsmith, P. F. 2002, ApJ, 565, L43
Viti, S., Collings, M. P., Dever, J. W., McCoustra, M. R. S., & Williams, D. A. 2004, MNRAS, 354, 1141
Watanabe, N., Nagaoka, A., Shiraki, T., & Kouchi, A. 2004, ApJ, 616, 638
Werner, M. W., et al. 2004, ApJS, 154, 1
Whittet, D. C. B., et al. 1996, A&A, 315, L357
Widicus Weaver, S. L. 2005, PhD thesis, California Institute of Technology
Willacy, K. & Langer, W. D. 2000, ApJ, 544, 903
Wu, C. Y., Judge, D. L., Cheng, B.-M., Shih, W.-H., Yih, T.-S., & Ip, W. H. 2002, Icarus, 156, 456
Yanti, M. & Snyder, L. 1997, ApJ, 480, L67
Young, C. H., et al. 2004, ApJS, 154, 396
Young, C. H., Shirley, Y. L., Evans, N. J., & Rawlings, J. M. C. 2003, ApJS, 145, 111

Christine Chen explains a difficult concept to Dan Jaffe.

Frank N. Bash Symposium 2005: New Horizons in Astronomy
ASP Conference Series, Vol. 352, 2006
S. J. Kannappan, S. Redfield, J. E. Kessler-Silacci, M. Landriau, and N. Drory

Dust and Gas Debris around Main Sequence Stars

Christine H. Chen[1]

National Optical Astronomy Observatory, Tucson, AZ, USA

Abstract.   Debris disks are dusty, gas-poor disks around main sequence stars (Backman & Paresce 1993; Lagrange, Backman & Artymowicz 2000; Zuckerman 2001). Micron-sized dust grains are inferred to exist in these systems from measurements of their thermal emission at infrared through millimeter wavelengths. The estimated lifetimes for circumstellar dust grains due to sublimation, radiation and corpuscular stellar wind effects are typically significantly smaller than the estimated ages for the stellar systems, suggesting that the grains are replenished from a reservoir, such as sublimation of comets or collisions between parent bodies. Since the color temperature for the excess emission is typically $T_{gr} \sim 110$–120 K, similar to that expected for small grains in the Kuiper Belt, these objects are believe to be generated by collisions between parent bodies analogous to Kuiper Belt objects in our solar system; however, a handful of systems possess warm dust, with $T_{gr} \geq 300$ K, at temperatures similar to the terrestrial planets. I describe the physical characteristics of debris disks, the processes that remove dust from disks, and the evidence for the presence of planets in debris disks. I also summarize observations of infalling comets toward β Pictoris and measurements of bulk gas in debris disks.

1.   Introduction

The observation of periodic variations in the radial velocities of nearby late-type stars has led to the discovery of ~150 giant planets (Marcy, Cochran, & Mayor 2000), including 14 multiple planet systems, suggesting that planetary systems may exist around >5% of solar-like main sequence stars. Giant planets, like Jupiter, are believed to form in proto-planetary disks made of gas and dust. As grains grow into larger bodies, the bulk of circumstellar gas dissipates from the system (e.g., via accretion onto the star or planetary objects) until the disk contains predominately large bodies, such as planets and/or planetesimals and dusty debris. Studying the dynamics of dust and gas in disks around main sequence stars reveals the presence of planets and small bodies, analogous to asteroids and comets in our solar system. Therefore, studying debris disks may allow us to determine the processes that shape the final architectures of planetary systems including our own solar system.

The lifetimes of dust grains in debris disks are believed to be significantly shorter than the ages of the systems, suggesting that the particles are replenished from a reservoir such as collisions between parent bodies. Our solar system shows evidence for collisions within the main asteroid belt that produce micron-sized dust grains. In 1918, Hirayama discovered concentrations of asteroids in

---

[1]Spitzer Fellow.

three regions of $a - e - i$ (a is the osculatory orbital semi-major axis; e is the eccentricity, and i is the inclination) space which he named the Themis, Eos, and Koronis families. The clumping of these asteroids is widely attributed to the break up of larger parent bodies (Chapman et al. 1989). *Infrared Astronomy Satellite* (*IRAS*) observations of the zodiacal dust discovered the α, β, and γ dust bands which have orbital properties identical to the Themis, Koronis, and Eos families, suggesting that the dust in each band was created by collisions between asteroids in each family. Non-equilibrium processes may be responsible for the generation of dust bands. Gravitational perturbations by Jupiter and other planets in our solar system are expected to cause the apsides and nodes of asteroid orbits to precess at different rates because of small differences in their orbital parameters. This precession leads to asteroid collisions that generate the small grains observed in the dust bands.

Observed debris disks typically possess luminosities 3–5 orders of magnitude higher than our zodiacal disk, suggesting that they are significantly more massive. The dust in these systems may be produced in steady state at earlier times when the planetesimal belt was more massive or during an epoch of intense collisions that produced especially large quantities of dust, triggered by the formation and/or migration of planets. Simulations of planetesimal disks, in which planets are forming from pairwise collisions, suggest that formation of massive planets may trigger collisional cascades between the remaining nearby planetesimals (Kenyon & Bromley 2004). Alternately, the migration of giant planets, soon after their formation, may also trigger collisions. In our solar system, the moon and terrestrial planets experienced an intense period of cratering 3.85 Gyr ago known as the Period of Late Heavy Bombardment. Two pieces of evidence suggest that the impactors during this period may have been asteroids. (1) Lunar impact melts, collected during Apollo missions, suggest that the composition of impactors is similar to asteroids. (2) The size distribution of impactors, inferred from the lunar highlands, is similar to that of the main asteroid belt. The migration of the Jovian planets during the Late Heavy Bombardment may have caused gravitational resonances to sweep through the main asteroid belt sending asteroids into the inner solar system (Strom et al. 2005).

2.   Gross Properties

The debris disks around Vega, Fomalhaut, ϵ Eridani, and β Pictoris were initially discovered from the presence of strong *IRAS* 60 μm and 100 μm fluxes, 10–100 times larger than expected from the photosphere alone (Backman & Paresce 1993). Studies comparing the *IRAS* fluxes with predictions for the photospheric emission of field stars subsequently discovered more than 100 debris disk candidates (e.g., Walker & Wolstencroft 1988; Oudmaijer et al. 1992; Backman & Paresce 1993; Sylvester et al. 1996; Mannings & Barlow 1998; Silverstone 2000). The launch of the *Spitzer Space Telescope* (Werner et al. 2004), with unprecedented sensitivity in the far-infrared, is making the discovery and statistical study of large numbers of debris disks possible. The first exciting results from this mission are determining the mean properties of debris disks and measuring the characteristic timescales for dust and gas decay. *IRAS* and MIPS (Multiband Imaging Photometer for *Spitzer*) surveys of main sequence stars now suggest that ~10–15% of A- through K-type field stars possess 10–200 μm excesses indicative of the presence of circumstellar dust (Bryden et al. 2006; Decin et al. 2003). However, surveys of M-type stars suggest that the fraction of dwarfs with infrared excesses is lower (Gautier et al. 2004; Song et al. 2002); only 2 out of 150 *IRAS/Spitzer* detected M-dwarfs possess far-infrared excess: Hen 3-600 in the TW Hydrae Association and AU Mic in the β Pic Moving Group, with estimated ages $\sim$10 Myr.

Fractional Infrared Luminosity: The average fraction of stellar luminosity reprocessed by the circumstellar dust grains in debris disks discovered by *IRAS* and *ISO* (the *Infrared Space Observatory*), L_{IR}/L_*, is typically between $\sim 10^{-5}$ and $\sim 5 \times 10^{-3}$ (Decin et al. 2003). As a result, the majority of debris disks discovered using these satellites are located around high luminosity A-type stars. The improved sensitivity of *Spitzer* may drive the discovery of debris disks at 70 μm and 160 μm with L_{IR}/L_* as faint as 10^{-8}. For comparison, the zodiacal dust in our solar system reprocesses, $L_{IR}/L_* = 10^{-7}$ (Backman & Paresce 1993), and the dust in the Kuiper Belt is estimated to reprocess, $L_{IR}/L_* = 10^{-7}$–10^{-6} (Backman, Dasgupta, & Stencel 1995).

Grain Size: The similarity between black body grain distances, inferred from spectral energy distribution (SED) modeling of *IRAS* fluxes, and the measured radii of debris disks in resolved systems implies that the circumstellar dust grains in debris disks are large. For example, a black body fit to the *IRAS* HR 4796A excess SED implies an estimated dust grain temperature, $T_{gr} = 110$ K, corresponding to a distance of 35 AU if the grains are large (the absorption coefficient, $Q_{abs} \propto (2\pi a/\lambda)^0$; Jura et al. 1998). However, if the grains are small ($2\pi a < \lambda$), then they radiate less efficiently ($Q_{abs} \propto (2\pi a/\lambda)$) and are expected to be located at a distance of 280 AU from the star. High resolution mid-infrared and coronagraphic imaging has resolved a narrow ring of dust around HR 4796A at a distance of 70 AU from the star, more consistent with large grains (Schneider et al. 1999; Telesco et al. 2000). In addition, *Spitzer* Infrared Spectrograph (IRS) observations (at 5–35 μm) of 60 main sequence stars with *IRAS* 60 μm excesses (including HR 4796A) do not detect silicate emission features at 10 μm and 20 μm, suggesting that the grains have radii, $a > 10$ μm (Jura et al. 2004; Chen et al. 2006, in preparation).

Disk Mass: Scattered light and far-infrared imaging observations are sensitive to dust grains with radii <100 μm; however, the majority of the disk mass is probably contained in larger objects that do not possess as much surface area. Therefore, measurements of disk mass are made at submillimeter wavelengths where the grains are more likely to be optically thin. The typical dust mass in debris disks, is 0.01–0.25 $M_\oplus$ (Zuckerman & Becklin 1993; Holland et al. 1998; Najita & Williams 2005), a tiny fraction of the 10–300 $M_\oplus$ measured toward pre-main sequence T-Tauri and Herbig Ae/Be stars (Natta, Grinin, & Mannings 2000) and a tiny fraction of the dust mass expected in a minimum mass solar nebula with an interstellar gas:dust ratio ($\sim$30 $M_\oplus$). The decline in submillimeter flux, as objects age from Herbig Ae stars to main sequence stars, may be the result of accretion of grains onto the central star or grain growth into larger bodies. After stars reach the main sequence, the decline in submillimeter flux may be the result of parent body grinding. The upper envelope of disk masses for A-type stars is higher than that of later-type stars, suggesting that high mass stars may initially possess more massive disks (Liu et al. 2004a). For comparison, the estimated total mass in the main asteroid belt is $\sim$0.0003 $M_\oplus$;

the asteroid belt may have contained as much as 1000× more mass prior to the period of Late Heavy Bombardment. The estimated total mass in the Kuiper Belt is 0.1 $M_\oplus$.

Gas:Dust Ratio Current (uncertain) models suggest that Jovian planets form either via rapid gravitational collapse through disk instability within a few hundred years (Boss 2003) or via coagulation of dust into solid cores within the first ~1 Myr and accretion of gas into thick hydrogen atmospheres within the first ~30 Myr (Pollack et al. 1996). At present, the timescales on which giant planets form and accrete their atmospheres have not been well constrained observationally. CO surveys suggest that the bulk of molecular gas dissipates within the first ~10 Myr (Zuckerman, Forveille, & Kastner 1995). Since the bulk disk gas is expected to be composed largely of H_2, recent surveys have focused on searching for this molecule. New high-resolution (R = 600) *Spitzer* IRS observations place upper limits on the H_2 S(0) and S(1) emission toward β Pictoris, 49 Ceti, and HD 105 that suggest that <1-15 $M_\oplus$ H_2 remains in these systems (Chen et al. 2006; Hollenbach et al. 2005). The gas:dust ratio in debris disks is probably <10:1. UV absorption line studies have placed upper limits on the gas:dust ratios two nearly edge-on systems. Observations using *FUSE* (the *Far Ultraviolet Spectroscopic Explorer*) and *STIS* (the *Space Telescope Imaging Spectrograph*) constrain the gas:dust ratio in the AU Mic disk (<6:1) by placing upper limits on the H_2 absorption in the O VI $\lambda\lambda1032$, 1038 emission lines and in the C II $\lambda\lambda1036$, 1037 and $\lambda1335$ emission lines (Roberge et al. 2005). *FUSE* and *STIS* observations constrain the gas:dust ratio in the HR 4796A disk (<4:1) by placing upper limits on H_2 and $\lambda2026$ Zn II absorption (Chen & Kamp 2004).

3.   Debris Disks as Solar System Analogs

Parallels are often drawn between parent body belts in debris disks and small body belts in our solar system. These analogies are based on the similarity of observed grain temperatures to those inferred for the main asteroid belt and the Kuiper Belt.

3.1.   Dust in Extra-Solar Kuiper Belts

Numerous small bodies have been discovered in our solar system at distances of 30–50 AU from the Sun. These bodies are collectively referred to as the Kuiper Belt and are the likely source for short-period comets in our solar system. Kuiper Belt objects are expected to collide and grind down into dust grains that may be detected at far-infrared wavelengths (Backman, Dasgupta, & Stencel 1995). If these dust grains are large, then they are expected to have grain temperatures, T_{gr} = 40–50 K; if they are small, then T_{gr} = 110–130 K. The bulk of the energy radiated by grains with these temperatures should emerge at 30 μm to 90 μm, consistent with the properties of debris disks discovered using *IRAS* (Backman & Paresce 1993); therefore, the majority of discovered debris disks are envisioned to be massive analogs to the Kuiper Belt. *Spitzer* IRS spectroscopy of 60 *IRAS*-discovered debris disks suggests that the 5–35 μm spectra of *IRAS* 60 μm excess sources can be modeled using a single temperature black body (Chen et al. 2006; Jura et al. 2004). The peak of the IRS inferred grain temperature distribution lies at T_{gr} = 110–120 K; although, the lack of data beyond 35 μm makes this study insensitive to grains with T_{gr} < 65 K. *Spitzer* MIPS surveys of main sequence FGK stars have discovered a number of solar-like stars with 70 μm excess but no 24 μm excess, consistent with $T_{gr} < 100$ K (Bryden et al. 2006; Chen et al. 2005b; Kim et al. 2005).

The low luminosity of M-type stars makes detecting thermal emission from circumstellar dust around these stars challenging compared to detecting thermal emission from dust around high luminosity A-type stars. For example, dust located 50 AU from an M2V star is expected to possess $T_{gr} = 17$ K, if $L_* = 0.034\ L_\odot$ and the grains are black bodies. Since the dust around M-type stars is so cool, disks around M-dwarfs may be detected best at submillimeter wavelengths. At the distance of the closest stars (50 pc), a disk that reprocesses $L_{IR}/L_* = 2 \times 10^{-3}$ would produce an undetectable flux at 100 μm, $F_\nu(100\ \mu\text{m})$ <10 mJy. However, it would produce a robust excess at submillimeter wavelengths, $F_\nu(850\ \mu\text{m}) = 30$ mJy. One star has been detected at submillimeter wavelengths despite the lack of *IRAS* excess. Photometry of TWA 7 at 850 μm from SCUBA (the Submillimetre Common-User Bolometer Array) on the James Clerk Maxwell Telescope (JCMT) detects thermal emission from dust, with $F_\nu(850\ \mu\text{m}) = 15.5$ mJy (Webb 2000), even though the source does not appear in the *IRAS* catalog. A survey of three young M-dwarfs in the β Pic Moving Group (with an estimated age of 12 Myr) and the Local Association Group (with an estimated age of 50 Myr) discovered 450 μm and/or 850 μm excesses associated with two of the stars: AU Mic and Gl 182 (Liu et al. 2004a). The disk around AU Mic is warm enough and massive enough that its disk is bright at 70 μm ($F_\nu(70\ \mu\text{m}) = 200$ mJy) while the disk around Gl 182 is not detected at either 24 μm or 70 μm (Chen et al. 2005b).

3.2. Dust in Extra-Solar Asteroid Belts

The zodiacal dust in our solar system possesses a grain temperature, $T_{gr} = 150$–170 K, suggesting that the bulk of the thermal energy is radiated at 20–25 μm. The Earth has a temperature, $T_{gr} \sim 300$ K, suggesting that the bulk of its thermal energy is radiated at $\sim$10 μm. Searches for 10 μm excesses around main-sequence stars have revealed that 300 K dust around A–M dwarfs is rare. *IRAS* surveys of main sequence stars suggest that <5% of debris disks possess 12 μm excesses. In a survey of 548 A–K dwarfs, Aumann & Probst (1991) were able to identify *IRAS* 12 μm excesses only with β Pictoris and ζ Lep. Identifying objects with 12 μm excess is challenging because the photosphere usually dominates the total flux at this wavelength. High resolution mid-infrared imaging using the Keck Long Wavelength Spectrometer (LWS) confirms that the debris disk around ζ Lep is compact. The disk is at most marginally resolved at 17.9 μm, suggesting that the dust probably lies within 6 AU although some dust may extend as far as 9 AU away from the star, consistent with the 230–320 K color temperature inferred 10–60 μm photometry (Chen & Jura 2001). Similarly, 12 μm excess around M-dwarfs appears to be rare. Follow-up Keck LWS 11.7 μm photometry of nine late-type dwarfs with possible *IRAS* 12 μm excesses is unable to detect excess thermal emission from any of the candidates (Plavchan et al. 2005).

Warm silicates with grain temperatures, $T_{gr} \sim 300$ K, may produce emission in the 10 μm Si-O stretching mode, if the dust grains are small, $a < 10\ \mu$m. These features can provide insight into not only the composition but also the size of circumstellar dust grains. More than a decade ago, ground-based spectroscopy of β

Pictoris revealed a broad 9.6 μm amorphous silicate and a weaker 11.2 μm crystalline olivine emission feature, similar to that observed toward comets Halley, Bradford, and Levy (Knacke et al. 1993), suggesting that the parent bodies may be similar to small bodies in our solar system. Several systems with $T_{gr} > 300$ K have now been discovered that possess silicate emission features. *Spitzer* IRS spectra of HD 69830, a K0V star with an age of 2 Gyr, show mid-infrared emission features nearly identical to those observed toward Hale-Bopp but with a higher grain temperature, $T_{gr} = 400$ K instead of $T_{gr} = 207$ K (Beichman et al. 2005b). Gemini Michelle spectroscopy of BD+20 307 (HIP 8920), a G0V star with an estimated age of 300 Myr, suggests that amorphous and crystalline silicates are present; models of the 9–25 μm SED imply $T_{gr} = 650$ K and a remarkably high $L_{IR}/L_* = 0.04$ for its age (Song et al. 2005). More warm dust systems with spectral features may soon be identified using *Spitzer* MIPS. For example, α^1 Lib (a F3V star in the 200 Myr old Castor Moving Group) and HD 177724 (an A0V field star) possess such strong 24 μm excess that their 12 μm, 24 μm, and 70 μm fluxes can not be self-consistently modeled using a modified black body, suggesting that their strong 24 μm excess may be the result of emission in spectral features (Chen et al. 2005b).

Since SED modeling is degenerate, high resolution imaging is needed to determine definitively the location of the dust and to search for structure in the disk. Nulling interferometers, now operational at Keck and the Multiple-Mirror Telescope (MMT), will allow dust in exo-zodiacal disks to be directly resolved. By placing the central star in a null, nulling interferometry destructively interferes stellar emission, obviating the need for accurate models of the stellar atmosphere. Without the bright central core, nulling observations can not only seek faint exo-zodiacal emission but can also apply the full diffraction limit of the telescope to resolve a source. For example, high-resolution Keck LWS imaging of the Herbig Ae star AB Aur at 18.7 μm (with a FWHM 0.5″) struggles to resolve faint extended emission in the wings of the PSF at 0.4″ from the star (Chen et al. 2003) while MMT nulling observations at 10.6 μm suppress all but 10–20% of the flux; fits of the percentage null versus rotation of the interferometer baseline suggest that the mid-infrared emitting component possess an angular diameter $\sim$0.2″ with a position angle, 30°±15°, and an inclination, 45°–65°, from face-on (Liu et al. 2005). Nulling observations of Vega at 10.6 μm do not detect resolved emission at >2.1% (3σ limit) of the level of the stellar photospheric emission, suggesting that Vega possess <650× as much zodiacal emission as our solar system (Liu et al. 2004b).

4.   Dust Removal Processes

Infrared spectroscopy and SED modeling suggest that the majority of debris disks possess central clearings. They may be generated by planets that gravitationally eject dust grains that are otherwise spiraling toward their orbit centers under Poynting-Robertson and stellar wind drag. However, a number of other processes may also contribute to presence of absence of central clearings:

Sublimation: If the grains are icy, then ice sublimation may play an important role in the destruction of grains near the star and may provide a natural explanation for the presence of central clearings in debris disks (Jura et al. 1998) implied from black body fits to *Spitzer* IRS spectra (Chen et al. 2006). The peak in the measured T_{gr} distribution, estimated from black body fits to *Spitzer* IRS spectra, suggest that grains in debris disks typically have $T_{gr} = 110$–120 K (Chen et al. 2006), near the sublimation temperature of water ice in a vacuum, $T_{sub} = 150$ K. Since sublimation lifetimes are sensitively dependent on grain temperature, cool grains may possess ices. For example, 3.5 μm grains with $T_{gr} = 70$ K, have a sublimation lifetime, $T_{sub} = 1.3 \times 10^7$ Gyr while 16 μm grains with $T_{gr} = 160$ K, have a sublimation lifetime, $T_{sub} = 7.4$ minutes! Since the majority of debris disks have $T_{gr} < 120$ K, they could possess icy grains that may be detectable with *Spitzer*. Crystalline water ice possesses an emission feature at 61 μm. Low resolution (R $= \lambda/\Delta\lambda = 15$–$25$) *Spitzer* MIPS SED mode observations may be able to detect emission from ices in debris disks.

Radiation Effects: The initial discovery of debris disks around high luminosity main sequence B- and A-type stars led to speculation that radiation pressure and the Poynting-Robertson effect may govern grain dynamics. The force due to radiation pressure acting on grains around A-type stars with radii <few μm is larger than the force due to gravity; therefore, small grains are expected to be effectively removed from the circumstellar environment on timescales $\sim 10^4$ years (Artymowicz 1988). Larger particles are subject to the Poynting-Robertson effect in which dust grains lose angular momentum through interactions with out-flowing stellar photons. As a result, larger grains spiral in toward their orbit center on timescales typically <1 Myr (Burns, Lamy, & Soter 1979). If a debris disk is composed of large black body grains which spiral inward under PR drag, then the dust should be contained in a continuous disk with constant surface density and an infrared spectrum, $F_\nu \propto \lambda$, at short wavelengths (Jura et al. 1998, 2004). *Spitzer* IRS spectroscopy of *IRAS*-discovered debris disks may have revealed one object that possess an excess spectrum better fit by $F_\nu \propto \lambda$ than a black body (HR 6670; Chen et al. 2006). Wyatt (2005) estimates that the grain density in *IRAS*-discovered debris disks is more than an order of magnitude too high for grains to migrate inward under Poynting-Robertson drag without suffering destructive collisions.

Corpuscular Stellar Wind Effects: The recent discovery of debris disks around solar-like and M-type stars has led to speculation that corpuscular stellar winds may contribute to grain removal in a manner analogous to radiation pressure and the Poynting-Robertson effect (Plavchan, Jura, & Lipscy 2005): (1) An out-flowing corpuscular stellar wind produces a pressure on dust grains which overcomes the force due to gravity for small grains; however, the corpuscular stellar wind in only important for low luminosity stars (M-dwarfs) with strong stellar winds ($\sim 100\ \dot{M}_\odot$; Chen et al. 2005b). (2) Large particles orbiting the star are subject to a drag force produced when dust grains collide with protons in the stellar wind. These collisions decrease the velocities of orbiting dust grains and therefore their angular momentum, causing them to spiral in toward their orbit center. Stellar wind drag may explain the observed anti-correlation between *Spitzer* 24 μm excess and *ROSAT* fluxes toward F-type stars in the 3–20 Myr Sco-Cen (Chen et al. 2005a) and the lack of 12 μm excesses observed toward nearby, >10 Myr-old, late-type M-dwarfs (Plavchan et al. 2005). Recently, Strubbe & Chiang (2006) have reproduced the radial brightness profile of the AU Mic disk assuming that dust grains, produced in collisions between parent bodies on circular orbits at 43 AU, generate the resolved scattered light.

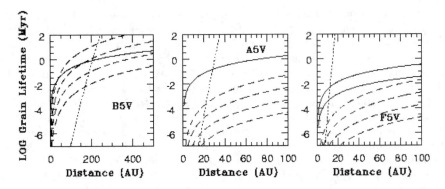

Figure 1.   Left: The grain lifetimes are plotted as a function of distance around a B5V star. The Poynting-Robertson Drag/Stellar Wind drag lifetime is shown with a solid line, the sublimation lifetime is shown with a dotted line, and the collisional lifetime is shown with a dashed line, assuming $M_{dust} = 0.001, 0.01, 0.1,$ and $1\ M_{\oplus}$ (from top to bottom). Center: same as the left panel for an A5V star. No stellar wind drag is assumed. Right: same as the left panel for a F5V star. The corpuscular stellar wind drag lifetime is shown with a solid line, assuming that $\dot{M}_{wind} = 100, 1000\ \dot{M}_{\odot}$ (from top to bottom).

In their model, large grains produce a surface density, $\sigma \propto r^0$, at $r < 43$ AU, under corpuscular and Poynting-Robertson drag modified by collisions, while small grains, that are barely bound under corpuscular stellar wind and radiation pressure, produce a surface density, $\sigma \propto r^{-5/2}$, in the outer disk.

Collisions: If the particle density within the disk is high, then collisions may shatter larger grains into smaller grains that may be removed by radiation pressure and/or corpuscular stellar wind pressure. *Spitzer* MIPS imaging has recently resolved symmetric extended emission from the face-on disk around Vega with radii >330 AU, >540 AU, and >810 AU at 24 μm, 70 μm, and 160 μm, respectively, that may be explained by small grains that are radiatively driven from the system (Su et al. 2005). Comparison of the 24 μm emission with the 70 μm emission suggests that the system possesses 2 μm grains, well below the blow-out limit of 14 μm. Statistical studies of the decline in fractional infrared luminosity may as a function of time also shed light on the processes by which dust grains are removed. The fractional infrared luminosity of a debris disk is expected to decrease inversely with time, $L_{IR}/L_* \propto 1/t_{age}$, if collisions are the dominant grain removal process and is expected to decrease inversely with time squared, $L_{IR}/L_* \propto 1/t_{age}^2$, if corpuscular stellar wind and Poynting-Robertson drag are the dominant grain removal processes (Dominik & Decin 2003). *Spitzer* MIPS and IRS and submillimeter studies of thermal emission from debris disks are consistent with a $1/t$ decay and a characteristic timescale, $t_o \sim 100$–200 Myr (Chen et al. 2006; Najita & Williams 2005; Liu et al. 2004a).

The dominant grain removal process within a disk is dependent not only on the luminosity of the central star and the strength of its stellar wind but also on grain distance from the central star. For example, Backman & Paresce (1993) estimate that collisions to small grain sizes and radiation pressure remove grains at 67 AU around Fomalhaut while the Poynting-Robertson effect removes grains at 1000 AU. In Figure 1, we plot the sublimation lifetime, the Poynting-Robertson (and corpuscular stellar wind) drag lifetime, and the collision lifetime for average-sized grains around typical B5V, A5V, and F5V stars. Sublimation may quickly remove icy grains in the innermost portions of the disk. At larger radii, collisions dominate grain destruction, and at the largest radii, where the disk has the lowest density, Poynting-Robertson and corpuscular stellar wind drag may dominate grain destruction. For typical A5V and F5V stars, the collision lifetime is shorter than the drag lifetime if the disk has a dust mass between 0.001 $M_\oplus$ and 1 $M_\oplus$, even if the F5V star has a stellar wind with a mass loss rate as high as $\dot{M}_{wind} = 1000 \ \dot{M}_\odot$. However, for a typical B5V star, the Poynting-Robertson and stellar wind drag lifetime may be shorter than the collision lifetime, especially at large radii, if the disk has a dust mass $M_{tot} < 0.1 \ M_\oplus$.

Gas-Grain Interactions: In disks with gas:dust ratios between 0.1 and 10, gas-grain interactions are expected to concentrate the smallest grains in the disk, with radii just above the blow-out size, at the outer edge of the disk, creating a ring of bright thermal emission. The presence of gas has been used to explain the central clearing in the HR 4796A disk (Takeuchi & Artymowicz 2001).

5. The Planet/Debris Disk Connection

The detection of asymmetries in the azimuthal brightness of debris disks can distinguish between whether the central clearings are dynamically sculpted by a companion or generated by other mechanisms. For example, a planet may create brightness peaks in a disk by trapping grains into mean motion resonances (Liou & Zook 1999; Quillen & Thorndike 2002). High resolution submillimeter imaging of ϵ Eri has revealed the presence of brightness peaks, that may be explained by dust trapped into the 5:3 and 3:2 exterior mean motion resonances of a 30 $M_\oplus$ planet, with eccentricity $e = 0.3$ and semimajor axis $a = 40$ AU (Ozernoy et al. 2000; Quillen & Thorndike 2002). Comparison of the 1997–1998 850 μm SCUBA map with the 2000–2002 850 μm SCUBA map, in Figure 2, indicates that three of the brightness peaks in the ring around ϵ Eri are orbiting counterclockwise at a rate of $1°$ yr^{-1}, consistent with that expected from planetary resonance models (Greaves et al. 2005; Ozernoy et al. 2000). The presence of two brightness peaks in submillimeter and millimeter maps of the face-on disk around Vega (Wilner et al. 2002) may also be explained by dust trapped, this time, into the 3:2 and 2:1 exterior mean motion resonances of a Neptune-mass planet that migrated from 40 AU to 65 AU over a period of 56 Myr (Wyatt 2003). However, the observation of orbital motion, the detection of the putative planet, and the observation of lower-level brightness asymmetries are needed to confirm this model.

If a planet in a debris disk has an eccentric orbit, then the planet may force circumstellar dust grains into elliptical orbits. Since dust grains at pericenter will be closer to the star and therefore warmer than grains at apocenter, disks with eccentric planets may possess brightness asymmetries (Wyatt et al. 1999). High resolution thermal infrared and submillimeter imaging of Fomalhaut has revealed a 30–15% brightness asymmetry in its disk ansae (Stapelfeldt et al. 2004; Holland et al. 2003). The dust grains in this disk may have experienced sec-

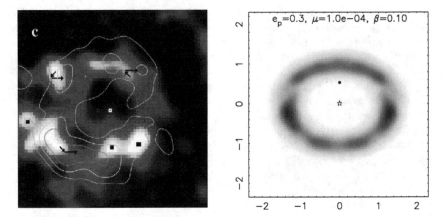

Figure 2.   Left: The 1997–1998 JCMT SCUBA 850 μm map of ϵ Eridani
is shown in grey scale while the 2000–2002 850 μm map is shown with overlaid contours at 30%, 50%, and 70% of the peak surface brightness. Black
squares are suggested background features and black arrows show the orbiting motion of the brightness peaks. (Figure taken from Greaves et al. 2005.)
Right: Simulated intensity distribution of the ϵ Eridani disk, assuming that
the brightness peaks are generated by dust trapped in the 5:3 and 3:2 exterior
mean motion resonances of a 30 $M_\oplus$ planet, with eccentricity, $e = 0.3$, and
semimajor axis, $a = 40$ AU. (Figure taken from Quillen & Thorndike 2002.)

ular perturbations of their orbital elements by a planet with $a = 40$ AU and
$e = 0.15$, which forces grains into an elliptical orbit with the star at one focus
(Wyatt et al. 1999; Stapelfeldt et al. 2004). Recent *Hubble Space Telscope* (*HST*)
ACS (Advanced Camera for Surveys) scattered light observations of Fomalhaut,
shown in Figure 3, have confirmed that the star is not at the center of the dust
grain orbits. Kalas, Graham, & Clampin (2005) measure an offset of ∼15 AU
between the geometric center of the disk and the position of the central star. A
15%–5% brightness asymmetry is observed toward HR 4796A at near-infrared
and mid-infrared wavelengths (Weinberger, Schneider, & Becklin 2000; Telesco
et al. 2000), which may be explained if the orbit of HR 4796B has an eccentricity
$e = 0.13$ or if there is a >0.1 M_{Jup} mass planet at the inner edge of the disk at
70 AU (Wyatt et al. 1999).

There is tantalizing evidence to suggest that giant planets and debris disks are
correlated. A near-infrared coronagraphic survey of six FGK stars with radial
velocity planets resolved disks around three objects (Trilling et al. 2000). However, NICMOS observations were unable to confirm the presence of a disk around
one of the objects: 55 Cnc (Schneider et al. 2001). *Spitzer* MIPS observations
of 26 planet-bearing FGK stars have discovered six objects that possess 70 μm
excesses, corresponding to a disk fraction of 23% (Beichman et al. 2005a), higher
than the ∼15% observed toward field stars. The MIPS observations suggest that
both the frequency and the magnitude of dust emission are correlated with the
presence of known planets. Since the 70 μm excess is generated by cool dust
($T_{gr} < 100$ K) located beyond 10 AU, well outside the orbits of the discovered
planets, the process that correlates the radial velocity planets at <5 AU with

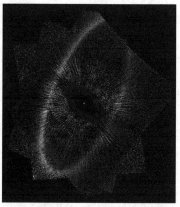 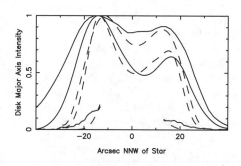

Figure 3.   Left: ACS scattered light observations of Fomalhaut with North up and East to the left. The star is offset from the geometric center of the disk by 15 AU, with the inner edge of the disk 133–158 AU from the star (courtesy of P. Kalas, University of California, Berkeley). Right: The solid lines are line cuts, along the disk axis, through *Spitzer* MIPS 24 μm, 70 μm, and 160 μm images (from bottom to top) of Fomalhaut. The dashed lines are the profiles expected if a planet with $a = 40$ AU, $e = 0.15$ forces the eccentricity of dust grains in the disk. (Figure taken from Stapelfeldt et al. 2004.)

cool dust beyond 10 AU is not understood. By contrast, a submillimeter survey of eight planet-bearing stars found no thermal emission toward any of the stars surveyed (Greaves et al. 2004). Radial velocity studies of main sequence stars that possess giant planets find a correlation between the presence of an orbiting planet and the metallicity of the central star (Fischer & Valenti 2005). If correlations exist between metallicity and the presence of a planet, and between the presence of a planet and a 70 μm excess, there might also be a correlation between the presence of infrared excess and stellar metallicity. However, no correlations between the presence of 70 μm excess and stellar metallicity have been found thus far (Bryden et al. 2006; Beichman et al. 2005a).

6.   Infalling Bodies and the Stable Gas Around β Pictoris

Ultraviolet (UV) and visual spectra of β Pictoris possess time-variable, high velocity, red-shifted absorption features, initially discovered in Ca II H and K and Na I D and later in a suite of metal atoms (including C I, C IV, Mg I, Mg II, Al II, Al III, Si II, S I, Cr II, Fe I, Fe II, Ni II, Zn II). These features vary on timescales as short as hours and are non-periodic (Vidal-Madjar, Lecavelier des Etangs, & Ferlet 1998). The excitation of the atoms, chemical abundance, high measured velocities, and rapid time-variability of the gas suggest that the material is circumstellar rather than interstellar. The velocity of the atoms, typically 100–400 km sec^{-1}, is close to the free fall velocity at a few stellar radii, suggesting that the absorption is produced as stellar photons pass through the coma of infalling refractory bodies at distances <6 AU from the star (Karmann, Beust, & Klinger 2001; Beust et al. 1998). At these distances, refractory materials may sublimate from the surface of infalling bodies and collisions may produce highly ionized species such as C IV and Al III. If infalling bodies generate the observed features, then the fact that the features are preferentially red-shifted (rather than equally red- and blue-shifted) suggests that some process aligns their orbits. Scattering of bodies by a planet on an eccentric orbit (Levison, Duncan, & Wetherill 1994) and bodies in mean motion resonances with an eccentric planet, the orbits of which decay via secular resonant perturbations by the planet (Beust & Morbidelli 1996), have been used to explain the preferentially red-shifted features; however, both models have difficulties.

The β Pictoris disk also possesses a stable component of gas at the velocity of the star. Many of the species that possess high velocity features also possess very low velocity features that vary slightly over timescales of hours to years. Spatially resolved visual spectra of β Pic have revealed the presence of a rotating disk of atomic gas, observed via emission from Fe I, Na I, Ca II, Ni I, Ni II, Ti I, Ti II, Cr I, and Cr II. Estimates of the radiation pressure acting on Fe and Na atoms suggest that these species should be accelerated to terminal velocities $\sim$100–1000 km sec^{-1}, significantly higher than is observed (Brandeker et al. 2004). One possible explanation for the low velocity of the atomic gas is that the gas is ionic and that Coulomb interactions between ions reduce the effective radiation pressure on the bulk gas. Fernández, Brandeker, & Wu (2006) suggest that ions in the disk couple together into a fluid that is bound to the system and that brakes the gas if the effective radiation pressure coefficient, $\beta_{eff} < 0.5$. In particular, they suggest that atomic carbon may be important for reducing the effective radiation pressure coefficient if the carbon abundance is $>10\times$ solar because the expected ionization fraction of atomic carbon is 0.5 and $F_{rad}/F_{grav} \approx 0$. Measurements of the line-of-sight abundance of ionized carbon, inferred from C II absorption in the O VI λ1038 emission using *FUSE*, confirm that carbon is overabundant by approximately a factor of 10 compared with measurements of the stable gas from the literature (Roberge et al. 2006, in preparation). Estimates of the total gas mass, inferred from measured elemental abundances and the gas density radial profile from scattered emission, suggest that the β Pic disk contains $\sim$0.004 $M_{\oplus}$ of gas or a gas:dust ratio $\sim$0.1 (Roberge et al. 2006).

HST STIS and GHRS (Goddard High Resolution Spectrograph) observations have also detected stable CO and C I (^{3}P) absorption at the velocity of the star (Jolly et al. 1998; Roberge et al. 2000). Since CO and C I are expected to be photodissociated and photoionized by interstellar UV photons on timescales of $\sim$200 years, these gases must be replenished from a reservoir. One possibility is that the CO is produced by slow sublimation of orbiting comets at several tens of AU from the star. However, the observed CO possesses a low ^{12}CO:^{13}CO ratio ($R = 15 \pm 2$) compared to solar system comets ($R = 89$). The overabundance of ^{13}CO may be explained if it is produced from ^{12}CO in the reaction

$$^{13}\text{C}^{+} + {}^{12}\text{CO} \rightleftharpoons {}^{13}\text{CO} + {}^{12}\text{C}^{+} + 35 \ \ \text{K} \qquad (1)$$

at temperatures below 35 K (Jolly et al. 1998). The order of magnitude difference in the measured column densities of C I ($N(\text{C I}) = (2\text{–}4)\times10^{16}$ cm^{-2}; Roberge et al. 2000) and CO ($N(\text{CO}) = 2.5\pm0.5\times10^{15}$ cm^{-2}; K. H. Hinkel, private communication) suggest that C I is not produced by photodissociation of CO. Similarly, the disparity in the measured excitation temperatures for CO and

C I, T_{ex} = 20–25 K and 80 K, suggests that the observed CO and C I are not cospatial (Roberge et al. 2000). One possibility is that the observed stable C I (^{3}P) is produced directly by sublimation of infalling bodies; if there are 100 bodies per year, then Roberge et al. (2000) estimate that each infalling object generates a C I (^{3}P) column density, $N_{comet} \sim 10^{11}$ cm^{-2} if C I is only destroyed via ionization by interstellar photons with $\Gamma = 0.004$ yr^{-1}, consistent with evaporation of kilometer-sized objects.

7. Future Work

The age $\sim$10 Myr appears to mark a transition in the gross properties of circumstellar disks. Pre-main sequence objects, such as T-Tauri and Herbig Ae/Be stars, possess optically thick, gaseous disks while young, main sequence stars possess optically thin, gas-poor disks. Recent studies of the 10 Myr old TW Hydrae association (Weinberger et al. 2004) and the 30 Myr Tucana-Horologium association (Mamajek et al. 2004) find that warm circumstellar dust ($T_{gr} = 200$–300 K), if present around young stars in these associations, re-radiates less than 0.1–0.7% of the stellar luminosity, $L_{IR}/L_* < (1$–7)$\times 10^{-3}$. If the majority of stars possess planetary systems at an age of $\sim$10 Myr, why do some stars, such as TW Hydrae, Hen 3-600, HD 98800, and HR 4796A in the TW Hydrae association, still possess large quantities of gas and/or dust? To investigate the origin of the dispersion in disk properties around $\sim$10 Myr old stars, I have begun a multi-wavelength study of solar-like stars in the 5–20 Myr old (Preibisch et al. 2002; Mamajek et al. 2002) Scorpius-Centaurus OB Association, located at a distance of 118–145 pc away from the Sun.

One component of my study is a *Spitzer* MIPS 24 μm and 70 μm search for infrared excess around $\sim$100 F- and G-type stars in Sco-Cen; the first results suggest that fractional 24 μm excess luminosity and fractional x-ray luminosity may be anti-correlated (Chen et al. 2005a). Observations of the first 40 targets detect strong 24 μm and 70 μm excess around six stars, corresponding to $L_{IR}/L_* = 7\times 10^{-4}$–1.5$\times 10^{-2}$, and weak 24 μm excess around seven others, corresponding to $L_{IR}/L_* = 3.5\times 10^{-5}$–4.4$\times 10^{-4}$. Only one of the 24 μm and 70 μm bright disks is detected by *ROSAT* while four of the weak 24 μm excess sources are. The presence of strong stellar winds, $\sim$1000$\times$ larger than our solar wind, may explain the depletion of dust in disks around x-ray emitting sources. Follow-up *Spitzer* IRS 5–35 μm spectroscopy may allow us to determine whether corpuscular stellar wind effectively removes dust grain in x-ray emitting systems. Since the inward drift velocity of dust grains under corpuscular stellar wind and Poynting-Robertson drag is $1 + \dot{M}_{wind}c^2/L_*$ larger than under Poynting-Robertson drag alone, a disk influenced by strong stellar wind is expected to possess a constant surface density distribution. If the grains are large, then the infrared spectra will be well modeled by $F_\nu \propto \lambda$. However, if collisions and ejection by radiation pressure is the dominant grain removal mechanism, then the infrared spectra may be better modeled by a black body.

Other components of my study include *Spitzer* IRS 5–35 μm spectroscopy and MIPS SED mode observations of MIPS-discovered excess sources to search for emission from silicates and water ice and to constrain the composition and grain size. The minimum grain size, due to blow-out from radiation pressure, around stars in this sample is $a_{min} \sim 0.5$–2.0 μm, small enough to produce spectral features. Ground-based 10 μm spectroscopy of HD 113766, an F3 binary member of the Sco-Cen subgroup Lower Centaurus Crux, suggests that the dust around this object is highly processed by an age of $\sim$16 Myr. Fits to the 10 μm silicate feature suggest that >90% of the amorphous silicate mass is contained in large grains (with radii $a \sim 2$ μm) and that $\sim$30% of the mass is contained in crystalline forsterite (Schutz, Meeus, & Sterzik 2005).

Acknowledgments. I would like to thank the Astronomy Department at the University of Texas at Austin for giving me the opportunity to write this review and M. Jura, P. Kalas, W. Liu, J. Najita, and A. Roberge for stimulating conversations and correspondence during the preparation of this manuscript. Support for this work was provided by the NASA through the Spitzer Fellowship Program under award NAS7-1407.

References

Artymowicz, P. 1988, ApJ, 335, L79
Aumann, H. H. & Probst, R. G. 1991, ApJ, 368, 264
Backman, D. E. & Paresce, F. 1993, in Protostars and Planets III, ed. E. Levy & J. Lunine (Tuscon: University of Arizona Press), 1253
Backman, D. E., Dasgupta, A., & Stencel, R. E. 1995, ApJ, 450, L35
Beichman, C. A., et al. 2005a, ApJ, 622, 1160
Beichman, C. A., et al. 2005b, ApJ, 626, 1061
Beust, H. & Morbidelli, A. 1996, Icarus, 120, 358
Beust, H., et al. 1998, A&A, 338, 1015
Boss, A. P. 2003, ApJ, 599, 577
Brandeker, A., Liseau, R., Olofsson, G., & Fridlund, M. 2004, A&A, 413, 681
Bryden, G., et al. 2005, ApJ, 636, 1098
Burns, J. A., Lamy, P. L., & Soter, S. 1979, Icarus, 40, 1
Chapman, C. R., Paolicchi, P., Zappala, V., Binzel, R. P., & Bell, J. F. 1989, in Asteroids II, ed. R. Binzel, T. Gehrels & M. Matthews (Tucson: University of Arizona Press), 386
Chen, C. H. & Jura, M. 2001, ApJ, 560, L171
Chen, C. H. & Jura, M. 2003, ApJ, 591, 267
Chen, C. H. & Kamp, I. 2004, ApJ, 602, 985
Chen, C. H., et al. 2005a, ApJ, 623, 493
Chen, C. H., et al. 2005b, ApJ, 634, 1372
Decin, G., Dominik, C., Waters, L.B. F. M., & Waelkens, C. 2003, ApJ, 598, 626
Dominik, C. & Decin, G. 2003, ApJ, 598, 626
Fernández, R., Brandeker, A., & Wu, Y. 2006, ApJ, in press (astro-ph/0601244)
Fischer, D. A. & Valenti, J. 2005, ApJ, 622, 1102
Gautier, T., et al. 2004, BAAS, 205, 5503
Greaves, J. S., et al. 2004, MNRAS, 348, 1097
Greaves, J. S., et al. 2005, ApJ, 619, L187
Holland, W. S., et al. 1998, Nature, 392, 788
Holland, W. S., et al. 2003, ApJ, 582, 1141
Hollenbach, D., et al. 2005, ApJ, 631, 1180
Jolly, A., et al. 1998, A&A, 329, 1028
Jura, M., et al. 1998, ApJ, 505, 897
Jura, M., et al. 2004, ApJS, 154, 453
Kalas, P., Graham, J. R., & Clampin, M. 2005, Nature, 435, 1067
Karmann, C., Beust, H., & Klinger, J. 2001, A&A, 372, 616
Kenyon, S. J. & Bromley, B. C. 2004, AJ, 127, 513
Kim, J. S., et al. 2005, ApJ, 632, 659

Knacke, R. F., et al. 1993, ApJ, 418, 440
Lagrange, A.-M., Backman, D. E., & Artymowicz, P. 2000, in Protostars and Planets IV, ed. V. Mannings, A. Boss, & S. Russell (Tuscon: University of Arizona Press), 639
Levison, H. F., Duncan, M. J., & Wetherill, G. W. 1994, Nature, 372, 441
Liou, J. & Zook, H. A. 1999, ApJ, 118, 580
Liu, M. C., et al. 2004a, ApJ 608, L526
Liu, W. M., et al. 2004b, ApJ, 610, L125
Liu, W. M., et al. 2005, ApJ, 618, L133
Mamajek, E. E., Meyer, M. R., & Liebert, J. 2002, AJ, 124, 1670
Mamajek, E. E. et al. 2004, ApJ, 612, 496
Mannings, V. & Barlow, M. J. 1998, ApJ, 497, 330
Marcy, G. W., Cochran, W. D., Mayor, M. 2000, in Protostars and Planets IV, ed. V. Manning, A. Boss, & S. Russell (Tuscon: University of Arizona Press), 1285
Najita, J. & Williams, J. P. 2005, ApJ, 635, 625
Natta, A., Grinin, V., & Mannings, V. 2000, in Protostars and Planets IV, ed. V. Mannings, A. Boss, & S. Russell (Tuscon: University of Arizona Press), 559
Oumaijer, R. D., ven der Veen, W. E. C. J., Waters, L. B. F. M., Trams, N. R., Waelkens, C., & Engelsman, E. 1992, A&AS, 96, 625
Ozernoy, L. et al. 2000, ApJ, 537, L147
Plavchan, P., Jura, M., & Lipscy, S. J. 2005, ApJ, 631, 1161
Pollack, J. B., et al. 1996, ApJ, 124, 62
Preibisch, T. et al. 2002, AJ, 124, 404
Quillen, A. C. & Thorndike, S. 2002, ApJ, 578, L149
Roberge, A., et al. 2000, ApJ, 538, 904
Roberge, A., Weinberger, A. J., Redfield, S., & Feldman, P. D. 2005, ApJ, 626, L105
Schneider, G. et al. 1999, 513, L127
Schneider, G. et al. 2001, 121, 525
Schutz, O., Meeus, G., & Sterzik, M. F. 2005, A&A, 431, 165
Silverstone, M. D. 2000, Ph.D. thesis, UCLA
Song, I., Weinberger, A., Becklin, E., Zuckerman, B., & Chen, C. 2002, AJ, 124, 514
Song, I., Zuckerman, B., Weinberger, A. J., & Becklin, E. E. 2005, Nature, 436, 363
Stapelfeldt, et al. 2004, ApJS, 154, 458
Strom, R. G., et al. 2005, Science, 309, 1847
Strubbe, L. E. & Chiang, E. I. 2006, ApJ, submitted (astro-ph/0510527)
Su, K. Y. L., et al. 2005, ApJ, 628, 487
Sylvester, R. J., Skinner, C. J., Barlow, M. J., & Mannings, V. 1996, MNRAS 279, 915
Takeuchi, T. & Artymowicz, P. 2001, ApJ, 557, 990
Telesco, C. M., et al. 2000, ApJ, 530, 329
Trilling, D. E. 2000, ApJ 529, 499
Vidal-Madjar, A., Lecavelier des Etangs, A., & Ferlet, R. 1998, Planet. Space Sci., 47, 629
Walker, H. & Wolstencroft, R. D. 1988, PASP, 100, 1509
Webb, R. A. 2000, Ph.D. thesis, UCLA
Weinberger, A., Schneider, G., & Becklin, E. E. 2000, in ASP Conf. Series, 219, 329
Weinberger, A. J. et al. 2004, AJ, 127, 2246
Werner, M. W., et al. 2004, ApJS, 154, 1
Wilner, D. J., et al. 2002, ApJ, 569, L115
Wyatt, M. C., et al. 1999, ApJ, 527, 918
Wyatt, M. C. 2003, ApJ, 598, 1321
Wyatt, M. C. 2005, A&A, 433, 1007
Zuckerman, B. & Becklin, E. E. 1993, ApJ, 414, 793
Zuckerman, B., Forveille, T., & Kastner, J. H. 1995, Nature, 373, 494
Zuckerman, B. 2001, ARA&A, 39, 549

Seth Redfield relates his work on the local ISM to astrobiology during a panel discussion.

Frank N. Bash Symposium 2005: New Horizons in Astronomy
ASP Conference Series, Vol. 352, 2006
S. J. Kannappan, S. Redfield, J. E. Kessler-Silacci, M. Landriau, and N. Drory

The Local Interstellar Medium

Seth Redfield[1]

Department of Astronomy and McDonald Observatory, University of Texas, Austin, TX, USA

Abstract. The Local Interstellar Medium (LISM) is a unique environment that presents an opportunity to study general interstellar phenomena in great detail and in three dimensions. In particular, high resolution optical and ultraviolet spectroscopy have proven to be powerful tools for addressing fundamental questions concerning the physical conditions and three-dimensional (3D) morphology of this local material. After reviewing our current understanding of the structure of gas in the solar neighborhood, I will discuss the influence that the LISM can have on stellar and planetary systems, including LISM dust deposition onto planetary atmospheres and the modulation of galactic cosmic rays through the astrosphere — the balancing interface between the outward pressure of the magnetized stellar wind and the inward pressure of the surrounding interstellar medium. On Earth, galactic cosmic rays may play a role as contributors to ozone layer chemistry, planetary electrical discharge frequency, biological mutation rates, and climate. Since the LISM shares the same volume as practically all known extrasolar planets, the prototypical debris disks systems, and nearby low-mass star-formation sites, it will be important to understand the structures of the LISM and how they may influence planetary atmospheres.

1. Introduction

The interstellar medium (ISM) is a critical component of galactic structure. Its role in the lifecycle of stars, mediating the transition from stellar death to stellar birth, evokes a sense of a "galactic ecology" (Burton 2004). The ISM provides a platform for the recycling of stellar material, by transferring and mixing the remnants of stellar nucleosynthesis and creating environments conducive for the creation of future generations of stars and planets. It also transfers energy and momentum, absorbing flows from supernovae blasts and strong winds from young stars and coupling these peculiar motions with galactic rotation and turbulence. Ultimately, when a dense interstellar cloud collapses, it is the conservation of momentum from the parent cloud that leads to the formation of a protostellar disk from which stars and planets are formed.

The local interstellar medium (LISM) is the interstellar material that resides in close ($\lesssim 100\,\mathrm{pc}$) proximity to the Sun. For a discussion on more distant ISM structures, see McClure-Griffiths, in this volume. Proximity is a special characteristic that drives much of the interest in the LISM. First, proximity provides an opportunity to observe general ISM phenomena in great detail, and in three

---

[1]Hubble Fellow.

dimensions. ISM structures and processes are repeated almost *ad infinitum* in our own galaxy (Dickey & Lockman 1990), and beyond in other galaxies (Mc-Cray & Kafatos 1987), even at high redshift (Rauch et al. 1999). Knowledge of general ISM phenomena in our local corner of the galaxy, discussed in §2, can be applied to more distant and difficult to observe parts of the universe.

Second, proximity implies an interconnectedness. The relationship between stars and their surrounding interstellar environment will be discussed in §3, with particular attention paid to the interaction of the Sun with the LISM. In §4, the consequences of the relationship between stars and the LISM on planetary atmospheres are discussed, and the LISM-Earth, or more generally, the ISM-planet connection is explored in more detail.

This manuscript should not be considered a comprehensive review of the subject of the LISM, but an individual, and biased, thread through a rich research area. I certainly will not be able to explore many LISM topics to the level they deserve, nor will I be able to highlight all the work done by the large number of researchers in this area. Hopefully, this short review will introduce you to some new ideas, and the references provided can escort you to even more work that was not specifically mentioned in this manuscript.

2.   Properties of the Local Interstellar Medium

Measuring the morphological and physical characteristics of the nearest interstellar gas has long been of interest to astronomers. Observations of the general ISM via interstellar extinction of background stars (Neckel et al. 1980), the 21 cm H I hyperfine transition (Lockman 2002), or foreground interstellar absorption in optical resonance lines (Cowie & Songaila 1986) typically focus on more distant ISM environments due to the observational challenges inherent in measuring the properties of the LISM (see §2.2). Recent reviews that focus specifically on the LISM by Ferlet (1999) and Frisch (1995, 2004) also provide some discussion on the history of the field.

2.1.   Our Cosmic Neighborhood

The LISM is a diverse collection of gas. The outer bound of what I consider to be "local" in the context of the LISM, is the edge of Local Bubble (LB), coincident, by definition, with the location of the nearby dense molecular clouds, such as Taurus and Ophiuchus (Lallement et al. 2003). Figure 1 is a schematic illustration of the volume populated by the LISM, adapted from a similar figure by Mewaldt & Liewer (2001). Within the LISM volume, our cosmic neighborhood so to speak, resides the nearest 10^4 to 10^5 stars, including almost all known planetary systems (see Ford, in this volume) and the prototypical debris disk stars (see Chen, in this volume). Among these stars drifts interstellar gas known as the LISM. Several warm partially ionized clouds, such as the Local Interstellar Cloud (LIC) and the Galactic (G) Cloud, are observed within the Local Bubble.

The LIC is the material that directly surrounds our solar system. The internal pressure and ram pressure of the LIC, functions of its density and velocity relative to the Sun, balance the force of the outward-moving solar wind to define the boundary and shape of the heliosphere (see §3.1). The heliospheric structure is not unique to the Sun, but is observed around other stars, including the near-

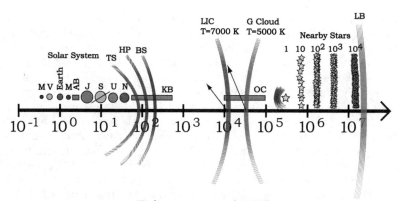

Distance (AU)

Figure 1.   Our cosmic neighborhood, shown on a logarithmic scale. The Solar System includes the major planets, the asteroid belt (AB), the Kuiper belt (KB), and the Oort cloud (OC). As of October 2005, Voyager 1 was 97.0 AU from the Sun. The heliospheric structure consists of the termination shock (TS), the heliopause (HP), and the bow shock (BS). The Local Interstellar Cloud (LIC) is the warm, partially ionized cloud that directly surrounds our solar system and currently determines the size and shape of the heliosphere. Our nearest neighboring cloud is the Galactic (G) Cloud which directly surrounds the α Cen stellar system. Analogous to the heliosphere, the α Cen system includes an astrosphere. The LISM resides in a larger ISM structure, the Local Bubble (LB). Within the Local Bubble, which extends $\sim$100 pc from the Sun, lie the nearest 10^4–10^5 stars. This figure is inspired by a diagram in Mewaldt & Liewer (2001).

est stellar system, α Cen (Wood et al. 2001). Nor is the heliospheric structure static. The heliosphere will expand and contract, in response to the density of the LISM material surrounding the Sun (see §3.2).

The boundary between the LISM and the solar system is not an obvious one. The Oort cloud contains the most distant objects that are gravitationally bound to the Sun, which reside at a third of the distance to α Cen. The Oort cloud is completely enclosed by LISM material, and due to the short extent of the LIC in the direction of α Cen (Redfield & Linsky 2000), Oort cloud objects in that direction may be surrounded by a different collection of gas (G Cloud material) than surrounds our planetary system (LIC material). Currently, even much of the Kuiper belt (KB) extends beyond the bow shock (BS) into pristine LISM material (see Sheppard, in this volume). Some neutral LISM material can penetrate well into the solar system. This material is utilized to make *in situ* measurements of properties of the LIC (see §2.2). The interconnectedness of our solar system with our surrounding interstellar medium could have important consequences for planetary atmospheres (see §4).

Hot Gas: Local Bubble   The Local Bubble refers to the apparent lack of dense cold material within approximately 100 pc of the Sun. Therefore, the distance to the edge of the Local Bubble cavity is equal to the distance at which cold dense gas is first observed. Lallement et al. (2003) were able to map out the contours of the Local Bubble by tracing the onset of the detection of foreground

Na I absorption lines in the spectra of ∼1000 early type stars. In general, the interstellar material within the Local Bubble is too hot for neutral sodium, and therefore none is detected until the edge of the Local Bubble is reached. The edge of the Local Bubble can be as close as 60 pc, and as far as ∼250 pc, or even unbound, as toward the north and south galactic poles (Lallement et al. 2003).

The hot gas within the Local Bubble is notoriously difficult to observe. Early direct detections of million degree gas came from diffuse soft X-ray emission (Snowden et al. 1998), although part of this emission is now thought to be caused by charge exchange reactions in the heliosphere, the same process that causes comets to emit X-rays (Lallement 2004). Absorption lines of highly ionized elements are generally weak or not detected, as Oegerle et al. (2005) found when looking for O VI absorption toward nearby white dwarfs. Detecting emission from highly ionized atoms has also been more difficult than expected. Using the *Cosmic Hot Interstellar Plasma Spectrometer* (*CHIPS*), Hurwitz et al. (2005) do not detect the array of extreme ultraviolet emission lines that are predicted from the "standard" Local Bubble temperature and density. Canonically, it is thought that the Local Bubble is filled with $T \sim 10^6$ K, $n \sim 5 \times 10^{-3}$ cm^{-3} gas that extends about $R \sim 100$ pc. However, much work remains to be done to understand the nature of the hot Local Bubble gas.

Warm Gas: Local Interstellar Clouds Within the hot Local Bubble substrate are dozens of individual accumulations of diffuse gas that are warm and partially ionized ($T \sim 7000$ K, $n \sim 0.3$ cm^{-3}, $R \sim 0.5$–5 pc). It is most commonly this material that is being referenced with the term "local interstellar medium." It is warm, partially ionized material that directly surrounds our solar system, and which can be measured with *in situ* observations and high resolution optical and ultraviolet (UV) spectroscopy (see §2.2). The warm LISM will dominate the remainder of this review, because it is the best studied of the different phases of LISM material, and the most significant with regards to interaction with stars.

Typical properties of the local warm interstellar clouds are given in Table 1. Absorption line spectroscopy and *in situ* observations often provide independent measurements of the same quantity (e.g., T, v_0, l_0, b_0). In addition, the two techniques are often complementary, as when parameters derived from absorption spectra of nearby stars (e.g., $N_{\rm HI}/N_{\rm HeI}$) are combined with *in situ* measurements (e.g., $n_{\rm HeI}$) to determine a third physical quantity that would be difficult or impossible to determine using either technique alone (e.g., $n_{\rm HI}$). The LISM is a diverse collection of gas. For example, individual temperature determinations of LISM material can be made to the precision of ±200 K, although temperatures ranging from 2000–11000 K are observed (Redfield & Linsky 2004b). The value given for the LISM in Table 1 is a weighted LISM mean.

Despite the diversity, there are several observational clues that indicate a coherence in the LISM. First, LISM absorption features in high resolution spectra can almost always be fit by one to three individual, symmetric, well-separated, Gaussian profiles, as opposed to a broad asymmetric feature that would result from multiple absorption features with a gradient of velocities. Second, the projected velocities of LISM features toward nearby stars can be characterized by a single bulk flow (Lallement & Bertin 1992; Frisch et al. 2002) that matches the observed ISM flow into our solar system (Witte 2004). However, small deviations from this bulk velocity vector are observed in the direction of the leading

Table 1. Properties of Warm Local Interstellar Clouds

Property	Value	Ref.	Comments[a]
Temperature (T)	6680 ± 1490 K	1	LISM, AL
	6300 ± 340 K	2	LIC, *IS*
Turbulent Velocity (ξ)	2.24 ± 1.03 km s^{-1}	1	LISM, AL
Velocity Magnitude (v_0)	25.7 ± 0.5 km s^{-1}	3	LIC, AL
	28.1 ± 4.6 km s^{-1}	4	LISM, AL
	26.3 ± 0.4 km s^{-1}	2	LIC, *IS*
Velocity Direction (l_0, b_0)	186°1, −16°4	3	LIC, AL
	192°4, −11°6	4	LISM, AL
	183°3 ± 0°5, −15°9 ± 0°2	2	LIC, *IS*
HI Column Density (log N_{HI})	17.18 ± 0.70 cm^{-2}	5	LISM, AL
HeI Volume Density (n_{HeI})	0.0151 ± 0.0015 cm^{-3}	6	LIC, *IS*
HI and HeI Ratio (N_{HI}/N_{HeI})	14.7 ± 2.0	7	LISM, AL
Electron Volume Density (n_e)	$0.11^{+0.12}_{-0.06}$ cm^{-3}	8	LIC, AL
HI Volume Density (n_{HI})	0.222 ± 0.037 cm^{-3}	6,7	$n_{HI} = n_{HeI} \times N_{HI}/N_{HeI}$
Thermal Pressure (P_T/k)	3180^{+1850}_{-1130} K cm^{-3}	1,6,7	$P_T = nkT$
Turbulent Pressure (P_ξ/k)	140^{+140}_{-130} K cm^{-3}	1,6,7	$P_\xi = 0.5\rho\xi^2$
Hydrogen Ionization (X_H)	$0.33^{+0.24}_{-0.13}$	6,7,8	$X_H = n_{HII}/(n_{HI} + n_{HII})$
Absorbers per sightline	1.8	9	LISM, AL
Cloud size (r)	~2.3 pc	10	LIC, AL
Cloud mass (M)	~0.32 $M_\odot$	10	LIC, AL

[a] LISM = average of several LISM sightlines; LIC = quantity for LIC material only; AL = derived from absorption line observations; *IS* = derived from *in situ* observations.

REFERENCES.—(1) Redfield & Linsky 2004b; (2) Witte 2004; (3) Lallement & Bertin 1992; (4) Frisch, Grodnicki, & Welty 2002; (5) Redfield & Linsky 2004a; (6) Gloeckler et al. 2004; (7) Dupuis et al. 1995; (8) Wood & Linsky 1997; (9) Redfield & Linsky 2002; (10) Redfield & Linsky 2000;

edge of the LIC, where the gas appears to be decelerated, possibly due to a collision of LIC material with neighboring LISM material (Redfield & Linsky 2001). Third, chaotic small scale structure in the LISM has not been detected. In one example, a collection of 18 Hyades stars, separated by only 1°–10°, shows a smooth slowly varying gradient in column density with angular distance, as opposed to a chaotic, filamentary geometry. However, an extensive database of observations is required to fully study the three dimensional structure of the LISM (see §2.3).

Several properties of warm LISM clouds are not well known. Except for the material currently streaming into the solar system, volume densities are very difficult to measure. Inherent in absorption line observations are: (1) the ignorance of a length scale to the absorbing material, other than the limit set by the distance of the background star, and (2) the ignorance of density variations within a single collection of gas, since only the total column density is observed.

Measurements of the magnetic field in the LISM have also been difficult to make. For lack of better measurements, observations of distant (several kpc) pulsars give a "local" galactic magnetic field strength of ~1.4 µG (Rand & Lyne 1994), although this value may have little to do with the magnetic fields entrained in the LISM. Polarization measurements of nearby (<35 pc) stars are weak, but seem to indicate an orientation parallel to the galactic plane (Tinbergen 1982; Frisch 2004). The same orientation is derived from heliospheric observations of 1.78–3.11 kHz radio emission by Voyager 1 (Kurth & Gurnett 2003). The orientation and strength of the local magnetic field will have important consequences on the structure and shape of the heliosphere (e.g., Gloeckler et al. 1997; Pogorelov et al. 2004; Florinski et al. 2004; Lallement et al. 2005). Additionally, since the magnetic pressure goes as B^2, the strength of the LISM magnetic field (B) could have important consequences for the relationship between the hot Local Bubble gas and the warm LISM clouds. The apparent pressure imbalance between the warm ($P_{tot}/k \sim 3300\,\mathrm{K\,cm^{-3}}$, see Table 1) and hot ($P_T/k \sim 10000\,\mathrm{K\,cm^{-3}}$, see "Hot Gas" section above) components of the LISM has been a persistent topic concerning the structure of our local interstellar environment (Jenkins 2002). Possible scenarios include: (1) the presence of a LISM magnetic field of $\sim$4.8 μG to match the hot Local Bubble pressure, (2) refinement of Local Bubble observations and models that may reduce the Local Bubble pressure, in particular refining the X-ray and extreme ultraviolet observations of nearby hot gas (Hurwitz et al. 2005; Lallement 2004), or (3) the hot and warm components of the LISM are not in pressure equilibrium. It will be important to resolve this issue in order to understand the interaction between the hot Local Bubble gas and warm local interstellar clouds.

Cold Gas? Although the volume of the Local Bubble is defined by the scarcity of cold dense gas, there are some indications that collections of cold gas may reside within the Local Bubble. Observations of Na I by Lallement et al. (2003), which define the morphology of the Local Bubble, also identify a number of isolated dense clouds, just inside the Local Bubble boundary. Magnani et al. (1996) identify several small molecular clouds that are thought to be relatively nearby ($\leq$200 pc), although their precise distances are not often known. These clouds have $T \sim 20\,\mathrm{K}$, $n \sim 30\,\mathrm{cm^{-3}}$, and sizes of about $R \sim 1.4\,\mathrm{pc}$. One such nearby molecular cloud, MBM 40 ($\sim$100 pc), contains molecular cores, although they are not massive enough for star formation (Chol Minh et al. 2003). MBM 40 is likely an example of a dense cloud that resides within the Local Bubble boundary.

2.2.   How do we Measure the Properties of the LISM?

The proximity of the LISM presents unique challenges and opportunities for measuring the properties of nearby interstellar gas, which is too sparse to cause measurable reddening or be detected in atomic hydrogen 21 cm emission. One unique observational technique is *in situ* measurements of ISM particles, which stream directly into the inner solar system. These observations complement the traditional ISM observational technique of high resolution absorption line spectroscopy. Observing the closest ISM provides many advantages, such as simple absorption spectra, well known distances to background stars, and large projected areas that allow multiple observations through different parts of a single cloud, enabling a probe of its properties in three dimensions.

In Situ Measurements A powerful observational technique, absent in the vast majority of astrophysical research, is the ability to send instruments to physically interact with, collect, and measure the properties of the material of interest directly, instead of relying on photons. Due to the close proximity of the LISM, interstellar particles are continually streaming into the interplanetary medium. Neutral helium atoms, and helium "pick-up" ions (neutral helium atoms ionized as they approach the Sun and are "picked-up," or entrained, in the solar wind plasma), are observed by mass spectrometers onboard the *Ulysses* spacecraft. Möbius et al. (2004) review these measurements, together with He I UV-backscattered emission, collected and analyzed by several groups over many years. These observations give consistent measurements of the temperature, velocity, and density of the interstellar medium directly surrounding the solar system (see Table 1). LISM dust particles collected in the interplanetary medium are among the sample onboard the *Stardust* mission, scheduled to return to Earth in January 2006 (Brownlee et al. 2003). Laboratories will be able to analyze the collected particles in detail. Not only will this provide information about the nature of dust in the LISM, but may answer questions about the origin of ISM dust and its role in circumstellar and disk environments (see Chen, in this volume). Both Voyager spacecraft, launched in 1977, are still functioning and returning data as part of the Voyager Interstellar Mission (VIM). On 16 December 2004, Voyager 1 provided a long sought-after measurement of the distance to the termination shock, at 94.01 AU. With enough power to last until 2020, Voyager 1 should provide measurements of its encounter with pristine interstellar material once it crosses the heliopause around 2015 (Stone et al. 2005).

High Resolution Absorption Line Spectroscopy A standard technique used to measure the physical properties of foreground interstellar material along the line of sight toward a background star is high resolution absorption line spectroscopy. This kind of work has a long and rich history, but has typically been dominated by more distant ISM environments, with large column densities and strong absorption signatures (Cowie & Songaila 1986; Savage & Sembach 1996). The challenge inherent in absorption line spectroscopy of the LISM is the low column density along sightlines to nearby stars. This limits the number of diagnostic lines to only the strongest resonance line transitions. In Figure 2, the hydrogen column density sensitivities are shown for the 100 strongest ground-state transitions at wavelengths from the far-ultraviolet (FUV), through the ultraviolet (UV), to the optical. The lower sensitivity limit indicates a 3σ detection in a high signal-to-noise observation with modern high resolution instruments. The upper sensitivity limit marks the column density at which the transition becomes optically thick and leaves the linear part of the curve of growth, where, although absorption is detected, limited information can be obtained from the saturated absorption profile. Ionization structure typical for warm LISM clouds is incorporated (Slavin & Frisch 2002; Wood et al. 2002), although typical LISM depletion is not; only solar abundances are assumed (Asplund et al. 2005). The range of LISM absorbers ($16.0 \leq \log N_{\mathrm{H}}\,(\mathrm{cm}^{-2}) \leq 17.7$) is such that less than 100 transitions are available to study the LISM. Taking into account such issues as blending or continuum placement, which can limit the diagnostic value of an individual transition, reduces the number of useful transitions even more. Most of the transitions lie in the FUV and UV, with only a few transitions available in the optical, most importantly Ca II resonance lines at $\sim$3950 Å. (Other notable optical transitions, such as Na I and K I, probe more distant and higher column density ISM environments, see Lallement et al. 2003 and Welty & Hobbs 2001.) Recent LISM absorption line observations of these transitions have been made in the FUV with the *Far Ultraviolet Spectroscopic Explorer* (*FUSE*) (e.g., Lehner et al. 2003; Wood et al. 2002), in the UV with the *Hubble Space Telescope* (*HST*) (e.g., Redfield & Linsky 2004a, 2002), and in the optical with ultrahigh resolution spectrographs, such as those at McDonald Observatory and the Anglo-Australian Observatory (e.g., Crawford 2001; Welty et al. 1994).

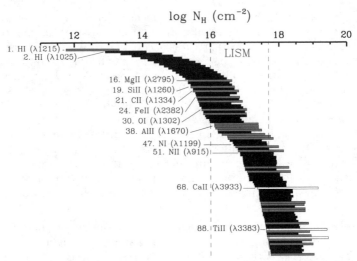

Figure 2.   The 100 strongest resonance lines, ranked in order of their hydrogen column density sensitivity. White bars indicate those that fall in the optical (3000-10000Å), gray for ultraviolet (1200-3000Å), and black for far-ultraviolet (900-1200Å). The top 15 lines are H I transitions from Lyman-α to Lyman-o, only Lyman series lines to Lyman-ω are shown. The vertical dashed lines indicate the typical range of hydrogen column densities observed for warm LISM clouds. Those transitions with sensitivities left of this range will be optically-thick and saturated, whereas those transitions to the right will not be sensitive enough to detect absorption from warm LISM clouds.

2.3.   Future Directions

Absorption line analyses, supported by *in situ* observations of the LIC, have resulted in numerous single sightline measurements of projected velocity, column density, and line width for several dozens of sightlines, leading to individual measurements of various physical properties of the LISM, including temperature and turbulence (Redfield & Linsky 2004b), electron density (Wood & Linsky 1997), ionization (Jenkins et al. 2000), depletion (Lehner et al. 2003), and small scale structure (Redfield & Linsky 2001). However, with the recent accumulation of significant numbers of observations, it is possible to go beyond the single sightline analysis and develop a global morphological and physical model of the LISM. Initial steps have been made toward this end with global bulk flow kinematic models of the LIC and G Clouds produced by Lallement & Bertin (1992), and a global morphological model of the LIC by Redfield & Linsky (2000).

Future LISM research will synthesize the growing database of LISM observations, taking advantage of the information contained in the comparison of numerous individual sightlines. A global morphological model would enable the development of global models of various physical properties, such as kinematics, ionization, depletion, density, etc. Ultimately, these global models are required to tackle larger issues that cannot be fully addressed by single sightline analyses, such as the interactions of clouds, the interaction of the warm LISM clouds with the surrounding hot Local Bubble substrate, the strength and orientation of magnetic fields, and the origin, evolution, and ages of clouds in the LISM.

Such work will not only be important in understanding the structure of gas in our local environment, but will be applicable to other, more distant and difficult to observe interstellar environments in our galaxy and beyond.

CLOSE: Ca II LISM Optical Survey of our Environment A global morphological model of the LISM requires high spatial and distance sampling. The development of a morphological model for the LIC by Redfield & Linsky (2000) was possible because LIC absorption is detected in practically every direction, since the material surrounds the solar system. More distant LISM clouds will subtend smaller angles on the sky, and will require higher density sampling of LISM observations in order to be morphologically characterized. The CLOSE (Ca II LISM Optical Survey of our Environment) project is a large scale, ultra-high resolution survey of ∼500 nearby stars that will enable a global morphological model of the LISM (work in collaboration with M.S. Sahu, B.K. Gibson, C. Thom, A. Hughes, N. McClure-Griffiths, P. Palunas). Previous Ca II surveys (Vallerga et al. 1993; Lallement et al. 1986; Lallement & Bertin 1992; Welty et al. 1996, and see summary, Redfield & Linsky 2002) only accumulated ∼50 nearby sightlines, but detected LISM Ca II absorption in 80% of the targets. Our survey will provide extensive coverage. The maximum angular distance between any two adjacent targets will be <10°, including more than 45 target pairs that will be <1° apart, providing an interesting study of small scale structure in the LISM. When combined with past and future observations, this survey will provide a significant baseline with which to search for long-term LISM absorption variation, as the sightlines to these high proper motion nearby stars vary over timescales of decades. Ultimately, the CLOSE project will provide a valuable database for the development of a global morphological model of the LISM.

3. Relationship Between Stars and their Local Interstellar Medium

As discussed in reference to Figure 1, the interaction between the LISM and solar/stellar winds is mediated by the heliospheric/astrospheric interface. This interface is defined by the balance between the solar/stellar wind and the LISM. Reviews of heliospheric modeling include Zank (1999) and Baranov (1990), and the detection of astrospheres around nearby stars is reviewed by Wood (2004).

3.1. The Heliosphere and Astrospheres

In the standard picture of the heliosphere, discussed by Zank (1999), Baranov (1990) and Wood (2004), the magnetized solar wind is shocked to subsonic speeds ("termination shock"), as is the ionized LISM material ("bow shock"). The interface in between ("heliopause") is where the plasma flows of the solar wind and LISM are deflected from each other. It was originally thought that neutral atoms from the LISM pass through the heliosphere unimpeded and therefore have a negligible influence on the structure of the heliosphere. However, charge exchange reactions between ionized hydrogen in the solar wind and neutral hydrogen in the LISM act to heat and decelerate LISM hydrogen atoms just prior to the heliopause. The resulting structure, referred to as the "hydrogen wall," is an accumulation of hot hydrogen between the heliopause and the bow shock.

At the same time that heliospheric simulations were indicating an enhancement of hydrogen just beyond the heliopause, it was becoming clear that observations of LISM absorption in H I Lyman α were discrepant with other LISM absorption lines along the same line of sight. In particular, excess H I absorption was required on the red and blue sides of the LISM H I absorption feature, in order to be consistent with the optically thin D I absorption profile, which is only $82\,\mathrm{km\,s^{-1}}$ to the blue of the H I absorption. The models predicted a column density for the "hydrogen wall" of log $N_\mathrm{H}\,(\mathrm{cm^{-2}}) \sim 14.5$. From Figure 2, it is clear that only the Lyman series hydrogen lines are sensitive enough to detect these low column densities. Indeed, heliospheric H I absorption was first detected using the Lyman-α profile, when Linsky & Wood (1996) measured excess absorption redshifted with respect to the LISM absorption. The blueshifted excess absorption is associated with an astrosphere. Because we observe the decelerated heliospheric hydrogen from the inside, the heliospheric absorption is redshifted, whereas the decelerated hydrogen in an astrosphere is observed exterior to the astrosphere, and therefore is blueshifted, see Figure 6 of Wood (2004).

It should be noted that if the interstellar absorption gets to be too large, log $N_\mathrm{H}\,(\mathrm{cm^{-2}}) \geq 18.7$, the saturated ISM absorption will obliterate any sign of a slightly offset heliospheric or astrospheric absorption. So, these measurements are only possible within the low column density volume of the LISM. Among a sample of nearby stars, heliospheric and astrospheric absorption has been detected for many stars (Wood et al. 2005b). Multiple heliospheric detections sample the structure of our heliosphere in three dimensions, while multiple astrospheric detections provide measurements of weak solar-like winds around other stars. More than 50% of stars within 10 pc that have high resolution UV Lyman-α spectra show signs of astrospheric absorption (Wood et al. 2005a).

3.2.   Heliospheric Variability

Short Term   The solar wind strength and distribution fluctuate with the 11-year solar cycle (Richardson 1997). These variations slowly propagate out to the heliospheric boundary and it is expected that the heliopause will expand and contract on a comparable timescale. The stochastic injection of energy into the solar wind in the form of flares and mass ejections leads to variability on even shorter timescales. This dynamic wind is constantly buffeting the magneto-spheres of planets in its path as well as the heliospheric boundary. Voyager 1 may have detected such short-term variability when over a 7-month period in 2003 the termination shock contracted inward, over Voyager 1, and then expanded back outward over Voyager 1 yet again, a year before Voyager 1 unambiguously crossed the termination shock (Krimigis et al. 2003; McDonald et al. 2003).

Long Term   Long-term variations in the solar wind strength are not well known, but observations of astrospheres around young solar analogs provide clues as to what kind of wind the Sun had in its distant past. The solar wind, 3.5 billion years ago, may have been $\sim 35\times$ stronger than it is today (Wood et al. 2005a). In contrast, density variations spanning 6 orders of magnitude are commonly observed throughout the general ISM. Since the Sun has likely encountered a number of different ISM environments with extreme variations in density, it seems quite intuitive to expect that the variation in density of our surrounding interstellar environment plays the dominant role in long-term variations of the heliosphere. Detailed models support this intuition. Zank & Frisch (1999) model the modern heliosphere surrounded by a LISM density $50\times$ the current value and find that the termination shock shrinks from $\sim 100\,\mathrm{AU}$ to $\sim 10\,\mathrm{AU}$. In the next section, we will explore the possible consequences for planets, such as Earth, caught in the midst of such a dramatic change in the structure of the heliosphere.

4. The LISM-Earth or ISM-Planet Connection

Discussion of a LISM-Earth connection has an incredibly long history. Shapley (1921) and Hoyle & Lyttleton (1939) are among the earliest references. Research has intensified as of late; observations and models have improved, geologic and climatic events remain unexplained, and it is becoming clear that the habitability of planets may be dependent on subtle astrobiological parameters (e.g., Fahr 1968; Begelman & Rees 1976; Thaddeus 1986; Frisch 1998; Florinski et al. 2003; Yeghikyan & Fahr 2004; Wallmann 2004; Pavlov et al. 2005; Gies & Helsel 2005).

4.1. Planetary Consequences of Heliospheric Variability

It appears inevitable that the heliosphere has and will continue to expand and contract as the Sun passes through different ISM environments. The Sun is now surrounded by a relatively modest density cloud, the LIC, and the heliopause stands well beyond the planets at $\sim 100\,\mathrm{AU}$. However, when the Sun encounters more dense ISM clouds, the heliosphere will shrink, perhaps to within the inner solar system. The question then arises: What are the consequences, on Earth and on other planets, of a compressed heliosphere/astrosphere?

Cosmic Rays The solar wind is magnetized and extends out to the heliopause. Energetic charged particles, or cosmic rays, interact with this magnetic field and if they are not too energetic ($\leq 1\,\mathrm{GeV}$), the cosmic rays are partially or completely prevented from penetrating far into the heliosphere. The heliosphere is one of three screens, together with the Earth's magnetic field and the Earth's atmosphere, that modulate the cosmic ray flux at the surface of the Earth. The loss of heliospheric modulation would lead to flux increases of 10–$100\times$ at energies of 10–$100\,\mathrm{MeV}$ at the top of the terrestrial magnetosphere (Reedy et al. 1983), and could have serious consequences for several planetary processes.

Cosmic rays are important sources of ionization in the upper atmosphere, creating showers of secondary particles, ultimately leading to muons, which dominate the cosmic ray flux in the lower atmosphere. Ions in the atmosphere may serve as cloud nucleation sites, increasing low altitude clouds and ultimately increasing the planetary albedo (Carslaw et al. 2002). A connection between cosmic rays and clouds was suggested by Svensmark & Friis-Christensen (1997) and a correlation between cosmic ray flux and low cloud cover over the course of a solar cycle was found by Marsh & Svensmark (2000), even though the global cosmic ray flux varied by only $\sim 15\%$. The 1–2 orders of magnitude variation that would result from the loss of the heliospheric screen could have a tremendous impact on the formation of clouds in the Earth's atmosphere. The ionization caused by cosmic ray secondaries may also trigger lightning production (Gurevich & Zybin 2001). Cosmic rays increase the production of NO and NO_2 in the upper stratosphere, which significantly influences ozone layer chemistry. Randall et al. (2005) tracked the enhancement of nitric oxide and nitrogen dioxide that resulted from an injection of low energy cosmic rays following a series of intense solar storms. In some locations within the polar vortex, the increased levels of NO and NO_2 led to 60% reductions in terrestrial ozone over several months.

Muons, electrons, and other cosmic ray products are significant natural radiation sources, accounting for 30–40% of the annual dose from natural radiation in the United States (Alpen 1998). Muons ionize atoms in our bodies, producing hydroxyl radicals, which can cause DNA mutations. The cosmic muon flux can be significant even at depths of 1 km below the Earth's surface, and is therefore a source of mutation even for deep-sea or deep-earth organisms. The loss of the heliospheric screen, and the subsequent increase in cosmic ray flux on the Earth's atmosphere, could have important implications for the long-term evolution of the Earth's climate, and for long-term mutation rates in terrestrial organisms.

Dust Deposition Passage through dense ISM environments will compress the heliosphere while also depositing a significant amount of interstellar dust onto the top of the Earth's atmosphere. Pavlov et al. (2005) presented atmospheric models in which large amounts of interstellar dust cause a reverse greenhouse effect, blocking or scattering incident visible light while being transparent to infrared thermal radiation. McKay & Thomas (1978) also find that increased deposition of interstellar H_2 onto Earth's atmosphere would decrease mesospheric ozone levels, decrease the mesospheric temperature, and cause high altitude noctilucent ice clouds, which would ultimately increase the planetary albedo. Dust deposition could be a natural trigger for "snowball" Earth episodes, periods of ~200,000 years in which the Earth is entirely glaciated (Hoffman et al. 1998).

4.2. Future Directions

Most of the work on the LISM-Earth or ISM-planet connection has been theoretical. The few observational tests of this relationship have focused on correlating the largest ISM fluctuations with the largest climatic or geological fluctuations. For example, Shaviv (2003) claimed a correlation between the passage through spiral arms, the glaciation period, and the cosmic ray flux derived from iron meteorites. The most significant hurdles that these kinds of empirical tests must overcome are the extreme systematic errors that plague long-term astronomical, climatic, and meteoritic timescales that are so vital to demonstrating a convincing correlation. Invoking passage through spiral arms and major glaciations requires accurate temporal calibration back more than a billion years.

Although it is intuitive to attach the largest fluctuations in the ISM to the largest fluctuations in the climatic record, with respect to the modulation of the cosmic ray flux, modest density clouds may have a significant influence. As demonstrated in the model by Zank & Frisch (1999), a modest density LISM has a dramatic effect on the structure of the heliosphere. Although higher density environments would compress the heliosphere still further, there will be a point of diminishing returns as the incident cosmic ray flux on Earth's atmosphere approaches the galactic cosmic ray flux level. Future work will explore the evolution of the heliosphere through a variety of ISM environments, besides our current LISM surroundings and intermediate to the extremes of giant molecular clouds in the heart of spiral arms.

Historical Solar Trajectory Linking the long-term timing of ice ages with spiral arm passages will continue to be an interesting test of the ISM-planet connection, but the LISM provides a provocative alternative empirical test (work in collaboration with J. Scalo and D.S. Smith). As discussed in §2, we know precisely the nature of the LISM that directly surrounds our solar system now from *in situ* measurements of the LIC. If we look at a very nearby star, in the direction of the historical solar trajectory (Dehnen & Binney 1998), the observed LISM absorption should provide information on the nature of the LISM that the Sun encountered only a short time ago. (The Sun travels 1 pc in approximately 73,000 years.) If we continue this exercise, looking out the rearview mirror, so to speak, it should be possible to reconstruct a deterministic history of the ISM that the Sun experienced in the not-too-distant past. The Sun's ISM history could then be converted into a cosmic ray flux history, based on the heliospheric response to the historical interstellar density profile. As we sample more distant environments, and therefore more distant times, the peculiar motions of the Sun and the ISM cease to make a true deterministic history of the Earth's cosmic ray flux possible, but would still represent a possible and plausible cosmic ray history of the Earth, or any other planet orbiting a nearby star. Although the most recently experienced ISM environments may not include dense molecular clouds, nor the most recent climatic history include dramatic periods of global glaciations, such a short-term test does not suffer the systematics that make long-term correlations difficult. Both the astronomical (*Hipparcos* distances to nearby stars) and climatic (Zachos et al. 2001) records of the most recent past are sampled at high temporal resolution and are well calibrated.

5.   Conclusions

The LISM is a unique conduit that connects large scale galactic and extragalactic structures with planetary atmospheres. Although every observation of an astronomical object outside our solar system (and even some "within" our solar system) peers through the LISM, we do not, as of yet, have a detailed three-dimensional global understanding of the morphological or physical properties of our own interstellar environment. The challenges of observing the properties of the LISM are slowly being met with larger databases of high resolution spectra taken from the FUV to the optical, along sightlines toward nearby stars. Due to the limited number of transitions that are sensitive to the low column densities of LISM material, it is critical to have access to high resolution spectrographs from the FUV to the optical. With the recent loss of the Space Telescope Imaging Spectrograph (STIS) on *HST*, no high resolution astronomical spectrograph ($R \equiv \lambda/\Delta\lambda \geq 50,000$) is currently operating in the ultraviolet. A new high resolution UV instrument is needed in order to preserve and expand our observational capabilities in the UV. Without such a facility, among other losses, we will no longer have the ability to observe gas in the LISM, and risk being ignorant of our most immediate interstellar surroundings. The LISM interacts with stellar and planetary systems and as we explore the subtle astrobiological parameters that control the degrees of habitability of planets, the role of the LISM may turn out to be significant.

Since the LISM, and the ISM in general, is an important part of a grand "galactic ecology," research on the LISM touches many different areas of astrophysics, often at a profound level. Carl Sagan once said that "we are made of star-stuff" (Sagan 1980). Created in stars, our "stuff" was mixed and transported from across the galaxy by the ISM and, eventually, it is likely that this "stuff" will return there to be recycled yet again.

Acknowledgments.   Support for this work was provided by NASA through Hubble Fellowship grant HST-HF-01190.01 awarded by the Space Telescope Science Institute, which is operated by the Association of Universities for Research in Astronomy, Inc., for NASA, under contract NAS 5-26555. I would like to thank John Scalo and Brian Wood for their helpful suggestions.

References

Alpen, E. L. 1998, Radiation Biophysics, 2nd edn. (San Diego: Academic Press)
Asplund, M., Grevesse, N., & Sauval, A. J. 2005, in ASP Conf. Ser. 336: Cosmic Abundances as Records of Stellar Evolution and Nucleosynthesis in Honor of David L. Lambert, ed. T. G. Barnes & F. N. Bash (San Francisco: ASP), 25
Baranov, V. B. 1990, Space Science Reviews, 52, 89
Begelman, M. C., & Rees, M. J. 1976, Nat, 261, 298
Brownlee, D. E., et al. 2003, J. Geophysical Research (Planets), 108, 1
Burton, M. 2004, in IAU Symp. 213, Bioastronomy 2002: Life Among the Stars, ed. R. Norris & F. Stootman (San Francisco: ASP), 123
Carslaw, K. S., Harrison, R. G., & Kirkby, J. 2002, Science, 298, 1732
Chol Minh, Y. C. Y., et al. 2003, New Astronomy, 8, 795
Cowie, L. L., & Songaila, A. 1986, ARA&A, 24, 499
Crawford, I. A. 2001, MNRAS, 327, 841
Dehnen, W., & Binney, J. J. 1998, MNRAS, 298, 387
Dickey, J. M., & Lockman, F. J. 1990, ARA&A, 28, 215
Dupuis, J., Vennes, S., Bowyer, S., Pradhan, A. K., & Thejll, P. 1995, ApJ, 455, 574
Fahr, H. J. 1968, Ap&SS, 2, 474
Ferlet, R. 1999, A&A Rev., 9, 153
Florinski, V., Pogorelov, N., Zank, G., Wood, B. E., & Cox, D. P. 2004, ApJ, 604, 700
Florinski, V., Zank, G. P., & Axford, W. I. 2003, Geophys. Res. Lett., 30, 5
Frisch, P. C. 1995, Space Science Reviews, 72, 499
Frisch, P. C. 1998, in Planetary Systems: The Long View, ed. L. M. Celnikier & J. Tran Than Van (Editions Frontiers), 1
Frisch, P. C. 2004, Advances in Space Research, 34, 20
Frisch, P. C., Grodnicki, L., & Welty, D. E. 2002, ApJ, 574, 834
Gies, D. R., & Helsel, J. W. 2005, ApJ, 626, 844
Gloeckler, G., Fisk, L. A., & Geiss, J. 1997, Nat, 386, 374
Gloeckler, G., et al. 2004, A&A, 426, 845
Gurevich, A. V., & Zybin, K. P. 2001, Uspekhi Fizicheskikh Nauk, 44, 1119
Hoffman, P. F., Kaufman, A. J., Halverson, G., & Schrag, D. 1998, Science, 281, 1342
Hoyle, F., & Lyttleton, R. A. 1939, in Proc. of the Cambr. Phil. Soc., 405
Hurwitz, M., Sasseen, T. P., & Sirk, M. M. 2005, ApJ, 623, 911
Jenkins, E. B. 2002, ApJ, 580, 938
Jenkins, E. B., et al. 2000, ApJ, 538, L81
Krimigis, S. M., et al. 2003, Nat, 426, 45
Kurth, W. S., & Gurnett, D. A. 2003, J. Geophysical Research (Space Physics), 108, 2
Lallement, R. 2004, A&A, 418, 143
Lallement, R., & Bertin, P. 1992, A&A, 266, 479
Lallement, R., Quémerais, E., Bertaux, J. L., Ferron, S., Koutroumpa, D., & Pellinen, R. 2005, Science, 307, 1447
Lallement, R., Vidal-Madjar, A., & Ferlet, R. 1986, A&A, 168, 225
Lallement, R., Welsh, B. Y., Vergely, J. L., Crifo, F., & Sfeir, D. 2003, A&A, 411, 447
Lehner, N., Jenkins, E. B., Gry, C., Moos, H. W., Chayer, P., & Lacour, S. 2003, ApJ, 595, 858
Linsky, J. L., & Wood, B. E. 1996, ApJ, 463, 254
Lockman, F. J. 2002, in ASP Conf. Ser. 276: Seeing Through the Dust, ed. A. R. Taylor, T. L. Landecker, & A. G. Willis (San Francisco: ASP), 107

Magnani, L., Hartmann, D., & Speck, B. G. 1996, ApJS, 106, 447
Marsh, N. D., & Svensmark, H. 2000, Physical Review Letters, 85, 5004
McCray, R., & Kafatos, M. 1987, ApJ, 317, 190
McDonald, F. B., Stone, E. C., Cummings, A. C., Heikkila, B., Lal, N., & Webber, W. R. 2003, Nat, 426, 48
McKay, C. P., & Thomas, G. E. 1978, Geophys. Res. Lett., 5, 215
Mewaldt, R. A., & Liewer, P. C. 2001, in COSPAR Colloq. Ser. 11, The Outer Heliosphere: The Next Frontiers, ed. K. Scherer, H. Fichtner, H. J. Fahr, & E. Marsch (Amsterdam: Pergamon Press), 451
Möbius, E., et al. 2004, A&A, 426, 897
Neckel, T., Klare, G., & Sarcander, M. 1980, A&AS, 42, 251
Oegerle, W., Jenkins, E., Shelton, R., Bowen, D., & Chayer, P. 2005, ApJ, 622, 377
Pavlov, A. A., Toon, O. B., Pavlov, A. K., Bally, J., & Pollard, D. 2005, Geophys. Res. Lett., 32, 3705
Pogorelov, N. V., Zank, G. P., & Ogino, T. 2004, ApJ, 614, 1007
Rand, R. J., & Lyne, A. G. 1994, MNRAS, 268, 497
Randall, C. E., et al. 2005, Geophys. Res. Lett., 32, 5802
Rauch, M., Sargent, W. L. W., & Barlow, T. A. 1999, ApJ, 515, 500
Redfield, S., & Linsky, J. L. 2000, ApJ, 534, 825
Redfield, S., & Linsky, J. L. 2001, ApJ, 551, 413
Redfield, S., & Linsky, J. L. 2002, ApJS, 139, 439
Redfield, S., & Linsky, J. L. 2004a, ApJ, 602, 776
Redfield, S., & Linsky, J. L. 2004b, ApJ, 613, 1004
Reedy, R. C., Arnold, J. R., & Lal, D. 1983, Science, 219, 127
Richardson, J. D. 1997, Geophys. Res. Lett., 24, 2889
Sagan, C. 1980, Cosmos (New York: Random House)
Savage, B. D., & Sembach, K. R. 1996, ARA&A, 34, 279
Shapley, H. 1921, J. Geology, 29, 502
Shaviv, N. J. 2003, New Astronomy, 8, 39
Slavin, J. D., & Frisch, P. C. 2002, ApJ, 565, 364
Snowden, S. L., Egger, R., Finkbeiner, D. P., Freyberg, M. J., & Plucinsky, P. P. 1998, ApJ, 493, 715
Stone, E. C., Cummings, A. C., McDonald, F. B., Heikkila, B. C., Lal, N., & Webber, W. R. 2005, Science, 309, 2017
Svensmark, H., & Friis-Christensen, E. 1997, J. Atmos. and Terrestrial Phys., 59, 1225
Thaddeus, P. 1986, in The Galaxy and the Solar System, ed. R. Smoluchowski, J. M. Bahcall, & M. S. Matthews (Tucson: University of Arizona Press), 61
Tinbergen, J. 1982, A&A, 105, 53
Vallerga, J. V., Vedder, P. W., Craig, N., & Welsh, B. Y. 1993, ApJ, 411, 729
Wallmann, K. 2004, Geochemistry, Geophysics, Geosystems, 5, 6004
Welty, D. E., & Hobbs, L. M. 2001, ApJS, 133, 345
Welty, D. E., Hobbs, L. M., & Kulkarni, V. P. 1994, ApJ, 436, 152
Welty, D. E., Morton, D. C., & Hobbs, L. M. 1996, ApJS, 106, 533
Witte, M. 2004, A&A, 426, 835
Wood, B. E. 2004, Living Reviews in Solar Physics, 1, 2
Wood, B. E., & Linsky, J. L. 1997, ApJ, 474, L39
Wood, B. E., Linsky, J. L., Müller, H.-R., & Zank, G. P. 2001, ApJ, 547, L49
Wood, B. E., Müller, H.-R., Zank, G., Linsky, J., & Redfield, S. 2005a, ApJ, 628, L143
Wood, B. E., Redfield, S., Linsky, J. L., Müller, H.-R., & Zank, G. P. 2005b, ApJS, 159, 118
Wood, B. E., Redfield, S., Linsky, J. L., & Sahu, M. S. 2002, ApJ, 581, 1168
Yeghikyan, A., & Fahr, H. 2004, A&A, 415, 763
Zachos, J., Pagani, M., Sloan, L., Thomas, E., & Billups, K. 2001, Science, 292, 686
Zank, G. P. 1999, Space Science Reviews, 89, 413
Zank, G. P., & Frisch, P. C. 1999, ApJ, 518, 965

Naomi McClure-Griffiths describes the energy balance of the Galactic
interstellar medium.

Frank N. Bash Symposium 2005: New Horizons in Astronomy
ASP Conference Series, Vol. 352, 2006
S. J. Kannappan, S. Redfield, J. E. Kessler-Silacci, M. Landriau, and N. Drory

Large Scale Structures in the Interstellar Medium

N. M. McClure-Griffiths

Australia Telescope National Facility, CSIRO, Epping, NSW, Australia

Abstract. Galactic interstellar medium (ISM) studies have undergone a renaissance in the past ten years due in part to new large-scale surveys of the Galactic ISM. New surveys of neutral hydrogen (H I), in particular the Canadian and Southern Galactic Plane Surveys, excel because of their exceptional spatial dynamic range, combining single-dish and interferometer data for arcminute resolution imaging over areas of many degrees. These surveys are allowing us to probe the physics of the global ISM in more detail than ever before. Among the largest and most energetic discrete objects observed in the ISM are H I supershells and chimneys. These objects play a pivotal, though not fully understood, role in the Galactic ecosystem. They heat and reshape the ISM on scales of parsecs to hundreds of parsecs and are a significant source of structure and energy input into the ISM. H I supershells also provide convenient locations for energy dissipation on small scales. For example, instabilities that develop along the dense swept-up walls may be sites of cooling and molecular cloud formation. Understanding this transition from the hot-ionized gas presumably filling the interior of shells, to the warm neutral medium of the outer walls, to the cold neutral medium along the small-scale instability-created features, and finally to the molecular gas is crucial to understanding the evolution of the ISM. In this review I show some examples of the detailed observations possible of large-scale supershells and also discuss their role in the interaction of the Galactic disk with the halo.

1. Introduction

Structure in the interstellar medium (ISM) of the Milky Way is observed over more than twelve orders of spatial scale, from AU scale features to kiloparsec scale features (Spangler 2001). Most of this structure can be described as turbulent, having no clear origins but following a power law of density fluctuations with size scale. Deterministic structure, or structure with clear origins, is also observed over a large range of size scales from sub-parsec to kiloparsec. Among the largest and most energetic of the deterministic structures are supershells, superbubbles and chimneys. These objects are detected at a variety of wavelengths, including atomic hydrogen (H I) emission, ^{12}CO emission, infrared radiation from dust, Hα, and soft X-rays.

Most of the supershells and chimneys in the Milky Way are detected in H I. These supershells are typically detected as voids in the interstellar H I with walls of swept-up emission (McClure-Griffiths et al. 2002). They are often detected as expanding objects with expansion velocities of 7–20 km s^{-1}. Although supershells are presumably filled with hot, ionized gas, soft X-ray emission has only been detected towards the closest shells. It is believed that most H I shells form through the combined effects of stellar winds and supernovae associated with massive stars. Clusters of massive stars are capable of releasing 10^{52-54} ergs of energy, ionizing the neutral medium and sweeping up a massive expanding shell.

In this review I discuss two topics related to supershells in the Milky Way: the role of supershells in the evolution of the ISM and the role of supershells in the interaction of the Galactic disk and halo. In § 2 I discuss in broad terms the role of supershells in the evolution of the ISM and follow this in § 3 by the discussion an instructive example object. In § 4 I discuss theories for the disk-halo interaction via supershells and the implications of recent observations. Finally, in § 5 I briefly explore some possible future directions of research in this field.

2. Role of H I Shells in the Evolution of the ISM

As a class of objects, supershells largely determine the structure and evolution of the ISM. In the Large Magellanic Cloud (LMC), the large scale structure of the ISM is dominated by over 100 H I supershells (Kim et al. 1999, 2003). The impact of supershells in the Milky Way is probably not as significant, but there are dozens of cataloged H I supershells (Heiles 1979, 1984; McClure-Griffiths et al. 2002). As a result of their large input energies, H I shells may be the large deterministic structures that drive the turbulent cascade of energy in the ISM. From soft X-ray studies of supershells, such as LMC2 in the LMC (Points et al. 2000), it is clear that supershells heat the ISM and ionize areas of hundreds of parsecs in scale. The walls of supershells may also play a role in energy dissipation on small scales by providing sites of cooling and molecular cloud formation. We do not know, however, how gas makes the transition from the hot, ionized state of the stellar outflows to the cool, neutral gas along the shell walls and finally how it breaks up into cold, molecular clouds. This question has dominated a large fraction of Galactic H I work in recent years, particularly in the Canadian and Southern Galactic Plane Surveys (Taylor et al. 2002; McClure-Griffiths et al. 2005).

Although we know that H I supershells play an important role in determining the structure of the ISM, there are a number of questions that remain: how important are H I shells as sites of cooling? how do H I shells evolve? how much of halo structure is formed from supershells? Theoretical simulations of evolving shells have addressed these questions at length (e.g., Mac Low & McCray 1988; Mac Low et al. 1989), even including magnetic fields (e.g., Tomisaka 1998) and extending to Galaxy-wide scales (e.g., de Avillez 2000; de Avillez & Berry 2001). Unfortunately, observations have lagged behind, owing largely to the poor resolution of single dish telescopes and the large angular sizes of supershells. In the past eight years new observational techniques have finally allowed us to approach these questions. By combining single dish and interferometric radio data it is now possible to image multi-degree sized H I supershells with arcminute resolution (e.g., McClure-Griffiths et al. 2003). The high spatial dynamic range of these data has re-opened the field of H I supershell studies, finally allowing us to study the physics of supershell evolution.

Velocity: 36.28 km/s

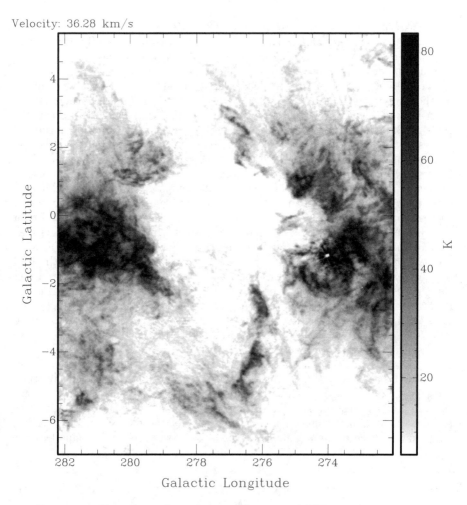

Figure 1. H I velocity channel grey-scale image of GSH 277+00+36 at $v =$ 36.3 km s^{-1}. The image shows the multiple chimney openings as well as the small-scale structure of the shells walls.

3. Physics of H I Shells: GSH 277+00+36 as an Example

Recent high resolution images of H I supershells have allowed us, for the first time, to resolve a large fraction of the ISM evolutionary life cycle as related to supershells. Figure 1 is a high-resolution image of a Galactic H I supershell, GSH 277+00+36 (McClure-Griffiths et al. 2003). This image has a resolution of $\sim3'$ across the entire 100 deg^2 field, corresponding to a physical resolution of ~6 pc at the distance of the shell. GSH 277+00+36 has a diameter of 610 ± 45 pc and an expansion velocity of 20 km s^{-1} (McClure-Griffiths et al. 2000) . In order to explain its size and rate of expansion, McClure-Griffiths et al. (2000) estimate an initial expansion energy of $\sim2.4 \times 10^{53}$ ergs. GSH 277+00+36 is a remarkable supershell because it appears to have broken at both sides of the Galactic disk, creating chimney-like conduits from the shell to the lower halo. It is also remarkable in the degree of detail visible within the shell; Figure 1 exhibits structure on all size scales.

The lifetime of an H I supershell like GSH 277+00+36 is ultimately limited by the shearing effects of Galactic differential rotation (Tenorio-Tagle & Bodenheimer 1988) and the onset of gas instabilities that can break down the walls of the shell (Dove et al. 2000). A variety of gas instabilities affect H I supershells. These instabilities are comprehensively reviewed by Dove et al. (2000), but quickly reviewing them here, the dominant types are the dynamical, Rayleigh-Taylor and gravitational instabilities. Dynamical instabilities tend to operate early in a shell's life and die down when expansion drops below Mach 3. Gravitational instabilities cause the shell walls to fragment into self-gravitating clumps with sizes much smaller than the radius of the shell. Rayleigh-Taylor (R-T) instabilities operate on the dense shell walls lying above the tenuous shell interior in the presence of a local acceleration. Late in a shell's life the walls should begin to collapse back on the shell's interior under the force of the ISM pressure and the Galactic gravitational field. The instability can also occur near the shell's polar caps as the dense walls accelerate along the density gradient towards the halo.

In GSH 277+00+36 there are many structures that resemble R-T instabilities. The chimney conduits are the most obvious examples but there are also interesting candidates on the smallest size scales. Some of these may be the scalloped appearance of the shell visible in Figure 2. In general R-T instabilities are observed at all size scales, with the smallest instabilities growing faster than the large ones. However, magnetic fields can act to place a lower limit on the size scale of the instability. The magnetic field effectively produces a surface tension that stabilizes against the development of the instability. For a given magnetic field, B, at an angle θ to the instability wave vector, the smallest size scale for the instability, λ_c, is

$$\lambda_c = \frac{B^2 \cos^2 \theta}{a \mu_0 (\rho_2/\rho_1)}, \qquad (1)$$

where ρ_2 and ρ_1 are the densities in the dense and light materials, respectively, a is the local acceleration and μ_0 is the magnetic permeability constant (Chandrasekhar 1961). In GSH 277+00+36 the smallest scale observed for the R-T instability is 15–20 pc. For a magnetic field of typical ISM value ~3 μG and $\rho_2/\rho_1 \sim$ 10–20, we expect that the critical scale will be 15–30 pc (McClure-Griffiths et al. 2003). This seems to agree well with the observed size. The timescale for these small features to develop is short, only ~1 Myr, which is much less than the estimated ~20 Myr lifetime of the shell.

The small structures we observe along the walls of GSH 277+00+36 are quite cold, with H I line widths indicating temperatures of only ~700 K. There is also some evidence that the instabilities in GSH 277+00+36 harbor cold, molecular gas. Figure 2 shows the H I at $v = 43$ km s^{-1} overlaid with contours of ^{12}CO emission from the NANTEN (Matsunaga et al. 2001) and Mopra (McClure-Griffiths & Dickey 2006b) telescopes. There are a number of CO clouds lying along the edges of the shell, many at z heights of 200–300 pc, far above the scale

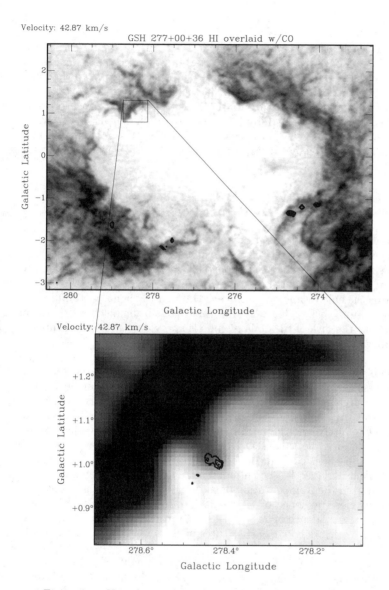

Figure 2.   H I velocity channel grey-scale image of GSH 277+00+36 at $v =$ 42.87 km s^{-1} overlaid with ^{12}CO contours from the NANTEN (upper panel, resolution 4 arcmin) and Mopra (lower panel, resolution 30 arcsec) telescopes.

height for the molecular disk. One of the most interesting molecular clouds is the very small cloud lying inside the Rayleigh-Taylor drip in the zoom panel of Figure 2. Though it is impossible with these data to tell whether this cloud formed within the R-T instability or was swept up by the expanding shell, it is tantalizing to consider that instabilities along shells walls may be sites of molecular cloud formation.

With these new observations we are able to image parsec-scale features on the kiloparsec-scale supershell. The data reveal a large number of potentially Rayleigh-Taylor instability-formed structures with size scales down to the lower limit implied by a Rayleigh-Taylor instability developing in a magnetic field. It seems that in this one object we are able to observe the ISM transition from the hot, ionized gas presumably filling the interior of the shell, to the warm diffuse gas along the shell walls, to the cold atomic gas of the instability-formed structures and finally the molecular gas lying along the shell walls and in the Rayleigh-Taylor instabilities.

4.   H I Shells and the Disk-Halo Interaction

One of the fundamental questions in the evolution of the Milky Way is: how does gas escape the disk to the halo? H I supershells may have an important role in this disk-halo interaction. If a supershell grows large enough to exceed the scale height of the Galactic H I disk, it will rapidly expand along the density gradient towards the halo forming a "chimney" through which hot gas and metals can stream away from disk. This provides a mechanism for distributing metal enriched gas about the disk of the Galaxy and also provides the thermal support necessary to maintain the halo against gravitational collapse.

There are two dominant models that describe the transfer of mass from the disk to the halo via H I shells. The "Chimney" model involves large pressure-driven H I supershells, like GSH 277+00+36, popping at high z heights. These supershells grow large enough to exceed the scale height of the disk and break at the caps, expelling their hot, enriched gas to the halo (e.g., Tomisaka & Ikeuchi 1986). The alternate, "Fountain" model involves supernova heated gas buoyantly rising towards the halo (e.g., Bregman 1980). Both theories speculate that the expelled gas will cool and condense into clouds that will rain back down on the Galactic plane. In the chimney model it is also possible that cool clouds may form from the fragmentation of the supershell caps at high latitudes. Although there is clear evidence of gas being forced out of the Galactic plane via chimneys and also evidence of gas falling back down on the plane it has been difficult to clearly associate cool gas structures in the lower halo with objects in the disk.

Recent work with the Green Bank Telescope by Lockman (2002) has shown that there is a population of cool, compact clouds in the lower Galactic halo with z heights of 1–2 kpc. These cloudlets appear to be dynamically linked to the Galactic disk, but because they are much denser than their surroundings they cannot be in a stable position; they should plummet towards the Galactic disk. The origin of these $\sim$10 pc, $\sim$50 $M_\odot$ clouds is not yet known. There is speculation that they are formed either from chimney- or fountain-expelled gas that has cooled and condensed or alternatively from fragmentation of supershell caps that have grown to large z heights.

Looking towards external galaxies as well as our own, there is mounting evidence for the disk-halo interaction operating via the chimney model of breaking shells. GSH 277+00+36 may be a prime example of that phenomenon in our own

Galaxy. In some edge-on galaxies, like NGC 4217, there are cold dust filaments that loop 1–2 kpc above the galactic plane (Thompson et al. 2004), suggesting that these are the caps of large supershells that have expanded into the lower halo. In our own Galaxy we have recently found another intriguing example of a broken supershell, GSH 242-03+37 (McClure-Griffiths et al. 2006a) that may provide direct evidence of cloud formation in the lower halo.

4.1. GSH 242-03+37

GSH 242-03+37, shown in Figure 3, is a remarkable supershell. It is one of the largest and most energetic in the Milky Way, with a radius of 565 pc and an expansion energy of $\sim 3 \times 10^{53}$ ergs (Heiles 1979). The supershell lies about 3.6 kpc from the Sun and 10.7 kpc from the Galactic Center. Although the shell was discovered in Heiles' seminal work on H I shells, it has only been with recent higher resolution data that we have noticed the shell is in the process of forming a chimney. Using data from the ongoing Galactic All-Sky Survey (McClure-Griffiths et al. 2006c) we have shown GSH 242-03+37 to be another chimney with very similar structure to GSH 277+00+36. As can be seen in Figure 3, GSH 242-03+37 appears to break both above and below the Galactic disk. These breaks result in a large, looping filament that extends from the shell up to 1.6 kpc above the disk. The filaments are roughly traced by the dashed lines in Figure 3.

If we look carefully at these filaments, particularly the filament at positive latitudes, we find that they appear to be breaking into clumps. These clumps have typical sizes on the order of tens of parsecs. The H I linewidth of the individual clumps is ~ 10 km s^{-1}, which implies thermal temperatures of $\sim 10^3$ K. The average column density of the clumps is $N_H = few \times 10^{19}$ cm^{-2}. If the clumps are roughly spherical then their average density is ~ 1 cm^{-3} and their average H I mass is ~ 100 M$_\odot$. The properties of the clumps are remarkably similar to the cloudlets discovered by Lockman (2002).

The typical thermal pressure of these clumps is $nT \sim 10^3$ cm^{-3} K. It is interesting to consider whether they are in pressure equilibrium with their surroundings and how long they can last. Unfortunately it is difficult to estimate the ambient pressure of the lower halo, estimates vary from 10 cm^{-3} K (Boulares & Cox 1990) to 10^3 cm^{-3} K (Shull & Slavin 1994). It is generally believed, however, that the pressure of the lower halo is dominated by outflows from chimneys like GSH 242-03+37. We can therefore estimate the ambient pressure by estimating the internal pressure in the evolved supershell. This leads to estimates of $nT \sim 10^{3-4}$ cm^{-3} K, suggesting that the clumps are in equilibrium or marginally pressure confined by the ambient medium (McClure-Griffiths et al. 2006a).

One factor that may limit the lifetime of the clumps is the possibility of evaporation due to heat flux from the ambient medium. We can get a rough handle on the timescale for evaporation by considering the classical evaporation limit (Cowie & McKee 1977). If the ambient medium is dominated by the hot, $\sim 10^6$ K, gas presumably filling the shell interior then the evaporation timescale for these clumps is ~ 35 Myr. Of course, this is an extreme oversimplification; the clumps are presumably not perfect spheres and may present significantly more surface area to the hot medium. On the other hand, the medium above the shell is probably not nearly as hot as inside the shell, so we can use the

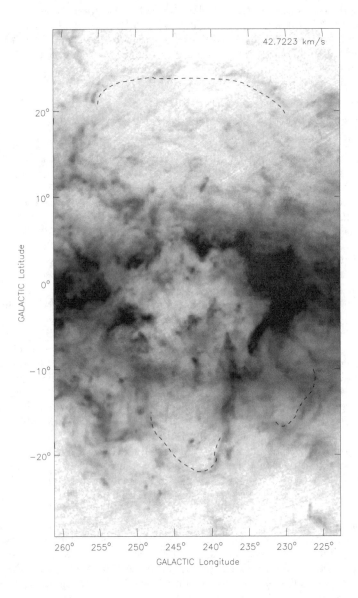

Figure 3.   H I velocity channel image at $v = 42.7$ km s^{-1} of GSH 242-03+37 from the Galactic All Sky Survey (McClure-Griffiths et al. 2006a). The grey scale in this image is logarithmic and scaled between 0 K (white) and 40 K (black). The caps of the chimney are marked with dashed lines. All caps are located $\sim$1.6 kpc from the Galactic plane.

evaporation timescale as a rough estimate of the clump lifetime. Given that the shell lifetime is only $\sim$20 Myr, it is possible that the fragments of the shell cap may outlast the shell structure in the Galactic plane. If this is true, the Lockman (2002) population of small halo clouds may result from broken H I supershells like GSH 242-03+37. Based on this one example it seems conceivable that some fraction of the structure in the lower halo comes not from clouds that have cooled and condensed out of hot outflows, but instead from fragmentation of very large, cool supershell caps.

5. Future Directions

The past five years have seen great advances in the observational aspects of Milky Way ISM research. By combining single dish and interferometric radio observations we are able to explore large scale structures in the ISM with high resolution, allowing us to study the detailed physics of supershell evolution. Whereas ten years ago simulations of individual shells far exceeded the level of detail observable, the reverse is now true. Recent simulations have focused on the global evolution of galaxy systems with supershells (e.g., de Avillez & Berry 2001) instead of focusing on the evolution of individual objects. Now that observational work on H I supershells has progressed so much I would encourage simulators to return to the important topic of how individual shells evolve. It should now be possible to compare the observational data with the numerical simulations.

Sophisticated MHD modeling has shown us that magnetic fields can have a significant impact on the evolution of a supershell, but these models are relatively unconstrained by observational details. The next observational advance will hopefully be the inclusion of observations probing the magnetic fields of supershells. In the near future we can expect to be able to compare H I imaging with imaging of the diffuse polarization of the radio continuum. This should allow us to explore some large scale properties of the magnetic field structure. When more sensitive radio telescopes, such as the SKA, become available, H I and possibly OH Zeeman measurements of shell instabilities will provide invaluable information about the magnetic field strengths in these objects.

6. Conclusions

H I supershells are among the largest and most energetic objects in the interstellar medium. In many galaxies, such as the LMC, H I supershells dominate the global structure of the galaxy. In the Milky Way H I supershells are also a major source of structure and may dominate the energy budget in the ISM. H I supershells play a major role in the evolution of the ISM ecosystem. They provide sites of both heating and cooling as well as structure on size scales ranging from parsecs to kiloparsecs. With recent high resolution observations of large supershells we are finally able to observe the detailed evolution of supershells, including the the fragmentation of the shell walls due to gas instabilities and the cold, small scale structures that result. Finally, H I supershells may play an important part in the interaction of the disk and halo, supplying heat and structure for the lower halo of the Galaxy. Supershells like GSH 242-03+37 seem to be forming relatively cold structures at z heights of $\sim$1–2 kpc and may contribute to the population of small, cool clouds observed by Lockman (2002) in the lower halo.

Acknowledgments. I would like to thank the Bash Symposium organizers for inviting me to attend the symposium and for providing financial support to enable my visit.

References

Boulares, A. & Cox, D. P. 1990, ApJ, 365, 544
Bregman, J. N. 1980, ApJ, 236, 577
Chandrasekhar, S. 1961, Hydrodynamic and Hydromagnetic Stability, International Series of Monographs on Physics (Oxford: Clarendon)
Cowie, L. L. & McKee, C. F. 1977, ApJ, 211, 135
de Avillez, M. A. 2000, MNRAS, 315, 479
de Avillez, M. A. & Berry, D. L. 2001, MNRAS, 328, 708
Dove, J. B., Shull, J. M., & Ferrara, A. 2000, ApJ, 531, 846
Heiles, C. 1979, ApJ, 229, 533
—. 1984, ApJS, 55, 585
Kim, S., Dopita, M. A., Staveley-Smith, L., & Bessell, M. S. 1999, AJ, 118, 2797
Kim, S., Staveley-Smith, L., Dopita, M. A., Sault, R. J., Freeman, K. C., Lee, Y., & Chu, Y.-H. 2003, ApJS, 148, 473
Lockman, F. J. 2002, ApJ, 580, L47
Mac Low, M. & McCray, R. 1988, ApJ, 324, 776
Mac Low, M., McCray, R., & Norman, M. L. 1989, ApJ, 337, 141
Matsunaga, K., Mizuno, N., Moriguchi, Y., Onishi, T., Mizuno, A., & Fukui, Y. 2001, PASJ, 53, 1003
McClure-Griffiths, N. M. & Dickey, J. M. 2006b, in preparation
McClure-Griffiths, N. M., Dickey, J. M., Gaensler, B. M., & Green, A. J. 2002, ApJ, 578, 176
—. 2003, ApJ, 594, 833
McClure-Griffiths, N. M., Dickey, J. M., Gaensler, B. M., Green, A. J., Haverkorn, M., & Strasser, S. 2005, ApJS, 158, 178
McClure-Griffiths, N. M., Dickey, J. M., Gaensler, B. M., Green, A. J., Haynes, R. F., & Wieringa, M. H. 2000, AJ, 119, 2828
McClure-Griffiths, N. M., Ford, A., Pisano, D. J., Gibson, B. K., Staveley-Smith, L., Calabretta, M. R., Kalberla, P. M. W., & Dedes, L. 2006a, ApJ, 638, 196
McClure-Griffiths, N. M., Ford, A., Pisano, D. J., Staveley-Smith, L., Calabretta, M. R., Gibson, B. K., Kalberla, P. M. W., Dedes, L., Bailin, J., & Kiessling, A. 2006c, in preparation
Points, S. D., Chu, Y.-H., Snowden, S. L., & Staveley-Smith, L. 2000, ApJ, 545, 827
Shull, J. M. & Slavin, J. D. 1994, ApJ, 427, 784
Spangler, S. R. 2001, Space Science Reviews, 99, 261
Taylor, A. R., Stil, J., Dickey, J., McClure-Griffiths, N., Martin, P., Rotwell, T., & Lockman, J. 2002, in ASP Conf. Ser. 276: Seeing Through the Dust: The Detection of HI and the Exploration of the ISM in Galaxies, ed. A. R. Taylor, T. L. Landecker, and A. G. Willis (San Francisco: ASP), 68
Tenorio-Tagle, G. & Bodenheimer, P. 1988, ARA&A, 26, 145
Thompson, T. W. J., Howk, J. C., & Savage, B. D. 2004, AJ, 128, 662
Tomisaka, K. 1998, MNRAS, 298, 797
Tomisaka, K. & Ikeuchi, S. 1986, PASJ, 38, 697

Frank N. Bash Symposium 2005: New Horizons in Astronomy
ASP Conference Series, Vol. 352, 2006
S. J. Kannappan, S. Redfield, J. E. Kessler-Silacci, M. Landriau, and N. Drory

Stellar Abundances: Recent and Foreseeable Trends

Carlos Allende Prieto

Department of Astronomy, University of Texas, Austin, TX, USA

Abstract. The determination of chemical abundances from stellar spectra is considered a mature field of astrophysics. Digital spectra of stars are recorded and processed with standard techniques, much like samples in the biological sciences. Nevertheless, uncertainties typically exceed 20% and are dominated by systematic errors. The first part of this paper addresses what is being done to reduce measurement errors, and what is not being done, but should be. The second part focuses on some of the most exciting applications of stellar spectroscopy in the arenas of galactic structure and evolution, the origin of the chemical elements, and cosmology.

1. Introduction

Stellar absorption lines, first discovered in the solar spectrum circa 1804, can be used to quantify the proportions of different chemical elements in stars. The strength of the lines depends on the abundance of absorbers, but also on the environment where the absorption takes place: the stellar atmosphere. The determination of chemical abundances requires an accurate knowledge of the physical conditions (e.g., temperature, density, radiation field) in the outer layers of the star. In other words, inferring chemical abundances from spectra is just a part of the problem of the physics of stellar atmospheres, and cannot be decoupled from it.

After Eddingon, Milne, and others established the theoretical basis, the field of quantitative stellar spectroscopy has evolved over time to become almost an industry. Stellar abundances have shed light on the origin of the elements, the early universe, what the interiors of the stars look like, and how galaxies form and evolve. As modeling is refined, so are the quality and reach of the observations, making it possible to use stellar abundances to tackle new problems.

Nowadays, a small army of some 10^3 researchers are *professionally* dedicated to measuring chemical abundances from stellar spectra in the entire world. Practitioners favor B-type stars on the warm side, and F- and G-type stars on the cool side. B-type stars can be observed at large distances, while the cooler F- and G-types live long, allowing us to study the past. Cooler stars have far more complex spectra, shaped by molecular absorption and dust.

In this paper, with no ambition of formally reviewing the field, I simply highlight recent accomplishments and tendencies, both in methods (§ 2) and applications (§ 3). The closing section is devoted to a personal (and surely biased) reflection on where the field is going, and where one wishes it would go.

2.   Stellar Abundances. How To?

Telescopes and spectrographs are used to obtain stellar spectra. The visible is
the window of choice for abundance measurements, because of the high trans-
parency of the Earth's atmosphere, limited line crowding, and simple continuum
opacities (basically H and H^-). The result of an observation is an account of the
detected stellar photons properly ordered by energy. The next step is removing
instrumental effects that distort the observed photon distribution.

The final stage consists of translating observed line strengths into chemical
abundances. This step involves many simplifications and assumptions that allow us to build an approximate description of the physical conditions in the
atmosphere of the star, given a small number of parameters (energy flux, surface gravity, and metal abundances[1]) to be inferred from observations. Libraries
of atmospheric models are computed by experts and then widely disseminated
(or not). The fact that one of the model parameters, the metal abundance, is
what we seek, makes this a problem that needs to be treated iteratively. In the
standard procedure, the equation of radiative transfer is solved to calculate the
spectra predicted by the models, which are then compared to an observation to
constrain the atmospheric parameters and estimate the metal abundances. This
information is then used to select a new model that will be used to refine the
abundances, going on until convergence is achieved.

Two critical parts of this process are the model atmospheres and the calculation of the emergent spectrum from the models. Obtaining the right observations
and performing a reliable analysis are other key elements involved.

2.1.   Model Atmospheres

The model atmospheres commonly used today are based on the concepts of
radiative and hydrostatic equilibrium. These models are one-dimensional: all
the relevant thermodynamical quantities depend only on height (plane-parallel
geometry) or radius (spherical models). Models for late-type stars also assume
local thermodynamical equilibrium (LTE), i.e., the source function is equal to
the Planck law and therefore only depends on the local temperature.

Spectral lines are most useful for deriving chemical abundances, but line absorption also affects the energy balance in the outer layers of a star. Iron is by
far the element that produces most of the observed lines. Metal line crowding
in the blue and UV spectra of late-type stars contributes significantly to the
total opacity, inducing extra warming in deep atmospheric layers and cooling in
the outer regions. Line blanketing becomes milder for hot stars, but even then
it is still quite significant. Hot star models that account for both departures
from LTE and line blanketing (see Fig. 1) have become available only recently
(Hubeny & Lanz 1995, Lanz & Hubeny 2003, Rauch 2003, Repolust, Puls, &
Herrero 2004). For the most massive stars, especially when dealing with UV
spectra, winds need to be considered (Pauldrach et al. 2001, Hillier 2003).

---

[1]Hereafter quantified in a logarithmic scale relative to hydrogen and solar values:

$$[\mathrm{X/H}] = \log \frac{N_X}{N_H} - \log \left(\frac{N_X}{N_H} \right)_\odot , \text{ where N represents number density.}$$

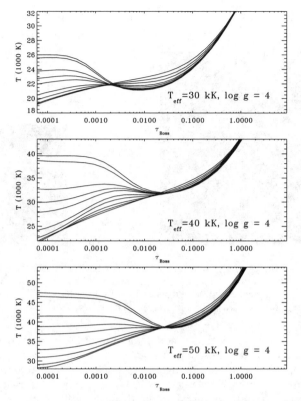

Figure 1.   Temperature structure as a function of the Rosseland optical depth from the O-star grid of Lanz & Hubeny (2003; reproduced with kind permission). The different curves for any given T_{eff} correspond to metallicities from 2× solar to zero. Departures from LTE are responsible for warming the outer layers of the more metal-poor models, where line cooling is not effective.

Models for late-type (F, G, and K-type) stars have been always based on the assumption that LTE holds. As stated above, line blanketing needs to be included. Unlike in more massive stars, convection develops in the envelopes of these stars to some degree. This is accounted for in the energy balance by a simplified treatment termed as mixing-length theory (Böhm-Vitense 1958). However, the impact of convection on the velocity fields at the surface of the star is usually ignored, i.e., hydrostatic equilibrium is still assumed. That this is a bad approximation is nowhere more evident than in a high-resolution image of the solar surface. Fig. 2 shows an area of the solar disk including a group of sunspots. The concentration of magnetic field in sunspots inhibits convection, which is responsible for the granular pattern apparent everywhere else, as evident in the section expanded in the right-hand panel of Fig. 2. The white areas correspond to bright granules: hot gas upflows. The dark lanes are the intergranules: cooler sinking gas. The temperature contrast in the solar photosphere ranges from several hundred to more than a thousand degrees, and the velocities reach a few kilometers per second. Typical granules have angular sizes of about 1 arcsecond, or a couple thousand kilometers on the solar surface, and evolve on time scales of the order of 10–15 minutes.

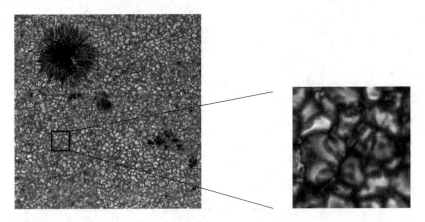

Figure 2.   Image of the solar surface obtained with the Swedish 1-m Solar Telescope at La Palma. The continuum image (436.4 nm) shows a group of sunspots on August 22, 2003. Observations by O. Engvold, J. E. Wiik, and L. R. van der Voort (Inst. of Theoretical Astrophysics of Oslo University).

Hydrodynamical models that account for the temperature and velocity fields associated with convection, oscillations, and shocks in the solar atmosphere were pioneered more than two decades ago (e.g., Massaguer & Zahn 1980, Nordlund 1980, Hurlburt, Toomre & Massaguer 1984) and have now reached maturity (Stein & Nordlund 1998, Asplund et al. 2000a). Unlike classical model atmospheres, which have a constant temperature at any given depth, these models exhibit temperature and velocity inhomogeneities that resemble solar observations. Fig. 3 compares the temperature at a given depth in the photosphere for one-dimensional and three-dimensional models — compare with Fig. 2.

Classical static models predict perfectly symmetric spectral lines. The line profiles produced in hydrodynamical models are asymmetric and blue-shifted relative to their rest wavelengths as a result of the correlation between temperature and velocity fields — very much like the observed line profiles (Asplund et al. 2000a, Allende Prieto et al. 2002a). In some cases, the strengths of spectral lines predicted by 3D and 1D models are significantly different. Occasionally, the ability of hydrodynamical models to reproduce closely the observed line shapes allows us to identify blends that otherwise would go unnoticed.

A recent revision of the solar photospheric abundances of several light elements based on 3D models has arrived at the conclusion that both oxygen and carbon (but also Ne, Ar, Fe and Si) are significantly less abundant, in some cases by as much as 40%, than previously thought (Allende Prieto et al. 2001, 2002b; Asplund et al. 2000b, 2004, 2005). The revised solar abundances show that hydrodynamics (and departures from LTE) are not mere refinements, but crucial ingredients. The updated abundances are in better agreement with other stars in the solar neighborhood, but they also ruin an earlier excellent agreement between helioseismic measurements and models of the solar interior (Bahcall et al. 2005a, Bahcall, Serenelli & Basu 2005c, Basu & Antia 2004, Antia & Basu

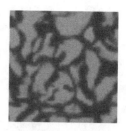

Figure 3.   The gray scale indicates temperature at a given photospheric depth for a classical model atmosphere (left-hand panel), and in a snapshot of the hydrodynamical simulation of Asplund et al. (2000a).

2005, Montalbán et al. 2004) — a conundrum that still needs a solution at the time of this writing (see Bahcall, Basu & Serenelli 2005b and Drake & Testa 2005 for possible light at the end of the tunnel, or Young & Arnett 2005 for a different source of light). Unfortunately, hydrodynamical models for stars other than the Sun are not widely available (and mostly non-existent). Moreover, the tools available to calculate spectra from time-dependent 3D hydrodynamical simulations are very limited, and unable, for example, to calculate emergent absolute fluxes with realistic (line and continuum) opacities over large spectral windows.

For stars with $T_{\text{eff}} < 4000$ K, molecules become very important both in the equation of state and in the radiative opacity (mainly H_2, H_2O, CH_4, CO, N_2, NH_3, FeH, CrH, TiO, and VO). In addition, for atmospheres cooler than $T_{\text{eff}} \sim 2500$ K, solid particles, such as silicates, which form clouds, are also an important source of opacity, with alkalis as the only atoms that still contribute significantly. The complicated opacities pose a challenge for modeling the coolest stars and brown dwarfs. Handling the formation and rainout of condensate clouds further complicates matters (e.g., Burrows, Sudarsky & Hubeny 2005, Tsuji 2005). Fig. 4 illustrates the change in the emergent flux between 700 K and 2100 K for models recently computed by Burrows et al. (2005) for a particular surface gravity, particle size, and cloud shape.

2.2. Spectrum Formation

Although the radiation field needs to be continuously evaluated as model construction proceeds, the detailed calculation of the emergent spectrum for comparison with observations is usually performed afterwards. At this time, higher spectral resolution and a more accurate modeling of the line profiles become affordable as the equation of radiative transfer needs to be solved just a few times. It is also possible to estimate departures from LTE for trace species, assuming that the atmospheric structure (e.g., temperature and electron density) is given by an LTE model atmosphere previously calculated.

Reliable opacities, as we have discussed in § 2.1, can be the limiting factor for obtaining realistic model atmospheres. For stars like the Sun, the optical and infrared continuum is shaped by bound-free and free-free absorption of H^-, a trace species nonetheless. In the UV window, however, atomic metal opacity (mainly due to Mg, Fe, and Al) becomes relevant, and even dominant at wave-

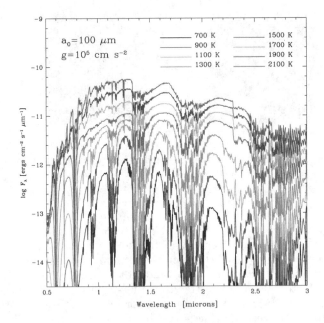

Figure 4.   Model fluxes for very low mass stars and brown dwarfs (dividing line at $T_{\rm eff} \sim 1700$ K) as they would be observed at a distance of 10 pc ($\log F \propto \log R + 2 \log T_{\rm eff}$, and the radii are adopted from Burrows et al. 1997). Adapted from Burrows, Sudarsky & Hubeny (2005) with kind permission.

lengths below 250 nm. Quantum mechanical calculations in the context of the Opacity Project (OP) have provided accurate photoionization cross-sections for light elements (Seaton 2005), and an extension of the calculations to iron ions is ongoing within the Iron Project (IP; Nahar & Pradhan 2005). These data are (slowly) being incorporated in the calculation of theoretical spectra (Cowley & Bautista 2003, Allende Prieto et al. 2003a,b). Fig. 5 illustrates the impact of the bound-free metal absorption on the UV solar flux.

Another important aspect of the calculation of stellar spectra that has been recently improved is related to the absorption line coefficients. Improved calculations of cross-sections for line broadening by elastic hydrogen collisions have been presented by Paul Barklem and collaborators (see Barklem, Anstee, & O'Mara 1998, Barklem & Aspelund-Johansson 2005, and references therein). Improved absorption coefficients are also now available for Balmer lines (see, e.g., Barklem, Piskunov & O'Mara 2000, Barklem et al. 2002). The wings of the Balmer lines, formed in deep layers very close to LTE conditions, are very sensitive indicators of the effective temperature of a star (see Fig. 6). Interestingly enough, the mixing-length parameter for convection (α) needs to be reduced to about half a pressure scale height in order to get consistent temperatures from Hα and Hβ for solar like stars — whereas the usual values inferred from standard solar models and multidimensional simulations of solar surface convection lead to $2 < \alpha < 2.1$ (Basu, Pinsonneault, & Bahcall 2000, Robinson et al. 2004). The new calculations of collisional broadening by atomic H have the largest impact

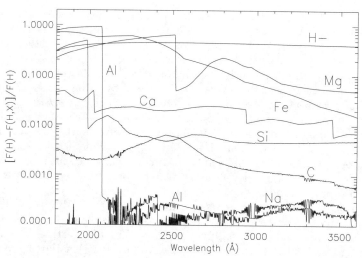

Figure 5.   Relative changes in the emergent flux from a solar-like atmosphere when continuum opacity from H^- and several metals are added to the atomic hydrogen opacity (Allende Prieto, Hubeny & Lambert 2003b).

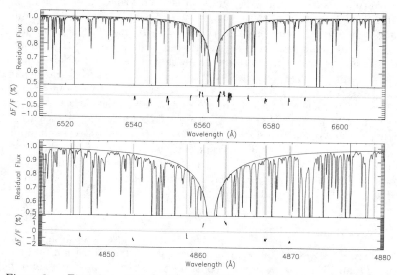

Figure 6.   Fittings to the solar $H\alpha$ and $H\beta$ line profiles with a MARCS model (Gustafsson et al. 1975, 2003) for a mixing-length parameter $\alpha = l/H_p = 0.5$, the Stark broadening profiles of Stehlé (1994), and the hydrogen collision broadening profiles of Barklem, Piskunov, & O'Mara (2000). The grey areas indicate the spectral windows used to evaluate the goodness-of-fit. Adapted from Barklem et al. (2002).

for metal-poor stars, where gas pressure is higher than at solar metallicity, and have made it possible to reconcile effective temperatures inferred from colors and Balmer lines.

The OP database at `http://vizier.u-strasbg.fr/topbase/home.html`, referred to as TOPBASE, includes other useful data in addition to photoionization cross-sections: energy levels and transition probabilities, as well as tools to compute opacities and radiative forces. The computed energies for atomic levels are not very accurate, and although the opacities include corrections for the ground level of each species, they can be improved by using observed energies, readily available from other sources such as NIST (`http://www.nist.gov`). OP and IP calculations assume LS coupling and neglect fine structure. For the line opacity, one commonly uses accurate observed line wavelengths. The most comprehensive atomic and molecular line lists have been compiled by Kurucz and collaborators (e.g., Kurucz 1993, `http://kurucz.harvard.edu`). Kurucz has also calculated and compiled wavelengths and weights for hyperfine structure (*hfs*) and isotopic components.

A noteworthy development is the recent availability of laboratory determinations of atomic parameters (transition probabilities, hyperfine and isotopic splitting constants, and partition functions) for many rare-earth elements (e.g., Den Hartog et al. 2005; Lawler, Sneden & Cowan 2004; Den Hartog, Wickliffe & Lawler 2002; Lawler, Wyart & Blaise 2001). Fig. 7 shows how important it is to consider *hfs* when deriving abundances for elements like holmium.

While LTE constitutes a very useful approximation that makes the calculation of model atmospheres and spectra simple, it is often wrong. Calculations of Non-LTE line formation in LTE model atmospheres for trace species show abundant examples where line strengths interpreted with equilibrium (LTE) populations lead to large systematic errors in the inferred abundances. NLTE models for hot stars are now available, but the same cannot be said for late-type stars. This shortcoming is most likely the result of the extended belief that departures from LTE are small in late-type stars. In fact, recent calculations by Short & Hauschildt (2005) suggest that there are significant changes in the temperature structure of a solar model when LTE is relaxed. The same investigation also finds that an NLTE solar model matches the observed fluxes worse than an LTE model — a puzzling result that suggests that this is still an unfinished business.

A more extensive discussion of some of the issues mentioned in this and the previous section can be found in the reviews by Gustafsson & Jørgensen (1994), Allard et al. (1997), Asplund (2005), and Kirkpatrick (2005).

2.3. Observation and Analysis

As modeling improves, so do some aspects of the data acquisition and analysis process. Quantities that have seen constant progress include telescope diameter, the spectral coverage of spectrographs, and their multiplexing capabilities. However, a victim of the difficulty of coupling large diameter telescopes with narrow slits, the resolving power has been neglected. Investigations of spectral line shapes, which are poorly matched by hydrostatic models and therefore nobody wants to see, have almost become faux pas.

Data reduction is done in most cases with packages such as IRAF or MIDAS. Pipelines for automatic reduction have been implemented for some facilities and large projects (e.g., Ritter & Washuettl 2004). For the most part, spectra are usually reduced interactively, which causes two problems: i) individual scientists, trying to move on as quickly as possible to the data analysis, do not dedicate the

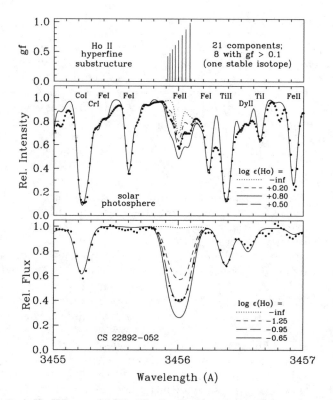

Figure 7.   Ho II line at $\lambda 3456$. The lines in the top panel show the strengths and positions of the hyperfine splitting (*hfs*) components due to ^{165}Ho, the only stable isotope of holmium. The middle and lower panels show the spectrum of the Sun and the metal-poor star CS 22892-052 (see § 3.1), respectively. The impact of *hfs* in the line shape is quite dramatic, and Ho abundances inferred would be seriously in error if *hfs* were neglected. Adapted from Lawler, Sneden & Cowan (2004) with kind permission.

necessary amount of time to make sure the reduction is the best possible, and ii) the knowledge on data reduction is not encapsulated into software. A more efficient and better scheme would involve dedicated research on the data reduction process specific for an instrument, with the results subsequently captured into a pipeline. For example, Piskunov & Valenti (2002) have recently reminded us that going the extra mile pays off.

As data sets grow in size — more stars and more frequencies per star — it becomes necessary to automate not only the data reduction, but also the data analysis. For example, a spectroscopic analysis typically starts with a search for the best estimates of the atmospheric parameters: effective temperature, surface gravity and overall metal content. Such a step is amenable to automation (Katz et al. 1998, Snider et al. 2001, Allende Prieto 2003, Erspamer & North 2003, Willemsen et al. 2005). Neural networks, genetic algorithms, self-organizing maps, and many other kinds of optimization methods can help to relieve the burden on the busy astronomer.

3.   Stellar Abundances. What For?

After some hard work, we get to the most exciting part: what can we do with the derived chemical abundances? A possible criterion to classify the applications of stellar abundances is whether or not they are based on the *golden rule*. The golden rule, a term introduced in this context by David Lambert, states:

The surface composition of a star reflects that of the interstellar medium at the time and location where the star formed.

As with any other good-looking law, the golden rule is often broken. But we will first see how to take advantage of those cases when it seems to hold.

3.1.   When the Golden Rule Applies

Only ^{1}H, ^{2}H, ^{3}He, ^{4}He, ^{7}Li and ^{6}Li were produced in significant amounts in the big bang, while heavier nuclei are mostly the result of stellar nucleosynthesis. When stars explode as supernovae, or lose mass from their envelopes by the action of stellar winds or thermal pulses on the asymptotic giant branch (AGB), the metals produced by stellar nucleosynthesis (explosive or not) enrich the interstellar medium. As low-mass stars have very long life spans (the hydrogen burning time scales $\propto M^{-5/2}$), if the golden rule applies, one can use stars of different ages to track the metal enrichment of the interstellar medium with time. An account of the state of this field is given in this volume by Yeshe Fenner, so I just briefly mention some recent observational developments that illustrate how stellar abundances can help to understand how galaxies form.

The disk of the Milky Way was first found to possess a *second* component with a larger scale height by Gilmore & Reid (1983). This component, which is known as the *thick disk*, contains more metal-poor and older stars than the thin disk. Thick disk stars are not only chemically, but also kinematically distinct from the thin disk population: they have a larger velocity ellipsoid, and an average galactic rotation velocity that lags the thin disk by some 30–40 km s^{-1}, as illustrated in Fig. 8 (see also Gratton et al. 1996; Fuhrmann 1998; Bensby et al. 2005; Mishenina et al. 2004; Reddy et al. 2003). A recent study that exploits calibration stars observed as part of the Sloan Digital Sky Survey (SDSS) suggests that the peak of the metallicity distribution of the thick disk is always at [Fe/H] ~ -0.7 between 4 and 14 kpc from the galactic center (Allende Prieto et al. 2006), while local thick disk members selected based on kinematics indicate a high chemical homogeneity in this population (Reddy, Lambert & Allende Prieto 2006). These and other observations seem consistent with the thick disk emerging in a period of intensive accretion of smaller gas-rich galaxies by the Milky Way at a redshift >1 or, equivalently, $>8 \times 10^9$ yr ago (Brook et al. 2005).

Some of the oldest stars in the Milky Way have extremely low metal abundances, much lower than any globular cluster in the Galaxy. Their compositions may resemble that of the early Galaxy, with Li/H ratios similar to the proportions produced in the big bang. This, in fact, was the interpretation given by Spite & Spite (1982) to the nearly constant values of Li/H found in the surface of turn-off field stars with very low metal abundances. Experts still debate whether or not the Li abundance is exactly the same in the most metal-poor turn-off stars (e.g., Ryan et al. 1999, Meléndez & Ramírez 2004), but for our purposes it suffices to say that big bang nucleosynthesis models yield different Li/H ratios depending on the universe's photon to baryon ratio, and therefore

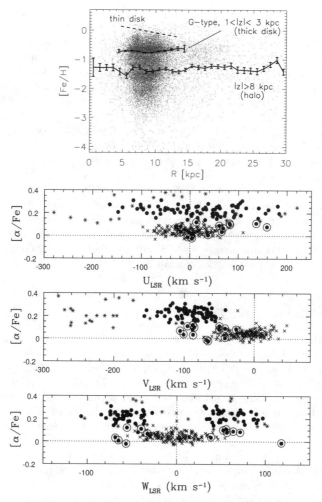

Figure 8.   Top panel: some 20,000 F- and G-type stars with $14 < V < 20$ mag spectroscopically observed as part of the SDSS are used here to trace the median metallicity of the thick disk and the halo as a function of distance from the Galactic center (R in cylindrical coordinates). In this figure z is the distance from the galactic plane. The flat values of the median iron abundances contrast with the gradients reported in the literature for the thin disk (the dashed line shows the mean metal abundances derived for B-type stars by Daflon & Cunha 2004). Adapted from Allende Prieto et al. (2006). Bottom panels: Chemical and kinematical separation of thick disk (filled circles), thin disk (crosses), and halo (asterisks) stars — plus the enigmatic stars with thick-disk kinematics and thin-disk abundances (filled circles surrounded by open circles). Here α indicates an average of Mg, Si, Ca and Ti. Adapted from Reddy et al. (2006).

the lithium abundances in stars can constrain such an important cosmological parameter.

When old very metal-poor halo stars formed, there were very few metals around to be incorporated in the stars' atmospheres. As massive stars died exploding as Type II supernovae, there must have been a period of high chemical heterogeneity in the halo. If a star happened to form at such early stages near the location where a supernova exploded, it could pick up the proportions of heavy elements produced before and during the explosion. That may have been the case for CS 22892-052, an extremely metal-poor ([Fe/H]= −3.1) and old (age > 11 × 10^9 yr) halo giant that is highly enriched in neutron-capture elements (Sneden et al. 1994, 2003), and also for several other well-known cases, such as BD +17 3248 (Cowan et al. 2002), or HD 115444 (Westin et al. 2000). As Fig. 9 illustrates, the abundance patterns of heavy neutron-capture elements in these stars are remarkably similar to the *r-process* contribution to the solar-system abundances of these elements (as inferred by two different methods: Burris et al. 2000 vs. Arlandini et al. 1999). The *r-process* nucleosynthesis operates when high neutron fluxes are available, and it is usually associated with Type II supernovae. Therefore measuring abundances in these stars allows us to study supernova yields. Ongoing active searches for more of these interesting objects are already producing results (Barklem et al. 2005).

3.2.   If the Golden Rule Does Not Apply

The chemical composition of the atmosphere of a star may have changed since the star formed. As research has already shown, this can happen by a myriad of different processes. An interesting example of the use of stellar abundances under these circumstances is very close to home. Chemical abundances derived for the solar photosphere have been found to provide valuable clues on the structure of the solar interior. Considering diffusion in the solar convection zone is necessary to match the p-mode frequencies from helioseismic observations (e.g., Demarque & Guenther 1988, Christensen-Dalsgaard 2002, Lodders 2003). Since the formation of the Sun 4.57×10^9 years ago, the surface abundance of He relative to H has seen a reduction of 18%, while the abundances of heavier elements have been reduced by 16%.

The surface abundances of Li or Be in late-type stars can change significantly when the material is transported by convection to more interior zones where the temperature reaches a few million degrees and these light nuclei are destroyed. This violation of the golden rule has been exploited to learn about stellar structure by studying abundances in clusters (e.g., Boesgaard et al. 2004, García López, Rebolo & Pérez de Taoro 1995).

We discussed above how supernova yields can be inferred from rare metal-poor stars when the golden rule applies, but some have also attempted to do the same when the rule is not respected. Binary systems that have as a primary a neutron star or a black hole offer a chance to detect supernova ejecta that may have been blocked by the companion. González-Hernández et al. (2005) detected unusual enhancements of Ti and Ni ratios in the K-type companion of the neutron star Cen X-4, and concluded by comparison with theoretical calculations that the observed abundances could be explained by a spherically symmetric supernova explosion. Nevertheless, the kinematics of the system supports an asymmetric supernova, leaving open an interesting puzzle.

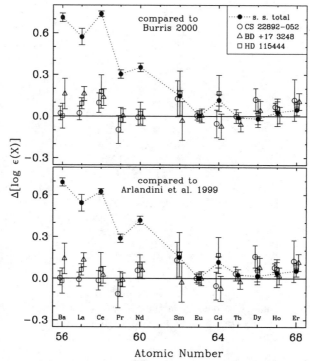

Figure 9.   Differences between the abundances of heavy neutron capture elements in three stars and the *r-process* component of the solar abundances. The *r-process* solar abundances were inferred by subtracting the *s-process* (slow neutron capture) abundances estimated by two different methods: an empirical approach (Burris et al. 2000) or AGB stars nucleosynthesis models (Arlandini et al. 1999). The patterns have been normalized to force a perfect agreement for europium. Adapted from Lawler et al. (2004) with kind permission.

Perhaps one of the worse cases of the golden rule being broken is that of the RV Tauri variables. The photospheres of these stars exhibit dramatic abundance anomalies that correlate with condensation temperature (T_c). Elements with a high T_c can be depleted by several orders of magnitude with respect to those with a low T_c. The elements depleted form dust grains, ending up in some sort of circumstellar reservoir or being driven away by a stellar wind. Fig. 10 shows the abundance pattern derived for UY Cam by Giridhar et al. (2005). Metal-deficient post-AGB stars such as HR 4049 (Van Winckel, Waelkens & Waters 1995) exhibit a photospheric iron abundance [Fe/H] $\simeq$ −5.0 but most likely started up with a near-solar Fe/H ratio. HR 4049 and its class seem to be all members of binary systems, and the same could be true for RV Tauris. We refer the reader to the recent review by Van Winckel (2003) for more on the amazing fauna of post-AGB stars.

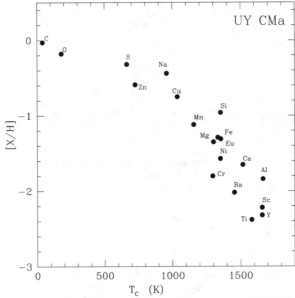

Figure 10.   Abundance pattern in the RV Tauri star UY CMa. Adapted from Giridhar et al. (2005) with kind permission.

4.   Conclusions

Recent noteworthy trends in research on stellar abundances include the development of improved model atmospheres for solar-like stars considering hydrodynamics, the availability of full-NLTE line-blanketed model atmospheres for hot stars, significant improvements in opacities and treatment of clouds for the coolest stars and brown dwarfs, the emergence of large spectroscopic surveys producing public data bases, significant progress toward the automation of spectroscopic analyses, and the availability of accurate atomic data for neutron-capture elements. A desirable short-term future would include 3D model atmospheres for late-type stars of all types, 1D models in full NLTE, NLTE line formation in 3D models, solar atlases observed at different positions on the disk to test line formation and model atmospheres, stronger efforts to measure/compute the necessary atomic and molecular data, stronger efforts to use the newly available atomic and molecular data, full analysis automation, a new generation of all-in-one archives for both multi-wavelength observations and models, and ultra-high resolution spectrographs on large telescopes. Let's make it happen!

Acknowledgments.   I am thankful to the organizers for a delightful meeting, and the opportunity to think (aloud) about the field of stellar abundance determination. Some of the figures in this paper were kindly provided by A. Burrows, S. Giridhar, I. Hubeny, T. Lanz, J. Lawler, B. Reddy, and C. Sneden. The Swedish 1-m Solar Telescope is operated on the island of La Palma by the Institute for Solar Physics of the Royal Swedish Academy of Sciences in the Spanish Observatorio del Roque de los Muchachos of the IAC. Support from NASA (NAG5-13057, NAG5-13147) is gratefully acknowledged.

References

Allard, F., Hauschildt, P. H., Alexander, D. R., & Starrfield, S. 1997, ARA&A, 35, 137
Allende Prieto, C. 2003, MNRAS, 339, 1111
Allende Prieto, C., Asplund, M., García López, R. J., & Lambert, D. L. 2002a, ApJ, 567, 544
Allende Prieto, C., Beers, T. C., Wilhelm, R., Newberg, H. J., Rockosi, C. M., Yanny, B., & Lee, Y. S. 2006, ApJ, 636, 804
Allende Prieto, C., Lambert, D. L., & Asplund, M. 2001, ApJ, 556, L63
Allende Prieto, C., Lambert, D. L., & Asplund, M. 2002b, ApJ, 573, L137
Allende Prieto, C., Lambert, D. L., Hubeny, I., & Lanz, T. 2003a, ApJS, 147, 363
Allende Prieto, C., Hubeny, I., & Lambert, D. L. 2003b, ApJ, 591, 1192
Antia, H. M., & Basu, S. 2005, ApJ, 620, L129
Arlandini, C., Käppeler, F., Wisshak, K., Gallino, R., Lugaro, M., Busso, M., & Straniero, O. 1999, ApJ, 525, 886
Asplund, M. 2005, ARA&A, 43, 481
Asplund, M., Grevesse, N., Sauval, A. J., Allende Prieto, C., & Kiselman, D. 2004, A&A, 417, 751
Asplund, M., Grevesse, N., Sauval, A. J., Allende Prieto, C., & Blomme, R. 2005, A&A, 431, 693
Asplund, M., Nordlund, Å., Trampedach, R., Allende Prieto, C., & Stein, R. F. 2000a, A&A, 359, 729
Asplund, M., Nordlund, Å., Trampedach, R., & Stein, R. F. 2000b, A&A, 359, 743
Bahcall, J. N., Basu, S., Pinsonneault, M., & Serenelli, A. M. 2005a, ApJ, 618, 1049
Bahcall, J. N., Basu, S., & Serenelli, A. M. 2005b, ApJ, 631, 1281
Bahcall, J. N., Serenelli, A. M., & Basu, S. 2005c, ApJ, 621, L85
Barklem, P. S., Anstee, S. D., & O'Mara, B. J. 1998, PASA, 15, 336
Barklem, P. S., & Aspelund-Johansson, J. 2005, A&A, 435, 373
Barklem, P. S., et al. 2005, A&A, 439, 129
Barklem, P. S., Piskunov, N., & O'Mara, B. J. 2000, A&A, 363, 1091
Barklem, P. S., Stempels, H. C., Allende Prieto, C., Kochukhov, O. P., Piskunov, N., & O'Mara, B. J. 2002, A&A, 385, 951
Basu, S., & Antia, H. M. 2004, ApJ, 606, L85
Basu, S., Pinsonneault, M. H., & Bahcall, J. N. 2000, ApJ, 529, 1084
Bensby, T., Feltzing, S., Lundström, I., & Ilyin, I. 2005, A&A, 433, 185
Boesgaard, A. M., Armengaud, E., King, J. R., Deliyannis, C. P., & Stephens, A. 2004, ApJ, 613, 1202
Brook, C. B., Veilleux, V., Kawata, D., Martel, H., & Gibson, B. K. 2005, in Island Universes, ed. R. de Jong, in press (astro-ph/0511002)
Böhm-Vitense, E. 1958, Zeitschrift fur Astrophysics, 46, 108
Burris, D. L., Pilachowski, C. A., Armandroff, T. E., Sneden, C., Cowan, J. J., & Roe, H. 2000, ApJ, 544, 302
Burrows, A., et al. 1997, ApJ, 491, 856
Burrows, A., Sudarsky, D., & Hubeny, I. 2005, ApJ, submitted (astro-ph/0509066)
Christensen-Dalsgaard, J. 2002, Reviews of Modern Physics, 74, 1073
Cowan, J. J., et al. 2002, ApJ, 572, 861
Cowley, C. R., & Bautista, M. 2003, MNRAS, 341, 1226
Daflon, S., & Cunha, K. 2004, ApJ, 617, 1115
Demarque, P., & Guenther, D. B. 1988, in IAU Symp. 123: Advances in Helio- and Asteroseismology, ed. J. Christensen-Dalsgaard & S. Frandsen, 91
Den Hartog, E. A., Herd, M. T., Lawler, J. E., Sneden, C., Cowan, J. J., & Beers, T. C. 2005, ApJ, 619, 639
Den Hartog, E. A., Wickliffe, M. E., & Lawler, J. E. 2002, ApJS, 141, 255
Drake, J. J., & Testa, P. 2005, Nature, 436, 525
Erspamer, D., & North, P. 2003, A&A, 398, 1121

Fuhrmann, K. 1998, A&A, 338, 161

García López, R. J., Rebolo, R., & Pérez de Taoro, M. R. 1995, A&A, 302, 184

Gilmore, G., & Reid, N. 1983, MNRAS, 202, 1025

Giridhar, S., Lambert, D., Reddy, B. E., Gonzalez, G., & Yong, D. 2005, ApJ, 627, 432

González Hernández, J. I., et al. 2005, ApJ, 630, 495

Gratton, R., Carretta, E., Matteucci, F., & Sneden, C. 1996, in ASP Conf. Ser. 92:
    Formation of the Galactic Halo, ed. H. L. Morrison & A. Sarajedini, 307

Gustafsson, B., Bell, R. A., Eriksson, K., & Nordlund, Å. 1975, A&A, 42, 407

Gustafsson, B., Edvardsson, B., Eriksson, K., Mizuno-Wiedner, M., Jørgensen, U. G.,
    & Plez, B. 2003, in ASP Conf. Ser. 288: Stellar Atmosphere Modeling, ed. I.
    Hubeny, D. Mihalas, & K. Werner (San Francisco: ASP), 331

Gustafsson, B., & Jørgensen, U. G. 1994, A&A Rev., 6, 19

Hillier, D. J. 2003, in ASP Conf. Ser. 288: Stellar Atmosphere Modeling, ed. I. Hubeny,
    D. Mihalas, & K. Werner (San Francisco: ASP), 199

Hubeny, I., & Lanz, T. 1995, ApJ, 439, 875

Hurlburt, N. E., Toomre, J., & Massaguer, J. M. 1984, ApJ, 282, 557

Katz, D., Soubiran, C., Cayrel, R., Adda, M., & Cautain, R. 1998, A&A, 338, 151

Kirkpatrick, J. D. 2005, ARA&A, 43, 195

Kurucz, R. 1993, CD-ROM No. 15. (Cambridge, Mass.: SAO)

Lanz, T., & Hubeny, I. 2003, ApJS, 146, 417

Lawler, J. E., Sneden, C., & Cowan, J. J. 2004, ApJ, 604, 850

Lawler, J. E., Wyart, J.-F., & Blaise, J. 2001, ApJS, 137, 351

Lodders, K. 2003, ApJ, 591, 1220

Massaguer, J. M., & Zahn, J.-P. 1980, A&A, 87, 315

Mishenina, T. V., Soubiran, C., Kovtyukh, V., & Korotin, S. A. 2004, A&A, 418, 551

Montalbán, J., Miglio, A., Noels, A., Grevesse, N., & di Mauro, M. P. 2004, in SOHO
    14 Helio- and Asteroseismology: Towards a Golden Future, ed. D. Danesy, 574

Meléndez, J., & Ramírez, I. 2004, ApJ, 615, L33

Nahar, S. N., & Pradhan, A. K. 2005, A&A, 437, 345

Nordlund, Å. 1980, Lecture Notes in Physics (Berlin: Springer Verlag), 114, 213

Pauldrach, A. W. A., Hoffmann, T. L., Lennon, M. 2001, A&A, 375, 161

Piskunov, N. E., & Valenti, J. A. 2002, A&A, 385, 1095

Rauch, T. 2003, A&A, 403, 709

Ryan, S. G., Norris, J. E., & Beers, T. C. 1999, ApJ, 523, 654

Reddy, B. E., Tomkin, J., Lambert, D.L., & Allende Prieto, C. 2003, MNRAS, 340, 304

Reddy, B. E., Lambert, D. L., & Allende Prieto, C. 2006, MNRAS, in press

Repolust, T., Puls, J., & Herrero, A. 2004, A&A, 415, 349

Ritter, A., & Washuettl, A. 2004, Astronomische Nachrichten, 325, 663

Robinson, F. J., Demarque, P., Li, L. H., Sofia, S., Kim, Y.-C., Chan, K. L., & Guenther,
    D. B. 2004, MNRAS, 347, 1208

Seaton, M. J. 2005, MNRAS, 362, L1

Short, C. I., & Hauschildt, P. H. 2005, ApJ, 618, 926

Sneden, C., Preston, G. W., McWilliam, A., & Searle, L. 1994, ApJ, 431, L27

Sneden, C., et al. 2003, ApJ, 591, 936

Snider, S., Allende Prieto, C., von Hippel, T., Beers, T. C., Sneden, C., Qu, Y., &
    Rossi, S. 2001, ApJ, 562, 528

Spite, F., & Spite, M. 1982, A&A, 115, 357

Stehlé, C. 1994, A&A, 104, 509

Stein, R. F., & Nordlund, Å. 1998, ApJ, 499, 914

Tsuji, T. 2005, ApJ, 621, 1033

Van Winckel, H. 2003, ARA&A, 41, 391

Van Winckel, H., Waelkens, C., & Waters, L. B. F. M. 1995, A&A, 293, L25

Westin, J., Sneden, C., Gustafsson, B., & Cowan, J. J. 2000, ApJ, 530, 783

Willemsen, P. G., Hilker, M., Kayser, A., & Bailer-Jones, C. A. L. 2005, A&A, 436, 379

Young, P. A., & Arnett, D. 2005, ApJ, 618, 908

Frank N. Bash Symposium 2005: New Horizons in Astronomy
ASP Conference Series, Vol. 352, 2006
S. J. Kannappan, S. Redfield, J. E. Kessler-Silacci, M. Landriau, and N. Drory

Evidence for Intermediate Mass Black Holes in Ultra-Luminous X-ray Sources

Jon M. Miller

Department of Astronomy, University of Michigan, Ann Arbor, MI, USA

Abstract. It is becoming clear that intermediate-mass black holes — black holes with masses in the 100–10,000 $M_\odot$ range — may have played an important role in early galactic evolution, and may currently be playing an important role in stellar cluster evolution. Here, I briefly highlight emerging evidence for intermediate mass black holes, in a subset of the most luminous non-nuclear X-ray point sources in nearby normal galaxies.

1. Introduction

Intermediate-mass black holes (IMBHs) may have played a very important role in the early universe. The first stars, so-called Population III stars, were likely extremely massive by present standards (few $\times 10^2$ $M_\odot$, Abel, Bryan, & Norman 2000). As such stars were necessarily metal-poor, they likely suffered minimal mass loss through stellar winds, and so would have retained much of their original mass in forming black holes (Heger et al. 2003). Initially, these IMBHs may have been a major source of ionizing radiation in young galaxies (Madau et al. 2004). Later, such IMBHs may have helped to build-up the mass of the central super-massive black hole in their host galaxy. It is possible that a fraction of these primordial IMBHs were not accreted, and might now accrete from the interstellar medium (ISM) or captured companion stars (Madau & Rees 2001).

Similarly, intermediate-mass black holes may have a strong impact on stellar cluster evolution. Studies of three-body interactions in globular clusters suggest that massive central black holes may grow through such interactions (Miller & Hamilton 2002). In extremely dense young stellar clusters, recent studies suggest that an IMBH may be quickly and naturally built-up through a run-away merger process (Portegies Zwart et al. 2004).

Supermassive black holes (somewhat arbitrarily, 10^6 $M_\odot$ and above) are well-known in active galactic nuclei (AGN), and in normal galaxies. Within the Milky Way and the Large Magellanic Cloud, optical radial velocity curves have identified nearly 20 binary systems which harbor stellar-mass black holes with masses up to 14 $M_\odot$ (McClintock & Remillard 2006). The theoretical considerations briefly mentioned above, may provide a basis for expecting that the broad divide between these super-massive and stellar-mass populations is not entirely barren.

Exploiting the $M - \sigma$ relationship, measurements of velocity dispersions (σ) in Seyfert AGN suggest that some of these galaxies may harbor $M \sim 10^5 M_\odot$ black holes (Barth, Greene, & Ho 2005). The same relationship may suggest that some globular clusters may harbor 10^4 $M_\odot$ black holes (Gebhardt, Rich,

& Ho 2005). Independently, a number of non-nuclear X-ray point sources have been discovered in nearby normal galaxies, which have luminosities in excess of the Eddington luminosity expected for $\sim$10 $M_\odot$ black hole X-ray binaries. Many studies of these sources have been aimed at understanding whether or not part of this population of "ultra-luminous" X-ray sources (ULXs) may harbor IMBHs.

2. Ultra-luminous X-ray Sources

ULXs are non-nuclear X-ray point sources in nearby normal galaxies. The luminosity at which a source qualifies as a ULX is somewhat arbitrary, but presently a widely-accepted and useful cut-off is $L > 2 \times 10^{39}$ erg s^{-1} (Irwin, Bregman, & Athey 2004). ULXs were first detected with the Einstein X-ray observatory. More detailed observations became possible with the Röntgensatellit (ROSAT) and the Advanced Satellite for Cosmology and Astrophysics (ASCA), and in many cases revealed variability properties and spectra consistent with accretion-powered sources. In the Chandra and XMM-Newton era, observations of ULXs have come of age, due to the spatial resolution and sensitivity of these observatories. For excellent reviews of the results of Chandra and XMM-Newton observations of ULXs and point source populations in nearby galaxies, please see Fabbiano & White (2006), Miller & Colbert (2004), and Swartz et al. (2004).

3. Searching for IMBHs in ULXs

The simplest method of searching for IMBH candidates, is to identify the most luminous sources and scale from the Eddington luminosity for a stellar-mass object, 1.3×10^{38} erg s^{-1}. Many ULXs only barely meet the luminosity criteria of $L_X \geq 2 \times 10^{39}$ erg s^{-1}, however, it is well-known that stellar-mass black holes can occasionally exceed the Eddington limit. To identify plausible IMBH candidates based only on luminosity, then, it is necessary to set a much higher luminosity threshold (at or near to 10^{40} erg s^{-1}, for instance).

Even with a high threshold, a luminosity cut is not necessarily a robust means of identifying IMBH candidates — the apparent luminosity of a source may not be an isotropic luminosity. It is possible that the flux from a given ULX may be relativistically beamed (Körding, Falcke, & Markoff 2002), or mechanically focused (e.g., by a funnel geometry at the inner disk; King et al. 2001). Fortunately, there are methods by which beaming and funneling can be addressed. Beaming tends to create flat spectra; radio upper limits on a number of strong IMBH-candidate ULXs appear to rule out beaming, and to be fully consistent with unbeamed stellar-mass black hole flux ratios (Miller et al. 2003). Funneling is more difficult to rule out, but one clever method has been devised: optical nebulae which surround many strong IMBH-candidate ULXs can be used as bolometers (Kaaret, Ward, & Zezas 2004), to address whether or not the given ULX is truly super-Eddington.

More sophisticated, and perhaps more reliable methods of identifying IMBH candidates in the ULX population, rely on scaling X-ray spectroscopic and timing properties from those commonly observed in stellar-mass black holes. (It should be noted that the ULXs in which multiple components can be required in energy spectra, and in which features can be seen in power spectra, are necessarily the most luminous sources within about 10 Mpc, given the sensitivity of Chandra and XMM-Newton. Thus, the set of sources for which such studies are possible is necessarily the same set given by a conservative luminosity threshold.) When stellar-mass black holes are observed at luminosities consistent with their Eddington luminosity, inner disk temperatures of $kT \simeq$ 1–2 keV are measured. For disks around stellar-mass black holes, $T \propto M^{-1/4}$. Scaling inner disk temperatures, then, provides a spectroscopic means of identifying IMBH candidates (Colbert & Mushotzky 1999, Miller et al. 2003). Similarly, characteristic "break" frequencies and quasi-periodic oscillation (QPO) frequencies (and relationships between breaks and QPOs) have been identified in stellar-mass black holes, and scaling from these frequencies to those seen in ULXs may provide a means of estimating ULX masses (Strohmayer & Mushotzky 2003, Cropper et al. 2004).

At present, the most secure IMBH candidate ULX is perhaps M82 X-1. This source has been observed at luminosities as high as 10^{41} erg s^{-1}. X-ray timing studies have revealed a break frequency and a QPO in its power spectra (Strohmayer & Mushotzky 2003), which may indicate that M82 X-1 harbors a black hole of 10^{3-4} $M_\odot$. Unfortunately, M82 X-1 resides in a region of diffuse soft X-ray emission, which makes it difficult to constrain the nature of the accretion disk in this system.

A number of ULXs have been identified with accretion disk temperatures of $kT \simeq$ 0.2 keV. Miller, Fabian, & Miller (2004) demonstrated that these sources lie well off the $L \propto T^4$ trend seen in a number of well-known stellar-mass black holes, and suggested that these particular ULXs could harbor 100–1000 $M_\odot$ black holes. Figure 1 demonstrates that some ULXs may represent a subpopulation of IMBHs within the broader ULX population, based on the position they occupy relative to stellar-mass black holes in a luminosity versus temperature diagram. This plot differs from that in Miller et al. (2004), in that Antennae X-37 has since been identified as a background AGN. Winter, Mushotzky, & Reynolds (2006) expanded on this work, and examined a large number of ULXs which were observed with XMM-Newton. In Figure 2, the ULXs considered by Winter et al. (2006) have been added to the ULXs and stellar-mass black holes considered by Miller et al. (2004). Only sources for which the error on the temperature diagnostic is not more than 50% of the measured value are included in the plot, and the luminosity values reported by Winter et al. (2006) were corrected downward by 20% to account for the broader energy range considered in that work. While some of the ULXs in the new sample do not smoothly join onto the luminosity versus temperature trend observed in stellar-mass black holes, they appear to be part of a different population than the ULXs with very low disk temperatures. Overall, Figure 2 serves to strengthen the possibility that a number of the brightest ULXs may harbor accreting IMBHs. (For an example of alternatives to the IMBH hypothesis for some ULXs and a discussion of the robustness of such alternatives, see e.g., Goad et al. 2006 and Miller, Fabian, & Miller 2006.)

It must be noted that scaling is an inexact means by which to estimate the mass of the black holes which may power ULXs, and a given mass estimate should only be regarded as representative. For a given IMBH candidate, scaling from the Eddington limit, the inner disk temperature, and timing parameters

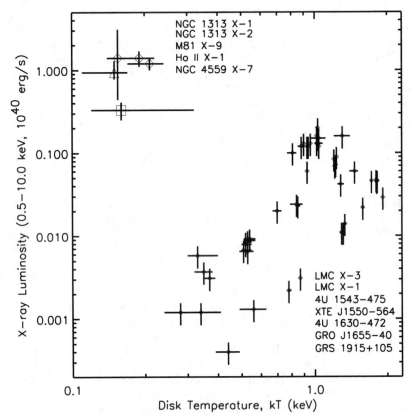

Figure 1.   The figure above plots X-ray luminosity versus apparent disk temperature, for a number of the most luminous ULXs (top left), and a number of well-known stellar-mass black hole X-ray binaries (bottom right). The stellar-mass black holes show a clear $L \propto T^4$ relationship, as expected for standard blackbody accretion disks. Another expectation for disks around black holes, that $T \propto M^{-1/4}$, suggests that the ULXs in the plot above may harbor 100–1000 $M_\odot$ black holes.

common for stellar-mass black holes, each give different mass estimates with large systematic errors. However, for the best IMBH candidates, independent scaling methods point in the direction of massive black hole primaries, and estimates generally lie in the 100–1000 $M_\odot$ range.

With a number of IMBH candidates now identified, it is possible to begin to examine IMBH formation mechanisms. This work is only beginning, but some trends are already emerging. The number of ULXs in elliptical galaxies, is consistent with the number of background AGN that could be mistaken for a galactic ULX source (Irwin et al. 2004). In contrast, the number and distribution of ULXs in spiral and starburst galaxies allows these sources to be confidently associated with a given host galaxy. This may indicate that most IMBHs which are visible in the present-day universe are not relics of Population-III stars, and

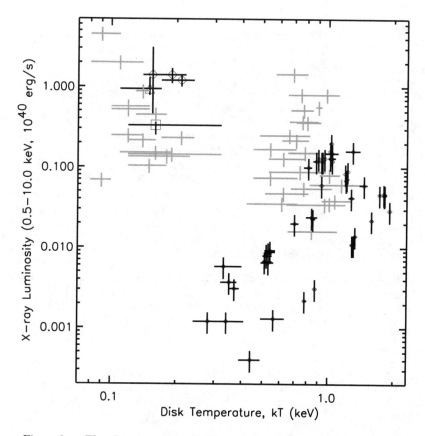

Figure 2. The plot above extends Figure 1 to include a much larger sample of ULXs. In this plot, the ULXs and stellar-mass black holes plotted in Figure 1 are shown in black, and ULXs considered by Winter et al. (2006) are shown in gray (please see the text for details). Though the ULXs with high disk temperatures do not appear to join onto the stellar-mass black hole sample seamlessly, the plot above does reinforce the idea that ULXs may be comprised of both stellar-mass and intermediate-mass black holes, typified by disjoint disk temperatures.

may indicate that IMBH formation in young stellar clusters is the more dominant formation process.

In searching for IMBHs in ULXs, there is an implicit assumption that the black hole is accreting from a companion star. Accretion from the ISM, or from a molecular cloud, is unlikely to drive a given IMBH to a high fraction of its Eddington luminosity, and IMBHs accreting in this way probably do not appear as ULXs. The need to fuel an IMBH with a companion star again supports models for IMBH formation in stellar clusters, as encounter rates in the field can be expected to be very low.

4. Future Studies

Multi-wavelength studies presently play an important role in understanding the nature of ULXs, and will continue to do so. However, it is unlikely that radial velocity curves can be extracted for mass constraints, even with 8–10 m telescopes. Radial velocity curves are very difficult to obtain when sources are actively accreting; in the few cases where this has succeeded in Galactic X-ray binaries, it has been due to a combination of excellent instrumental resolution and extremely high signal-to-noise ratios. Such studies may be possible in the era of 30 m telescopes, but even then it is likely to be extremely difficult.

In the near future, a very deep XMM-Newton stare at a galaxy harboring two or more strong IMBH-candidate ULXs can have a significant impact. An observation of approximately 500 ksec will allow for detailed spectroscopy and timing studies, that will strongly constrain the nature of the accreting compact objects in these systems. In the long run, the high effective area of planned missions such as Constellation-X will make it possible to study IMBH-candidate ULXs with a sensitivity approaching that which is presently achieved in studies of Galactic stellar-mass black hole binaries.

Acknowledgments. I would like to thank the organizers of the 2005 Bash Symposium, for making it a wonderful event. I would also like to acknowledge my frequent collaborators in studies of ULXs, Andy Fabian and Cole Miller.

References

Abel, T., Bryan, G. L., & Norman, M. L. 2000, ApJ, 540, 39
Barth, A. J., Greene, J. E., & Ho, L. C. 2005, ApJ, 619, L151
Colbert, E. J. M., & Mushotzky, R. F. 1999, ApJ, 519, 89
Cropper, M., Soria, R., Mushotzky, R. F., Wu, K., Markwardt, C. B., & Pakull, M. 2004, MNRAS, 349, 39
Fabbiano, G., & White, N. E. 2006, in Compact Stellar X-ray Sources, ed. W. H. G. Lewin & M. van der Klis (Cambridge: Cambridge University Press), in press
Gebhardt, K., Rich, R. M., & Ho, L. C. 2005, ApJ, 634, 1093
Goad, M. R., Roberts, T. P., Reeves, J. N., & Uttley, P. 2006, MNRAS, 365, 191
Heger, A., Fryer, C. L., Woosley, S. E., Langer, N., & Hartmann, D. H. 2003, ApJ, 591, 288
Irwin, J. A., Bregman, J. N., & Athey, A. E. 2004, ApJ, 601, L143
Kaaret, P., Ward, M. J., & Zezas, A. 2004, MNRAS, 351, L83
King, A. R., Davies, M. B., Ward, M. J., Fabbiano, G., & Elvis, M. 2001, ApJ, 552, L109
Körding, E., Falcke, H., & Markoff, S. 2002, A&A, 382, L13
Madau, P., & Rees, M. J. 2001, ApJ, 551, L27
Madau, P., Rees, M. J., Volonteri, M., Haardt, F., & Oh, S. P. 2004, ApJ, 604, 484
McClintock, J., & Remillard, R. 2006, in Compact Stellar X-ray Sources, ed. W. H. G. Lewin & M. van der Klis (Cambridge: Cambridge University Press), in press
Miller, J. M., Fabbiano, G., Miller, M. C., & Fabian, A. C. 2003, ApJ, 585, L37
Miller, J. M., Fabian, A. C., & Miller, M. C. 2004, ApJ, 614, L117
Miller, J. M., Fabian, A. C., & Miller, M. C. 2006, MNRAS, submitted (astro-ph/0512552)
Miller, M. C., & Colbert, E. J. M. 2004, International Journal of Modern Physics D, 13, 1
Miller, M. C., & Hamilton, D. P. 2002, MNRAS, 330, 232

Portegies Zwart, S. F., Baumgardt, H., Hut, P., Makino, J., & McMillan, S. L. W. 2004, Nature, 428, 724

Strohmayer, T. E., & Mushotzky, R. F. 2003, ApJ, 586, L61

Swartz, D. A., Ghosh, K. K., Tennant, A. F., & Wu, K. 2004, ApJS, 154, 519

Winter, L. M., Mushotzky, R. F., & Reynolds, C. S. 2006, ApJ, submitted (astro-ph/0512480)

Black hole accretion experts Jon Miller and Sera Markoff compare notes on a technical point before answering a question.

Frank N. Bash Symposium 2005: New Horizons in Astronomy
ASP Conference Series, Vol. 352, 2006
S. J. Kannappan, S. Redfield, J. E. Kessler-Silacci, M. Landriau, and N. Drory

Accretion and Jets in Microquasars and Active Galactic Nuclei

Sera Markoff

Astronomical Institute "Anton Pannekoek," University of Amsterdam, Amsterdam, The Netherlands

Abstract.   Black holes from stellar to galactic scales are observed to accrete material from their environments and, via an as yet unknown mechanism, produce jets of outflowing plasma. In X-ray binaries (XRBs), the systems display radically different radiative properties depending on the amount of captured gas reaching the event horizon. These modes of behavior (one of which includes "microquasars") correspond to actual physical changes in the environment near the black hole and can occur on timescales of days to weeks. Some of this behavior should hold true for active galactic nuclei (AGN) if the underlying physics scales with central mass and accretion power, as would be expected if black holes can be characterized mainly by their mass and local environment. However, the timescales on which changes occur should be inversely proportional to the mass. Recent studies support that this scaling applies in some cases, opening the way for comparisons of different stages of time-dependent behavior in microquasars to different classes of AGN zoology. In this distinctly jet-biased review, I will summarize our current understanding of accretion and outflow in these systems and present some of the newest progress addressing unanswered questions about the nature of the accretion flows, jet formation, and jet composition.

1.   Introduction

Although the data are pouring in, and computer simulations are increasingly able to tackle the necessary scenarios, many of the fundamental questions regarding the accretion/jet phenomenon are still largely unanswered. The partnership of infalling, rotating material and outflowing jets appears in objects, and on scales, as diverse as young stellar objects (YSOs), pulsars, and AGN, among others. Jets are obviously a natural byproduct of the accretion/infall process and are intimately related to the local angular momentum and magnetic fields. However, the exact relationship between inflow and outflow, as well as how jets are created and their internal composition, are still active areas of controversy and research on many fronts. In this summary, I will highlight some of the more recent advances and their related articles, with a distinctly jet-based bias because of my own research interests. A description of earlier groundwork can naturally be found in the quoted articles' references.

Because physical characteristics vary significantly depending on the class of accreting object, it has proven quite challenging to discern underlying trends common to all such sources. For instance, the weakly collimated outflows from symbiotic stars are thought to be predominantly comprised of non-relativistic, thermal plasma (e.g., Galloway & Sokoloski 2004), while the jets in AGN can be highly collimated, accelerated to Lorenz factors of $\Gamma \geq 10$ (Jorstad et al.

2005), and may even be dominated by electromagnetic fields rather than plasma (for recent simulations and discussions, see, e.g., De Villiers et al. 2005 and McKinney 2006). The most fruitful approach seems therefore to initially focus on one type of system, to look for clues about the underlying physics that may later turn out to be valid for all inflow/outflow systems.

Accreting black holes have one distinct advantage in the search for generic trends, in that they occur in systems ranging over orders of magnitude in central mass. While accretion disks are necessarily studied by mostly indirect means (they are usually too small or too distant to resolve), jets are now regularly observed in both stellar-mass accreting black holes (XRBs) and AGN, and may well exist in intermediate-mass black holes (IMBHs, see, e.g., Kaaret et al. 2003). If the physics governing accretion and its related outflows is generic to black holes regardless of size, then some observables should scale predictably between XRBs and AGN.

2.   Fundamental Questions

Much of the information necessary to determine what exactly is involved at the inflow/outflow interface, and thus in jet production, is still largely undetermined. A key factor is of course that this area cannot be directly imaged, except in the rarest of cases (e.g., there are hopes that submm very long baseline interferometry will resolve this region in our Galactic center supermassive black hole, Sgr A*). While the quality of data for emission originating close to black holes can be quite high, particularly in the X-ray band, it is mostly spectral and timing information, which does not usually unambiguously select a single physical picture.

We do at least know something about the content and general structure of accretion disks. Spectroscopy reveals both emission and absorption processes in the optical as well as X-ray bands, in many cases allowing detailed plasma diagnostics. And although the microphysics may prove much more complicated, interestingly rather simple analytical models perform fairly well for predicting some macroscopic accretion disk observables. Jets, however, are arguably far less determined. Their broadband emission comes from processes which can occur for a broad range of internal conditions, so fundamental parameters such as their total power, the distribution of this energy between plasma and electromagnetic fields, the type of plasma contained and its origins, and the physical structure all remain outstanding questions. These ambiguities make it quite difficult to draw conclusions about how jets form, and whether the jets in black holes differ in any significant way from jets in other types of accreting systems, such as neutron stars.

These questions are important not just for understanding accretion, but for gauging the effect of jets on their local environment. Even though we are used to thinking of gravity as the weakest force, it is surprisingly efficient in the case of accretion. While nuclear fusion in stars has $\sim 0.7\%$ efficiency in converting rest mass to energy, accretion around black holes has almost 10% efficiency! Jets carry away a fair fraction of the liberated energy (on the order of 10-50%, depending on the model), depositing it well beyond the sphere of influence of the central engine with obvious effects. The most visibly apparent of these are the "bubbles" full of plasma and fields which are created by XRB jets in the interstellar medium, and by AGN jets in the intergalactic or intercluster medium (Fig. 1). The size of the bubbles gives an estimate of the work necessary to inflate them, if the external pressure is known, and in the case of the supercavities in clusters of galaxies, this can be as large as 10^{61} ergs (McNamara et al. 2005)! If the lifetime of the source or outburst is known, this then provides an independent method of determining the total power in the outflows. So far the indications are that jets carry away a large fraction of the accretion energy, and this needs to be accounted for in dynamical models.

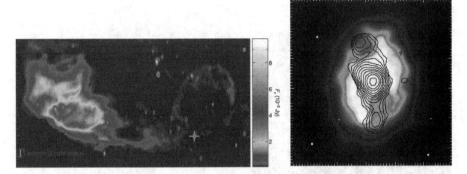

Figure 1.   The power of jets: significant fractions of energy gravitationally liberated by black holes are dumped far from the compact object into the local environment. a) 1.4 GHz radio image from the Westerbork Synthesis Radio Telescope showing the 15 light-year wide (in projection) bubble blown into the local ISM by the relativistic jet of the Galactic stellar black hole Cyg X-1 (figure reproduced with kind permission from Gallo et al. 2005). The wedge shows the (log) flux scale. b) Overlay of Very Large Array 1.4 GHz radio contours on a *Chandra* X-ray image of the MS0735.6+7421 cluster of galaxies. The center of the cluster harbors a cD galaxy whose radio jets have carved out the ~200 kpc diameter cavities, each requiring ~ 10^{61} ergs of work to inflate (figure reproduced with kind permission from McNamara et al. 2005).

One of the major issues being considered at the moment is the structure of accretion disks, particularly in the inner regions closest to the central object. The investigation of the roles of turbulence, magnetic fields and microphysics in general is still in progress, but the latest results are quite promising. Several approaches are being taken towards addressing questions about how magnetic fields are generated, maintained, and fed into (possibly ultimately helping to accelerate and collimate) the jets.

A few groups (e.g., McKinney & Gammie 2004, De Villiers et al. 2005, and Nishikawa et al. 2005), have succeeded in developing fully relativistic, three dimensional magnetohydrodynamical (MHD) simulations. For accretion disk simulations, turbulence clearly does play an important role (see Fig. 2a). However, one caveat is that it is still very difficult to include dissipation and radiative transfer (but see, e.g., Goldston et al. 2005). So for this and other reasons there may still be significant differences between the simulated disks and the real systems. For instance, MHD simulations generally show turbulence bubbling up into a corona, but this would predict higher levels of noise in power density spectra. Such a high level of rms noise is actually observed only in the XRB accretion states which are not in fact traditionally associated with bright accretion disk emission (for a review of timing behavior and XRB states, see van der Klis 1995). In fact, very high precision X-ray observations show that accretion-disk–dominated states such as the traditional "soft state" have rms noise on the order of only a few percent. Similarly, relatively simple analytical models treating the accretion disk as a quiet system, radiating as a smooth multicolor blackbody (e.g., Mitsuda et al. 1984), perform remarkably well against the data (Fig. 2b). These discrepancies will likely be resolved as more physics is included in the simulations.

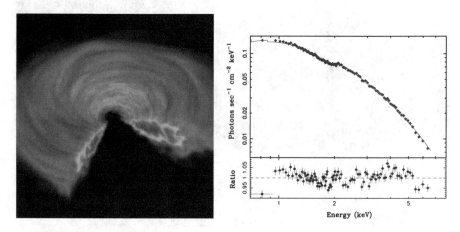

Figure 2.   Despite complicated simulations showing that disks should be quite turbulent, spectral data can be fit remarkably well by simple multicolor blackbody models. a) Snapshot of a 3D MHD simulation of the density in the inner regions of an accretion disk accreting onto a black hole. The disk evolves as a result of angular momentum transport, produced via the magnetorotational instability-generated MHD turbulence (figure reproduced with kind permission from Hawley & Krolik 2001). Such simulations predict significant turbulence. b) Comparison of *Chandra* HEG-1 data from a soft state (disk-dominated) black hole XRB to a simple, two parameter (normalization and inner radius temperature) multicolor black body model (Mitsuda et al. 1984). These ideal disks would be surprisingly rather "quiet" (figure reproduced with kind permission from Nowak et al. 2006).

When these types of open questions exist about inflows and outflows in their own right, it obviously inhibits our ability to understand which mechanisms govern outflow formation, and how. For instance, even if the jet power is determined, it is still not clear how the internal energy is distributed between particles and fields. Similarly, any matter present could either be purely leptonic (i.e., e^+e^- pairs) or hadronic (protons and/or ions and electrons), and obviously the difference in terms of kinetic energy and momentum transport will be important. Without knowing these parameters, it is challenging to work backwards to infer the conditions at work in jet formation, or even to properly interpret the radiative signatures of outflows. A very strong magnetic field with a small amount of hot ionized plasma can in principle give the same flux of synchrotron emission as a weak magnetic field with a larger amount of even hotter plasma. The differences will arise in details of the spectral modeling, and in particular, predictions for radiation in other wavebands. Therefore multiwavelength observations have become one of the best sources for information about internal conditions in the outflows, and their relationship to the accretion inflow.

3. Clues from Multiwavelength Observations

Multiwavelength observations are valuable for any accreting source, but the advantage in studying XRBs is twofold. First, they exist in several different accretion states, each with its own unique characteristics in the various wavebands and power density spectra, and they exhibit transitions between these states repeatedly on observable timescales. For instance the Galactic XRBs GX 339-4 and Cyg X-1 cycle between the various states on approximately yearly timescales (e.g. Pottschmidt et al. 2003; Homan et al. 2005). Thus theories can be developed for XRB behavior which can make testable predictions for the next outburst cycle of the same source. Second, and this is especially true in the last few years, there are high-quality broadband simultaneous data available for many sources.

The XRB short timescales and good data are particularly relevant for making comparisons of black hole accretion on all scales. If the processes governing accretion onto black holes are somewhat generic (as might be expected because despite differences in the local environment, the central black hole accretes via an accretion disk), then any trends found to hold for XRB black holes should also hold for AGN. The difficulty in testing this premise is that even though the underlying physics may be the same, the relevant timescales for systematic changes are determined to a great extent by the system size, which scales linearly with the black hole mass. Therefore a change which is observed to occur over a day in an XRB could take 10^8 days in a typical AGN. While this is obviously not something observable directly in our lifetimes, statistical studies are possible using large samples of AGN. Then the most difficult question becomes understanding which class of AGN corresponds to which "snapshot" in the life of an XRB, and which accretion state — a question complicated by orientation, beaming, and other effects.

There are a couple of examples of ongoing work where the premise of a consistent scaling in the physics of XRB and AGN black holes seems to hold. The first is not a multiwavelength project but is worth a quick mention in this context. The power density spectra (PDS) of accreting black holes show many unique identifiers, depending on the accretion state (again, see the review by van der Klis 1995). Several groups have been comparing the PDS of XRBs and AGN, a venture which has only recently been possible because for AGN it requires many years of observations to obtain a comparable range in frequencies. These groups are discovering convincing mappings between XRB states and AGN, including the discovery of trends suggesting a mass and accretion rate scaling in the PDS break frequency (e.g., Markowitz et al. 2003; Uttley & McHardy 2005; McHardy et al. 2005).

The example that I will focus more on here depends on broadband observations to learn more about processes close to the central black hole. This is in some ways the more radical approach, simply because for many years it was thought that the X-ray band alone would provide this type of information, and we are now finding that in combination with low frequencies, even more constraints are possible.

In the late 1990s, a nonlinear correlation was discovered between the radio and X-ray emission in the hard state of GX 339-4 (Hannikainen et al. 1998). The "hard" or "low hard" state and the "soft" or "high soft" state are the two most stable states, representing low and high disk luminosity, respectively (see a review of the various states in, e.g., McClintock & Remillard 2006). The hard state is the only state associated with weak, small but steady jets (revealed via their synchrotron emission and occasional imaging). Further joint radio/X-ray studies of the hard state in GX 339-4 ensued (Corbel et al. 2000; Wilms et al. 1999; Nowak et al. 2002; Corbel et al. 2003), with the last of these papers establishing that the correlation takes the form $L_{\rm R} \propto L_{\rm X}^{0.7}$ over orders of magnitude changes in the source luminosity with time. More recent studies have established that this correlation appears to be universal for all hard state XRBs (Fig. 3 and Gallo et al. 2003), though not always with the same normalization (Nowak et al. 2005; this reference also discusses possible infrared/X-ray correlations).

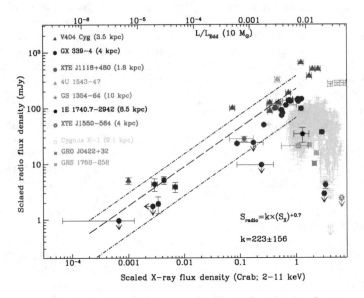

Figure 3.   The radio flux density vs. the X-ray flux density for a sample of 10 hard state black holes, scaled to a distance of 1 kpc and absorption corrected. The dashed line indicates the best fit to the observed radio/X-ray correlation $S_{\rm radio} \propto S_{\rm x}^{+0.7}$ (figure reproduced with kind permission from Gallo et al. 2003).

The 0.7 slope of the $\log(L_{\rm R})$-$\log(L_{\rm X})$ correlation follows directly from analytic predictions of scaling synchrotron jet models (Falcke & Biermann 1995; Markoff et al. 2003) and can be generalized for any X-ray emission process (disk or jet related) in terms of its dependence on the accretion rate (Heinz & Sunyaev 2003).

The correlation's normalization in the (L_X, L_R) plane also depends on the central engine mass. When this scaling is accounted for, unbeamed, extragalactic, supermassive black hole sources agree remarkably well with the same radio/X-ray correlation found for Galactic, stellar-mass sources (Fig. 4 and Merloni et al. 2003; Falcke et al. 2004). In general, one must be very careful in comparing samples of AGN against trends derived from individual XRB sources. One danger is that the correlation could be a distance-driven artifact (essentially plotting distance vs. distance) from a flux-limited sample. In Merloni et al. (2006), this and other statistical caveats are explored, such as by randomly scrambling all fluxes in one waveband (see, e.g., Fig. 5). Clearly the correlation seen in the original data is not apparent after scrambling, to high statical significance. These works strongly support the idea that the underlying physics governing accretion and broadband emission characteristics is fundamental to all accreting black holes regardless of mass.

One other factor that may influence this so-called "fundamental plane of accreting black holes" is the simultaneity of the observations. Most of the measurements in the AGN samples used in Figs. 4 & 5 were non-simultaneous. However for XRBs, flux measurements separated in time by more than days would result in an altogether different correlation coefficient reflecting the lag in observation time. In general, because AGN evolve much slower, non-simultaneity should not be a major factor. Nearby low-luminosity AGN (LLAGN), however, often have smaller central masses and show variations in both radio and X-ray fluxes of tens of percent over month-long timescales. An alternate approach to studying the fundamental plane using large samples is to use just a few sources with very well measured mass and distance, where any scatter is dominated by intrinsic variability. An initial test was performed in Markoff (2005) using data from GX 339-4 as well as our Galactic, underluminous supermassive black hole, Sgr A*, and two nearby LLAGN, M81 and NGC 4258. All of these sources have well-determined physical parameters such as mass and distance and are not highly beamed, allowing a detailed assessment of the fundamental plane coefficients as well as the scatter from variability. A linear regression fit was performed based on Monte Carlo simulations from decades of observations. The best fit plane is shown in Fig. 6, with contours in average scatter indicated. Interestingly Sgr A* in quiescence falls statistically off the correlation, while the flares increasingly approach it, suggesting a state transition at extreme ($<10^{-7}$ L_{Edd}) low luminosities. For comparison, the first simultaneous broadband observations of M81* were carried out this year (Markoff et al. 2006) and four data points corresponding exactly to those for XRBs can now be added to the fundamental plane projection. These are shown in Fig. 6 for comparison to the data point representing the average and rms variation based on all prior non-simultaneous observations. It is clear that simultaneity gives different results than using samples based on prior non-simultaneous observations.

The existence of the fundamental plane has placed significant limits on the processes that contribute to the hard (low luminosity) X-ray emission. Time variability indicates that this emission originates very close to the black hole, but leaves significant degeneracy in the process responsible for the emission. Only in combination with information from the lower frequencies can we begin to constrain the underlying physics of the extreme conditions at the interface of inflow and outflow.

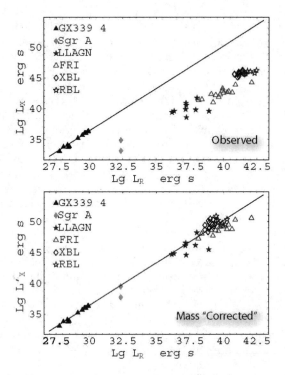

Figure 4.   The same radio/X-ray correlation shown in Fig. 3 for X-ray binaries, now including a sample of AGN. The solid line indicates the nonlinear
correlation predicted by a jet model (Markoff et al. 2003), normalized for the
Galactic XRB GX 339-4. The top panel shows the observed values of the
luminosities. The bottom panel shows the same data, except now plotting
L'_X, the "corrected" X-ray luminosity of each AGN after scaling by a theoretically predicted mass factor to the value if all sources had the same mass as
GX 339-4, 6 $M_\odot$ (figures reproduced from Falcke et al. 2004). An independent
investigation of this result can be found in Merloni et al. (2003). The correlation is a projection of a fundamental plane relating accreting black holes
across mass scales.

4.   Inflow/Outflow Connections: Spectral

There are various ways to interpret the intimate relationship between the radio and X-ray emission in the hard state. As this is the only state associated
with steady jet formation, it is also extremely relevant for learning about the
inflow/outflow relationship. Historically, however, this has not been how this
state has been approached. Until the discovery of the radio/X-ray correlations,
the hard X-ray continuum of this state was almost exclusively modeled in terms
of hot electrons in a compact accretion disk corona, which Compton upscatters
softer photons from the cooler accretion disk. Such a picture does not address
the relationship to the low-frequency data, but it is interesting to note that
the existence of energetic electrons near the compact object is also empirically
required by the jets which are imaged in the radio wavebands.

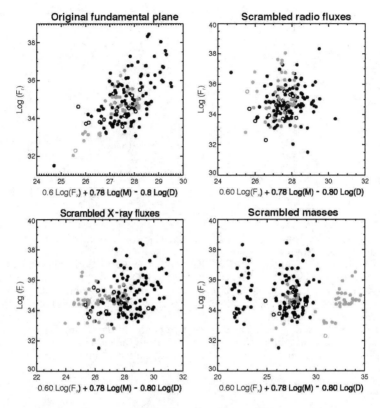

Figure 5.   An investigation of the fundamental plane for a different sample of AGN used by Merloni et al. (2003). The top left panel shows the original correlation, this time with the radio flux at 5 GHz plotted against the mass-factor scaled X-ray flux, also correcting for the measured distances. The other panels illustrate the loss of the correlation when the various input parameters are randomly scrambled, arguing against the fundamental plane being a distance-driven artifact (figures reproduced with kind permission from Merloni et al. 2006).

Some authors have suggested the relationship may be as simple as a correspondence between the X-ray flux and the rate at which radio blobs are emitted (Zdziarski et al. 2003). There is, however, no detailed model yet worked out for this scenario that shows how the spectra and correlations would follow. At the other extreme, it is interesting to investigate the possibility that the X-rays and radio emission both originate in the outflow. This is not as surprising as it may at first sound, since simulations that include magnetic fields in general find that the inner regions of the accretion disk become increasingly unbound, and that static coronal layers do not likely exist as have been envisioned for earlier models. Whether jets follow from the outflow or are enhanced or dominated by Poynting-flux flowing out from the central regions (see Fig. 7a) is not yet known. An important starting point is to investigate whether the base of the jets can

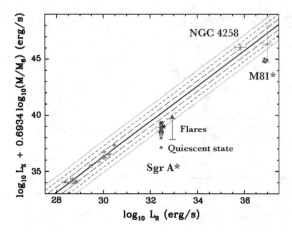

Figure 6.   The fundamental plane correlation for the three best-measured sources bracketing the Galactic center supermassive black hole, Sgr A*: the XRB GX 339-4, and the LLAGN NGC 4358 and M81. Both quiescent and flaring states of Sgr A* are indicated. The LLAGN data with the larger error bars represent the average and rms variation of all prior non-simultaneous observations. For M81, the four dark points indicate results from simultaneous radio/X-ray observations (Markoff et al. 2006). The solid line indicates the best fit correlation using linear regression from Monte Carlo simulations of the data, with contours in average scatter $<\sigma>$ from the correlation represented as increasingly finer dashed lines (figure modified from Markoff 2005).

"subsume" the role of the corona, providing the same spectral characteristics (both direct and reflected) for which coronae were conceived.

Fig. 7b shows such a model, in which the dominant contributor to the radio and X-ray continuum is from a continuous jet, along with a weak accretion disk and consistent reflection signatures (fluorescent iron line and reflection "hump"; see Markoff & Nowak 2004). The radio and soft X-ray emission (below $\sim$10 keV) are dominated by synchrotron emission by accelerated electrons in the jets, while the hard X-rays result from inverse Compton processes (by electrons in the jet base acting on seed photons from both the weak accretion disk and the synchrotron emission from the jet) very close to the black hole. The model has been statistically fit to an average of hard state, simultaneous radio/X-ray observations of Cyg X-1. While the figure shows the details of the X-ray fit, where the most data and higher statistics are, the entire fit including the radio band is shown as an inset. This model obtains the same goodness-of-fit, within the limits of the data, as a "standard" thermal Comptonizing accretion disk corona model. Clearly even the high statistics of *RXTE* X-ray data cannot distinguish between these models, so in this sense the base of the jets can be said to effectively subsume, at least spectrally, the role created for the corona. The main difference between the two pictures comes down to geometry, inflow versus outflow, and the relationship of the corona to the outer jets (see Markoff et al. 2005 for more details).

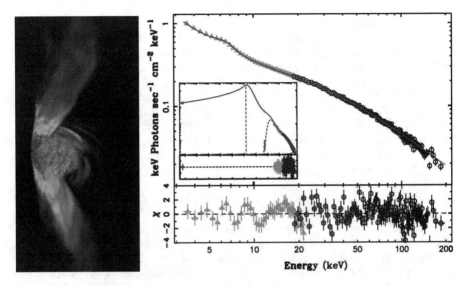

Figure 7.   a) Snapshot of a fully relativistic numerical simulation of an accretion disk in the Kerr metric, in a case of high spin a/M, where Poynting flux-dominated outflows are produced. This figure shows the Poynting flux on a logarithmic scale (simulations discussed in De Villiers et al. 2005; figure courtesy of J. Krolik, private communication). b) As the simulation shown in panel (a) suggests, it may be hard to distinguish between accretion disk corona and jet emission, if there are unbounded outflows from the disk. This panel shows a close up on the X-ray band of a fit of a jet-dominated model to a simultaneous radio/X-ray observation of the Galactic XRB Cyg X-1. Inset is the fit to the entire radio/X-ray broadband spectrum in the same units, with the flux axis extended up to 120 keV Photons s^{-1} cm^{-2} keV^{-1} and the energy axis down to 5×10^{-8} keV, in order to show the full model and all data. Note that the same model fits the radio data over 8 decades lower in frequency. The statistical goodness of the fit is $\chi^2_{\mathrm{dof}} = 186/179 = 1.04$. For comparison, a fit of the X-rays only with a thermal Comptonization model (as would be expected for a more static accretion disk compact corona) gives $\chi^2_{\mathrm{dof}} = 1.03$ (figure reproduced from Markoff et al. 2005).

5.   Inflow/Outflow Connections: Timing

Spectral modeling is only one part of the puzzle, however. There is an increasing amount of important information about the inflow/outflow relationship coming from monitoring of XRBs in outburst, particularly during state transitions. The time-dependent behavior, both spectral evolution and power density spectral changes, still needs to be integrated into any picture for the inflow/outflow relationship. Because of increased awareness of the importance of these types of observations, and plans for many extensive monitoring campaigns of existing and transient sources, this is likely the area where many of the answers about jet formation and inner disk configuration will ultimately be found.

As ever, GX 339-4's consistent outburst performances has made it the star of many studies. From multiwavelength monitoring during a 2002 outburst, Homan et al. (2005) explored the spectral evolution, adding important new information about the infrared and optical frequencies to the radio/X-ray picture. During this outburst, GX 339-4 transitioned through most of the possible accretion states, a representation of which can be effectively made in a hardness-intensity diagram (Fig. 8a and Homan & Belloni 2005). Through monitoring campaigns such as this, the definitions of states are being revised, and a new understanding of the role of jets and their relationship to timing features is being forged. For instance, while steady jets are effectively quenched between what are labelled the "hard intermediate" and "soft intermediate" states, there is a phase right in this transition "zone" where rms noise drops dramatically (Homan et al. 2006) and strongly beamed discrete radio ejecta events are observed. These transient ejecta may result from colliding internal shocks where more relativistic ejecta hit slower jet material, similar to models of gamma-ray bursts (Fender et al. 2004a). Another source, GRO J1655-4 is currently undergoing intensive monitoring (Homan et al. 2006), and the hardness-intensity diagram shown in Fig. 8b demonstrates the information now available for research. Each data point represents individual hours of observational time with *RXTE* from a campaign conducted over the good part of a year, and has a high-quality X-ray spectrum and power-density spectrum associated with it. Another Galactic black hole, GRS 1915+105 has been the target of many observations, in part because of its rather unique behavior. One recent result on this target is the discovery of a link between quasi-periodic oscillations (QPOs; seen in many sources and of as-yet unknown origin) and Fe K_α emission thought to arise from reflected hard X-ray continuum impinging on the disk, supporting proposals that QPOs originate in the inner disk (Miller & Homan 2005). These campaigns are generating new discoveries about connections between spectral hardness, noise and quasi-periodic oscillations, and low-frequency characteristics directly pertaining to jet formation.

6. Neutron Stars

Most of what I have been describing in this review has concerned black holes, because of their large range in mass. But one key question about jet formation is whether a black hole is necessary to power the high Lorentz factor jets we associate with AGN and transient black hole XRB outbursts. This question is particularly important for theoretical models because of two different mechanisms possibly involved in jet formation. In one, magnetic field lines from the accretion disk transfer angular momentum to the outflows (rotation centrifugally drives matter up and out along the fields lines; Blandford & Payne 1982). In the other, rotational energy from the black hole is extracted via magnetic field lines threading the ergosphere (Blandford & Znajek 1977). Obviously if jets could be shown to be the same in neutron star and black hole XRBs, it would argue strongly against the latter process playing a dominant role.

The jury is still out, but there are some interesting new results that seem on the surface to support opposing outcomes. For instance, a recent study to test whether the radio/X-ray correlation exists in neutrons star XRBs seems to indicate that these sources have possibly weaker jets than black holes do (Fig. 9a and Muno et al. 2005). On the other hand, recent observations of an outburst

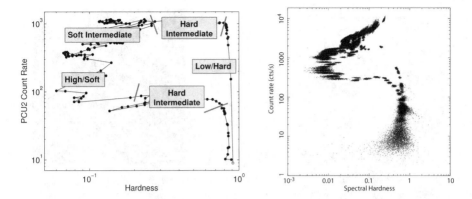

Figure 8.   State changes in black hole XRBs map out a pattern in spectral hardness and intensity space, called an HID (hardness-intensity diagram). While general patterns exist, each source and each outburst has a unique shape. a) HID for the 2002/2003 outburst of Galactic XRB GX 339-4, with labels for the various accretion states and their transitions (grey lines) superimposed (figure reproduced with kind permission from Homan & Belloni 2005). b) The most complete HID to date from a new, extensive monitoring campaign of the XRB GRO J1665-4 by Homan et al. (2006). Each data point represents typically a few hours of *RXTE* observation time, from a campaign conducted over about nine months (Homan et al. 2006; figure modified and reproduced with kind permission from J. Homan).

in the neutron star XRB Cir X-1 suggest the jets may be highly relativistically beamed, which would instead support that the jets are similar to those in black holes (Fig. 9b and Fender et al. 2004b).

7. Conclusions

The goal of this short article was to review some of the recent results that are at the forefront of our understanding of accretion in XRBs, including the microquasar states, and AGN, as well as the relationships between them. The observations clearly support several areas of correspondence between black holes on the stellar and galactic scales, and one of the challenges for the future is to further map out how XRB states may play into the AGN zoology.

The last few years have brought several surprises, to a great deal influenced by the success of coordinated observations from the radio to X-ray (and higher) frequencies. Clearly with the advent of the H.E.S.S. telescopes, γ-rays are also now an important waveband to exploit, as well as the "multi-messenger" approach, which will introduce information about high-energy cosmic rays and neutrinos using instruments such as LOFAR, Auger, AMANDA and ANTARES. If the significant advances from using new energy ranges in photons is any indication, the addition of other particles that are predicted in hadronic models of XRB and AGN jets will provide a critical piece of the puzzle. These latter projects will contribute greatly to our ability to finally resolve questions about jet matter and energy content.

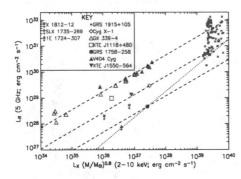 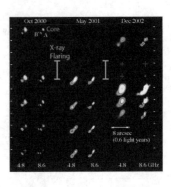

Figure 9.   A very important question for jet formation is whether the black hole contributes something "extra" to the system; i.e., are jets from accreting black holes stronger than from accreting neutron stars, both in otherwise fairly similar XRBs? a) So far, no radio/X-ray correlation has been established for neutron stars as shown in earlier figures for black holes. This figure shows a comparison of several hard state black-hole and three neutron-star (data points with small arrows) observations. Only upper limits are found in the radio for the neutron star sources, suggesting that perhaps the jets are weaker compared to those in black holes, or at the very least may follow a different correlation. Dashed lines indicate the $L_X^{0.7}$ correlation while the dotted line indicates a linear proportionality (figure reproduced with kind permission from Muno et al. 2005). b) An ultrarelativistic outflow is suggested for the neutron star XRB Cir X-1, based on the relationship inferred between this sequence of expanding radio blobs and the X-ray outburst. If confirmed, this would be a strong argument against the need for a black hole to power relativistic jets (figure modified and reproduced with kind permission from Fender et al. 2004b).

Acknowledgments.   This talk was prepared in part while supported by an NSF Astronomy & Astrophysics postdoctoral fellowship, under NSF Award AST-0201597.

References

Blandford, R. D., & Payne, D. G. 1982, MNRAS, 199, 883
Blandford, R. D., & Znajek, R. L. 1977, MNRAS, 179, 433
Corbel, S., Fender, R. P., Tzioumis, A. K., Nowak, M., McIntyre, V., Durouchoux, P., & Sood, R. 2000, A&A, 359, 251
Corbel, S., Nowak, M. A., Fender, R. P., Tzioumis, A. K., & Markoff, S. 2003, A&A, 400, 1007
De Villiers, J.-P., Hawley, J. F., Krolik, J. H., & Hirose, S. 2005, ApJ, 620, 878
Falcke, H., & Biermann, P. L. 1995, A&A, 293, 665
Falcke, H., Körding, E., & Markoff, S. 2004, A&A, 414, 895
Fender, R., Wu, K., Johnston, H., Tzioumis, T., Jonker, P., Spencer, R., & van der Klis, M. 2004a, Nature, 427, 222
Fender, R. P., Belloni, T. M., & Gallo, E. 2004b, MNRAS, 355, 1105
Gallo, E., Fender, R., Kaiser, C., Russell, D., Morganti, R., Oosterloo, T., & Heinz, S. 2005, Nat, 436, 819
Gallo, E., Fender, R. P., & Pooley, G. G. 2003, MNRAS, 344, 60

Galloway, D. K., & Sokoloski, J. L. 2004, ApJL, 613, L61

Goldston, J. E., Quataert, E., & Igumenshchev, I. V. 2005, ApJ, 621, 785

Hannikainen, D. C., Hunstead, R. W., Campbell-Wilson, D., & Sood, R. K. 1998, A&A, 337, 460

Hawley, J. F., & Krolik, J. H. 2001, ApJ, 548, 348

Heinz, S., & Sunyaev, R. A. 2003, MNRAS, 343, L59

Homan, J., & Belloni, T. 2005, ApSS, 300, 107

Homan, J., Buxton, M., Markoff, S., Bailyn, C. D., Nespoli, E., & Belloni, T. 2005, ApJ, 624, 295

Homan, J., Miller, J., Wijnands, R., Lewin, W., et al. 2006, in preparation

Jorstad, S. G., et al. 2005, AJ, 130, 1418

Kaaret, P., Corbel, S., Prestwich, A. H., & Zezas, A. 2003, Science, 299, 365

Markoff, S. 2005, ApJL, 618, L103

Markoff, S., & Nowak, M. A. 2004, ApJ, 609, 972

Markoff, S., Nowak, M., Corbel, S., Fender, R., & Falcke, H. 2003, A&A, 397, 645

Markoff, S., Nowak, M. A., & Wilms, J. 2005, ApJ, 635, 1203

Markoff, S., Nowak, M., Young, A., et al. 2006, in preparation

Markowitz, A., et al. 2003, ApJ, 593, 96

McClintock, J. & Remillard, R. 2006, in Compact Stellar X-ray Sources, ed. W. H. G. Lewin & M. van der Klis (Cambridge: Cambr. Univ. Press), in press (astro-ph/0306213)

McHardy, I. M., Gunn, K. F., Uttley, P., & Goad, M. R. 2005, MNRAS, 359, 1469

McKinney, J. C. 2006, ApJ, submitted (astro-ph/0506369)

McKinney, J. C., & Gammie, C. F. 2004, ApJ, 611, 977

McNamara, B. R., Nulsen, P. E. J., Wise, M. W., Rafferty, D. A., Carilli, C., Sarazin, C. L., & Blanton, E. L. 2005, Nature, 433, 45

Merloni, A., Heinz, S., & di Matteo, T. 2003, MNRAS, 345, 1057

Merloni, A., Körding, E., Heinz, S., Markoff, S., Di Matteo, T. & Falcke, H. 2006, NewA, submitted (astro-ph/0601286)

Miller, J. M., & Homan, J. 2005, ApJL, 618, L107

Mitsuda, K., et al. 1984, PASJ, 36, 741

Muno, M. P., Belloni, T., Dhawan, V., Morgan, E. H., Remillard, R. A., & Rupen, M. P. 2005, ApJ, 626, 1020

Nishikawa, K.-I., Richardson, G., Koide, S., Shibata, K., Kudoh, T., Hardee, P., & Fishman, G. J. 2005, ApJ, 625, 60

Nowak., M., Juett, A., Yao, Y., Schulz, N., Wilms, J. & Canizares, C., 2006, in preparation

Nowak, M. A., Wilms, J., & Dove, J. B. 2002, MNRAS, 332, 856

Nowak, M. A., Wilms, J., Heinz, S., Pooley, G., Pottschmidt, K., & Corbel, S. 2005, ApJ, 626, 1006

Pottschmidt, K., et al. 2003, A&A, 407, 1039

Sharma, P., Hammett, G. W., Quataert, E. & Stone, J. M. 2006, ApJ, submitted (astro-ph/0508502)

Uttley, P., & McHardy, I. M. 2005, MNRAS, 363, 586

van der Klis, M. 1995, in X-ray Binaries, ed. W. H. G. Lewin, J. van Paradijs, and E. P. J. van den Heuvel (Cambridge: Camb. Univ. Press), 252

Wilms, J., Nowak, M. A., Dove, J. B., Fender, R. P., & di Matteo, T. 1999, ApJ, 522, 460

Zdziarski, A. A., Lubiński, P., Gilfanov, M., & Revnivtsev, M. 2003, MNRAS, 342, 355

Mark Krumholz, Carlos Allende-Prieto, and Yeshe Fenner relax over a beer.

Frank N. Bash Symposium 2005: New Horizons in Astronomy
ASP Conference Series, Vol. 352, 2006
S. J. Kannappan, S. Redfield, J. E. Kessler-Silacci, M. Landriau, and N. Drory

Galactic Chemical Evolution

Yeshe Fenner[1]

Harvard-Smithsonian Center for Astrophysics, Cambridge, MA, USA

Abstract. Stars and gas in galaxies exhibit diverse chemical element abundance patterns that are shaped by their environment and formation histories. Many generations of stars forming, synthesizing new elements, and releasing their nuclear debris when they evolve and die, are largely responsible for the abundance of the elements throughout the cosmos. By investigating Galactic Chemical Evolution (GCE), one hopes to trace the distribution of the elements from the present-day back to the early universe and in doing so, gain insight into the lives of galaxies. This review approaches the field of GCE from a theoretical perspective, describing the design, results and future prospects for models tracing the production and distribution of the elements within galaxies.

1. Introduction

The study of galaxy formation and evolution is one of the most active fields in cosmology and extragalactic astronomy. From a theoretical perspective, understanding the evolution and distribution of chemical elements within stars and the interstellar medium (ISM) provides great insight into the behavior of galaxies and their interaction with the environment. Chemical abundance patterns reflect the cumulative history of star formation (SF) and gas exchange. Thus, galactic chemical evolution (GCE) models are powerful tools for interpreting the chemical signatures. They take advantage of the fact that chemical species have different characteristic production sites. Consequently, *absolute* and *relative* elemental abundances tell us about the role played by different types of chemical factories, which in turn, helps us decipher the history of galactic systems.

Each object in the cosmos has its own chemical fingerprint that contains clues about its nuclear history and evolution. The grand aim of GCE modelers is to interpret these chemical fingerprints and use them to piece together the formation and history of galaxies. Recent years have seen an explosion in the characterization of the chemical properties of stars and gas in the Milky Way, nearby dwarf galaxies and in the high-redshift universe (e.g., Barklem et al. 2005; Shetrone et al. 2003; Dessauges-Zavadsky et al. 2006). Such observational progress helps constrain the current GCE theories, while revealing a new collection of puzzles to be solved, ranging from detailed isotopic anomalies to broader issues concerning the relationship between objects at low- and high redshift.

Figure 1 shows a typical plot that GCE models strive to interpret. [Mg/Fe] and [Ca/Fe] vs [Fe/H] are plotted for stars in a variety of systems. A ratio like

---

[1]Institute for Theory and Computation (ITC) Fellow.

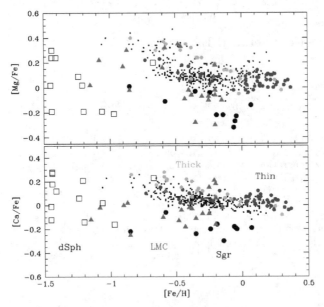

Figure 1.   [Mg/Fe] and [Ca/Fe] vs [Fe/H] for different stellar populations in the Milky Way and local group dwarf galaxies (figure taken from Venn et al. 2004). At a given metallicity, α/Fe is lower in the LMC, Sagittarius (Sgr) dwarf and dSphs with respect to thick and thin disk Galactic stars. This is indicative of slower chemical evolution, such that SNeIa contribute Fe at lower metallicity.

Mg to Fe is often used to gauge star formation timescales because Mg is created mainly in massive short lived stars, while Fe is largely produced by Type Ia supernovae (SNeIa) on longer characteristic timescales. Thus, the particular trends displayed by stellar populations belonging to the Milky Way thick and thin disk, and local group dwarf galaxies (Figure 1), reflect their unique nucleosynthetic and SF histories, as well as possible environmental effects. For instance, thin disk stars formed over a longer period of time than those of the thick disk (e.g., Bensby et al. 2005), allowing more time for SNeIa to contribute Fe. This causes [Mg/Fe] to decline at lower [Fe/H] than for thick disk stars. Less massive galaxies like the Large Magellanic Cloud (LMC), Sagittarius (Sgr) dwarf and dwarf spheroidals (dSphs) are thought to have accrued metals even more slowly, leading to lower [Mg/Fe] for a given metallicity (see Figure 1). Of course, SF timescales are not the sole factor influencing abundance patterns — the real situation is much more complicated owing to gas inflows and outflows and environmental effects.

2.   Basics of Galaxy Formation

The fundamental processes driving the chemical evolution of a Milky Way-like galaxy are illustrated in Figure 2. After the early universe had expanded and cooled sufficiently, gravitational effects pulled matter into denser protogalaxies,

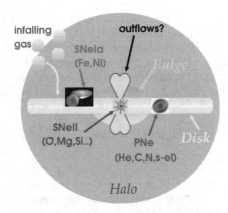

Figure 2.   Schematic diagram of processes effecting the chemical evolution of a Milky Way-like galaxy. SNeII are responsible for creating the bulk of the metals. Lower-mass stars, which end their lives as planetary nebulae (PNe), are more numerous but produce fewer metals. PNe winds are rich in He, C, N, s-process elements and certain isotopes. SNeIa have sufficiently long lifetimes that they are likely to have drifted from their birth place before their explosive death pollutes the ISM with Fe-peak elements. Infalling gas feeds ongoing star formation, while stars and SNe inject chemicals and energy into the surrounding gas.

in which molecular clouds began to form stars. The rate at which gas is converted into stars has been empirically estimated for a wide range of galaxies, and it appears to depend most sensitively on the surface density of gas (e.g., Elmegreen 2002). Each newly formed generation of stars has a range of masses described by the Initial Mass Function (IMF). Like the SF law, the IMF is best estimated empirically, and it is found to roughly follow a multicomponent power law (e.g., Kroupa et al. 1993). The relative and absolute elemental abundances are quite sensitive to the shape and limits of the IMF.

For a given stellar generation, the most massive stars die first because their larger gravitational potential energy leads to higher core temperatures and densities, which hastens the rate of nuclear reactions and fuel consumption. These stars end their lives exploding as energetic Type II SNe (SNeII), which are responsible for making most of the elements from O to Zn and the heavy r-process elements like Eu. Low- and intermediate-mass stars (LIMS) have longer lifetimes and don't explode as SNe. Instead, they release C, N, certain isotopes, and s-process elements like Ba and Pb through stellar winds and planetary nebulae. Lower-mass stars in binary systems may end their lives as Type Ia SNe (SNeIa). The mechanism responsible for SNeIa is thought to be the accretion onto a white dwarf (WD) of matter lost by its binary companion. If the WD accretes enough mass, it can no longer support itself through electron degenerate pressure, and its mass is violently dispersed in the form of metals — especially Fe (Iwamoto et al. 1999).

Not only do Type Ia and II SNe enrich the interstellar medium with newly synthesized metals, but they are sufficiently energetic to affect the structure of the ISM, carving holes, bubbles and chimneys (e.g., Wada et al. 2000; Marlowe et al. 1995). In lower mass galaxies, moreover, SNe can drive galactic-scale outflows of metal-enriched material (Martin 1998). The continual accretion of gas clouds fuels further SF, cycling the elements between the stellar and gas phases. At any point in time, the chemical abundances in stars, the ISM, and galactic winds, reflect the cumulative history of all the processes outlined above. Techniques for modeling these processes are described in the following section.

3.   Techniques for Modeling Galactic Chemical Evolution

Analytical Solution

The equations describing GCE can be solved analytically if one assumes that stars release their ejecta instantaneously at the time of their birth (see the seminal work of Tinsley 1980). This approximation is reasonable only for very massive short-lived stars and it precludes one from reproducing the relative trends of elements that are restored to the ISM by different mass stars on a variety of timescales.

Semi-analytical/numerical

In order to treat the nucleosynthesis in great detail, semi-analytic GCE models take into account the mass- and metallicity-dependent stellar lifetimes and yields and must therefore solve the equations governing GCE numerically. Such models have traditionally formed the cornerstone of the field, as they can explore a very wide range in parameter space. One weakness of many of these models is an inherent assumption of homogeneity, whereby the stellar ejecta from dying stars is instantly mixed back into the ambient interstellar medium. These semi-analytic models are the focus of the present review.

Chemodynamical N-body codes

N-body chemodynamic codes are the most computationally expensive modeling approach. They couple the kinematics and dynamics of stars and gas to their chemical properties. However, the kinematic information often comes at the expense of detailed nucleosynthetic information.

4.   Chemical Evolution in the Milky Way — A Template Model

The most thoroughly observed and best understood galaxy is the Milky Way, and in particular, the "local" solar neighborhood. So extensive is the Milky Way (MW) dataset that many model ingredients can be well-constrained empirically. Thus our own Galaxy is often the gauge by which chemical evolution models are calibrated. Indeed, studies of the cosmic evolution of disk galaxies often adopt scaling laws based on the Milky Way (e.g., Boissier & Prantzos 2000).

Key ingredients in a template MW model are the SF law and the IMF. Arguably, the most crucial feature of the models are the adopted stellar yields, because they describe the production of each isotope produced by stars as a function of their mass and metallicity. For the MW, a good match to the data is obtained assuming two phases of infall: a rapid phase, corresponding to the halo; and a more protracted episode associated with disk formation (Chiappini

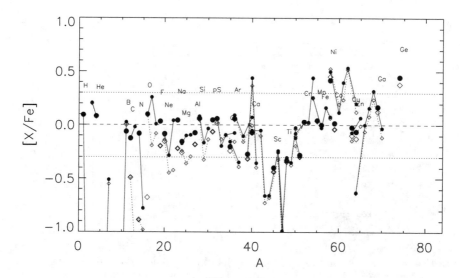

Figure 3. Predicted isotopic abundance pattern of the Sun with (*solid points*) and without (*open diamonds*) the contribution from low- and intermediate-mass AGB stars. Lines connect families of isotopes. Massive star yields are from Woosley & Weaver (1995) while AGB stellar yields are from Karakas & Lattanzio (2003).

et al. 2001). The model then predicts the spatiotemporal distribution of chemical elements, gas, stellar populations and supernova rates. A minimal set of observational constraints for GCE models should include: the solar abundance pattern; the metallicity distribution of long-lived stars; the evolution of diagnostic abundance ratios; SF and SN rates; the age-metallicity relationship; radial gas and metallicity gradients; and isotopic abundances. We compare predictions with some of these constraints in the following section.

4.1. Comparing with Observations

Solar Abundance Pattern

Figure 3 displays the predicted isotopic abundance pattern in the Sun from a standard dual-phase chemical evolution model (e.g., Fenner & Gibson 2003) with and without a chemical contribution from low- and intermediate-mass stars (solid dots and open diamonds, respectively). The yields for LIMS on the asymptotic giant branch (AGBs) were taken from recent calculations by Karakas & Lattanzio (2003). If we consider reasonable agreement to be predictions within a factor of two of solar, then the standard MW chemical evolution model incorporating Type II and Ia SNe and LIMS does a good job of matching the data. Discrepancies are evident for K, ^{15}N, Ca isotopes, Sc, Ti, Ni, and the Zn isotopes.

While the role of LIMS in producing considerable quantities of C and N has long been appreciated, these stars leave their signature in other important ways. For instance, they help explain the solar abundance of ^{17}O, F, Al and P, as well as the heavy isotopes of Ne, Mg and Si. AGB stars are also major production sites

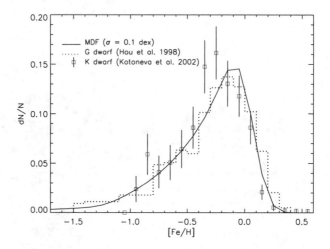

Figure 4.   The calculated metallicity distribution function (MDF) of long-lived nearby stars (*solid curve*), plotted against two observational datasets (*dotted histogram and open squares*). The predicted MDF has been convolved with a Gaussian of dispersion $\sigma = 0.1$ dex, consistent with the empirical uncertainties.

for certain *s*-process elements formed via neutron-capture (Busso et al. 1999). Not only do AGBs help reconcile the theoretical and observed solar abundance pattern, but they may prove vital to understanding the formation and evolution of objects from Galactic globular clusters through to high-redshift galaxies.

Metallicity Distribution Function

The distribution function of metallicities of long-lived stars, shown in Figure 4, can constrain both the star formation and gas infall assumptions. For instance, if all the gas is allowed to accumulate rapidly, then the models greatly overestimate the number of very metal-poor stars. A simple way to avoid this problem is by building the model galaxy via a rapid phase of halo formation plus a more drawn out disk formation phase. Figure 4 demonstrates that this dual-phase MW model fits the data well. The corresponding predicted ages for the stellar halo and thin disk of 12.5 and 7.4 Gyr, respectively, are in excellent agreement with empirically determined values (e.g., Hansen et al. 2002).

Radial Abundance Gradients

Since a key assumption in our MW model is that star formation is sustained via ongoing gas accretion, one may speculate about the composition of this infalling matter. We have investigated this issue by varying the metallicity of the disk-forming gas. Figure 5 shows that the radial gradient of oxygen in HII regions and OB stars is not reproduced when the infalling gas metallicity exceeds about 20% solar. This is consistent with observations of high velocity clouds (HVCs), whose metallicities typically range from 10% to 30% solar. Moreover, the predicted present-day infall rate agrees with the observed HVC infall rate of about 1 $M_{\odot}$/yr. Based on metallicity arguments, we can not conclusively

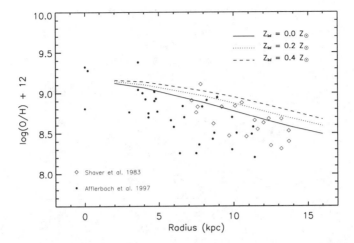

Figure 5.   Present-day Milky Way radial oxygen abundance profile predicted by our dual-infall phase GCE model. Observed abundances in HII regions are represented by symbols, while the curves indicate model predictions for disk-phase infalling gas with primordial (*solid line*), 20% solar (*dotted line*) and 40% solar (*dashed line*) metallicity.

rule out the scenario in which HVCs are fueling SF. However, some HVCs have metallicities >30% solar, which seems inconsistent with an extragalactic origin.

4.2.   Don't Ignore the Isotopes!

Detecting individual isotopic abundances in stars and gas is very challenging, yet the hard-to-detect species can provide valuable supplementary information about galaxy evolution. Magnesium is one such promising element whose isotopic composition can now be measured in stars. Figure 6 displays measurements of $^{26}Mg/^{24}Mg$ in nearby stars (diamonds and circles). Early GCE models assumed that the heavy isotopes ^{25}Mg and ^{26}Mg were created by massive stars alone, and consequently underestimated the $^{25,26}Mg/^{24}Mg$ ratios at low metallicity (Goswami & Prantzos 2000). The dotted line in Figure 6 shows that the expected ^{26}Mg abundance from SNeII alone falls well short of the observations. Recent calculations of nucleosynthesis in LIMS from Karakas & Lattanzio (2003) suggest that metal-poor intermediate-mass stars on the AGB are efficient at making the heavy Mg isotopes. Incorporating the new yields dramatically improves the match to the data (solid line in Figure 6).

4.3.   Homogeneous versus Inhomogeneous Models

Given the filaments and cavities that pervade the ISM, it is perhaps surprising that homogeneous models can reproduce Galactic abundance trends. They owe their success to the "smoothing" of the distribution of the elements that naturally occurs over time, as the ISM abundance pattern reflects the integrated effects of multiple stellar generations. But when one probes further back in time, the homogeneous approximation breaks down, as indicated by an increase in scatter in many abundance ratios (Ryan et al. 1996). The most metal-deficient

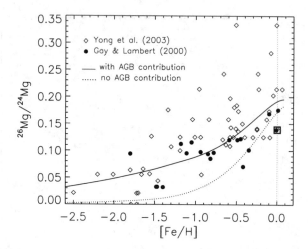

Figure 6.   Variation of Mg isotopic ratio with [Fe/H]. Circles show stellar abundances from Gay & Lambert (2000). Diamonds represent a sample of halo and thick disk stars from Yong et al. (2003). The square is the solar value. The predicted trend from the solar neighborhood model *with* Mg isotopic yields from AGBs (*solid line*) is shown against a model *without* the AGB contribution (*dotted line*). Both models arrive at similar present-day values, but only the AGB model matches the data at low metallicities. Figure from Fenner et al. (2003).

stars are believed to be composed of material that was polluted by just one or two stars. By their nature, homogeneous models are not equipped to interpret the lowest-metallicity data — to do so, one must carefully consider the way SN products are dispersed and the mechanisms triggering star formation.

Some of the most sophisticated attempts to address inhomogeneous GCE (iGCE) in the Milky Way are the models of Argast et al. (2000, 2002), which were designed to explain the apparent scatter in halo star abundances. The Argast et al. models start by imposing density fluctuations on a volume of gas representing a portion of the halo. Stars then form probabilistically, depending on local gas density. The metal-rich ejecta from SNeII is confined within the SN remnant shell, such that the only mode of mixing is via the overlapping of SN shells. These models therefore represent a "minimal mixing" case, at the opposite extreme from homogeneous models. Figure 7 shows a slice through a simulation at two epochs. After 170 Myr, corresponding to a mean [Fe/H] = −3 (*lefthand panel*), individual SN bubbles surrounded by dense shells are apparent. This scenario naturally gives rise to SN-triggered SF because stars preferentially form in shells, which contain the most compressed gas. After 1.5 Gyr, corresponding to a mean [Fe/H] = −2 (*righthand panel*), multiple overlapping SN have stirred the ISM so that the abundance pattern of a star born at this time reflects the integrated yields from many previous stellar generations. Thus, the initially large scatter decreases with time, until the ISM is no longer dominated by local chemical inhomogeneities but shows an averaged element abundance pattern.

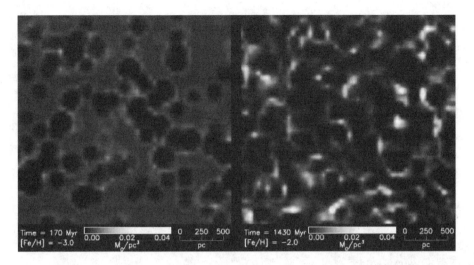

Figure 7.   A cut through the computed volume of an iGCE model, showing the density distribution of the halo ISM during the transition from the
unmixed to the well-mixed stage. Figure from Argast et al. (2000).

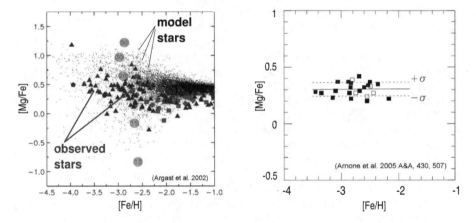

Figure 8.   (a) [Mg/Fe] ratios vs. metallicity [Fe/H] of metal-poor halo stars
(*squares and triangles*) and model stars (*small dots*). Circles depict [Mg/Fe]
ratios of SN II models of the given progenitor mass. Figure from Argast et al.
(2002). (b) Abundances of [Mg/Fe] vs. [Fe/H] in a sample of metal-poor stars
selected to be homogeneous in T_{eff} and $\log(g)$, so as to minimize systematic
uncertainties.   Thick line shows results of a linear fit and the dotted lines
depict the $\pm 1\sigma$ region.   Figure from Arnone et al. (2005).   The standard
deviation is about 0.06 dex — considerably less than the 0.4 dex standard
deviation predicted by the model shown in the lefthand panel.

The lefthand panel of Figure 8 compares [Mg/Fe] vs [Fe/H] predicted by
Argast et al. (2002, *small dots*) with observed halo stars (*triangles and squares*).
Judged against this dataset, a minimal mixing iGCE model appears to do a

good job of reproducing the scatter. However, the accepted notion that halo stars have large intrinsic chemical scatter was recently challenged by a study revealing *almost uniform* Mg/Fe in a carefully selected sample of metal-poor stars. Arnone et al. (2005) selected stars which were homogeneous in T_{eff} and $\log(g)$, in order to minimize systematic errors. By reducing systematic errors and ensuring internal consistency, they were able place an upper limit on the intrinsic dispersion of [Mg/Fe] of <0.06 dex — considerably less than the 0.4 dex dispersion predicted by the Argast et al. model. From the surprisingly small intrinsic scatter, shown in the righthand panel of Figure 8, Arnone et al. concluded that the early ISM was well mixed.

Despite the uniformity in [Mg/Fe], metal-poor stars still exhibit considerable star-to-star scatter in heavy neutron-capture elements (e.g., Sneden & Cowan 2003). A challenge for Galactic chemical evolution models is to simultaneously account for these observations.

5. Dwarf Galaxies

5.1. The Role of Feedback, Winds and Environment

Theoretical studies of the chemical evolution of *spiral* galaxies have a rich history, having benefited enormously from the wealth of data of our own Milky Way. The chemical evolution of *dwarf* galaxies (dGs) is less certain, but is a topic of great interest at the moment. Dwarfs are the most numerous galaxies and play a vital role as building blocks of larger galaxies in the hierarchical structure formation scenario (Tosi 2003). Moreover, they may lose a large fraction of their metals due to shallow potential wells, making them possible polluters of the intergalactic medium (IGM) (Murakami & Babul 1999). Dwarf galaxies may also give rise to some of the distant high column-density gas clouds, known as damped Lyman-α absorbers (DLAs), whose chemical content can be inferred from their absorption of a bright background light-source (e.g., Schaye 2001).

Unlike spiral galaxies, the low mass of dGs makes them vulnerable to feedback effects from starbursts and SNe. Hydrodynamical simulations suggest that the amount of gas and metals escaping from dwarf galaxies strongly depends on the depth of the dark matter dominated potential well, star formation history, merger history, and tidal and ram-pressure stripping (e.g., Marcolini et al. 2004; Fujita et al. 2004). Fragile et al. (2004) used 3D hydrodynamical simulations to explore the effects of SN feedback on the ISM of dwarf galaxies. One such simulation of an intense, centrally concentrated dwarf starburst is shown in Figure 9. The distribution of gas (*light gray*) and metals (*dark gray isosurface*) are plotted at $t = 25$, 50, 75, and 100 Myr. By the end of the simulation, less than a few percent of the gas has escaped the galaxy, but 90% of the metals are lost.

Figure 10 depicts results from a similar model, but with the SNe distributed across 80% of the disk. Individual SN bubbles are visible, but their metal-rich ejecta are now mostly confined by the gas disk. Only in the very center does a chimney spew forth metals. The escape fraction for gas in this model is again negligible, but the metal escape fraction has dropped to about 20% — greatly reduced from the centrally concentrated case.

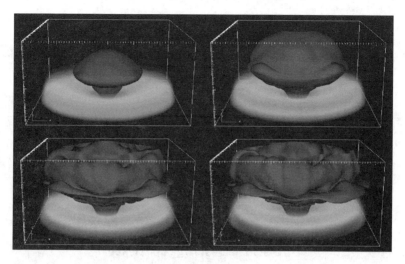

Figure 9. Volume rendering of gas density (*light gray*) plus an isosurface of metal tracer (*dark gray*) for a 3D hydrodynamic model of a dwarf disk galaxy at t = 25, 50, 75, and 100 Myr, from Fragile et al. 2004. The box size is 30 × 30 × 15 kpc³.

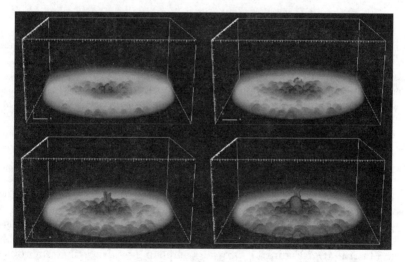

Figure 10. Same as Figure 9 but with SNe spread across the 80% of the disk, rather than being centrally concentrated.

The work of Fragile et al. (2004) illustrates how sensitive metal-loss is to details of the SF history and gas distribution. This directly impacts upon the chemical properties of the stars, gas, and winds in dwarf galaxies. Many questions about feedback in dGs still need to be answered. What is the escape efficiency of ejecta from SNeII, SNeIa and LIMS? What is the ultimate fate of winds — do they cool and fall back or are they lost forever? How much gas is lost via ram-pressure

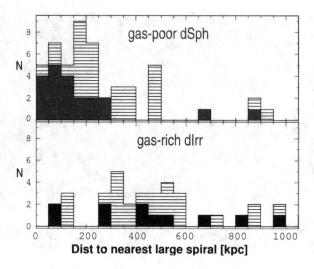

Figure 11.   Morphological segregation in the number distribution N of different types of galaxies in the Local Group (solid histograms) and in the M81 and Centaurus A groups (hashed histograms) as a function of distance D to the closest massive primary. Figure taken from Grebel (2005).

and tidal stripping? While future hydrodynamical simulations will provide some answers, in the meantime we may indirectly investigate feedback scenarios using the chemical properties of dwarf galaxies in the local group (LG). LG dGs offer a unique opportunity to study the evolution of low-mass systems. Their proximity enables the resolution of individual stars, which represent fossil evidence of prior star formation and feedback activity. The chemical properties of stars in nearby dGs, as revealed by recent observational initiatives, should shed light on the role of feedback and environment in the lives of dGs.

5.2.   The Local Group

At least 38 galaxies belong to the local group, with ~11 members being gas-rich higher-mass dwarf irregulars (dIrrs) and ~22 being gas-poor lower-mass dwarf spheroidals (dSphs) and dwarf ellipticals (dEs) (Grebel 2005). The smaller gas-poor galaxies show a striking tendency to be clustered around the closest massive spiral, whereas the larger gas-rich dIrrs are more widely distributed. These trends, illustrated in Figure 11, provide compelling evidence that environment plays a key role in the evolution of dwarf galaxies — perhaps through tidal and/or ram-pressure stripping triggered by interactions with other galaxies.

In addition to the clear morphology-distance relationship, dIrrs are distinguished from dSphs by having active SF within the past Gyr (Dolphin et al. 2005). Whereas dSphs harbor mostly old stellar populations, all dIrrs contain some young, as well as old, stars. To investigate the differences between dSphs and dIrrs, I have simulated the chemical evolution of local group dGs, including prescriptions for SN feedback, cooling and fallback of galactic winds, and stripping of the ISM due to galaxy-galaxy interactions. Star formation histories for individual galaxies were taken from those inferred by Dolphin et al. (2005)

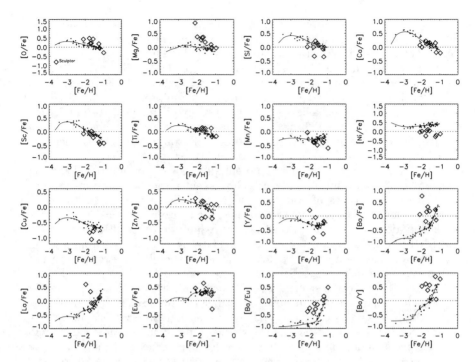

Figure 12.   Predicted abundances for a Sculptor dSph chemical evolution
model, incorporating SN feedback and galactic winds, plotted against stellar
observations from Shetrone et al. (2003) and Geisler et al. (2005, *diamonds*).
The curve shows predicted abundances in the ISM, while the small dots represent model stars (Fenner et al. in prep).

from color-magnitude diagrams. The rate of gas accretion can then be coupled
to the star formation rate (SFR) — under the usual assumption that changes
in SFR reflect changes in gas density — or may decline exponentially. Feedback
from each type of SNe is treated independently, with the choice of mass-loss effi-
ciencies being guided by hydrodynamical simulations and observations of dwarf
starburst winds (e.g., Martin et al. 2002).

Figure 12 shows results from a Sculptor dwarf spheroidal model, plotted
against stellar observations. The curve represents predicted abundance ratios as
a function of metallicity in the ISM, while small dots indicate abundances in 60
model stars picked randomly from the SF history (random uncertainty with 0.1
dex dispersion was imposed on model star abundances).

Sculptor's SF history is rather typical for a dSph. Over 90% of its stars are
older than 10 Gyr and SF has been essentially dormant for the last ∼7 Gyr.
Sculptor stars have an observed mean [Fe/H] ∼ −1.57 dex. This is well matched
by our predicted mean of ∼ −1.65. We also produce a significant metallicity
spread that agrees with the data. Observations reveal a decrease in [α/Fe]
with rising [Fe/H], where the α-element group includes O, Mg, Si, Ca and Ti.
This is typical of systems with low SF efficiency and consequently long metal
enrichment timescales. In our model, the loss of SN products via galactic winds

also hampers the growth of metals in the ISM. Note that the *subsolar* [α/Fe] values in metal-poor Sculptor stars contrasts dramatically with the *supersolar* ratios characterizing MW stars at the same metallicity.

Iron-peak elements such as Sc, Mn and Cu also show strikingly different behavior in Sculptor with respect to the MW. Whereas Sc tracks Fe closely in MW stars, [Sc/Fe] decreases steeply with rising Fe in Sculptor, in a manner akin to the α-elements. This behavior is well matched by our model. We predict subsolar [Mn/Fe] of about -0.25 dex, in rough agreement with the measured mean of -0.4, however we do not recover the observed drop in Mn/Fe with increasing Fe, which is opposite from the trend seen in MW stars.

Heavy neutron-capture elements potentially provide an additional independent probe of the evolution of dwarf galaxies, since they are presumed to originate in different types of stars via different mechanisms. For instance, the *s*-process element Ba is thought to come chiefly from low-mass stars with a metallicity-dependent yield. Based on our assumption that dGs preferentially lose SNII and SNIa products and retain relatively more ejecta from low- and intermediate-mass stars, one would expect Ba to be elevated relative to both α and iron-peak elements. Figure 12 shows hints of this behavior. Theories associating the *r*-process element Eu with massive stars may lead one to expect Eu to behave like the α-elements. Yet [Eu/Fe] is supersolar in most Sculptor stars. Our models only produce [Eu/Fe] ratios as high observations by including an additional Eu source whose nucleosynthetic products are *not* preferentially expelled in galactic winds. Candidates for a supplemental *r*-process site are lower mass stars (in the 8 to $10\,M_\odot$ range) or mergers between neutron-stars or neutron-star/black hole binaries. However, very little is presently known about the nucleosynthesis and energy associated with these phenomena. Nevertheless, we can use [Ba/Eu] vs [Fe/H] to gauge the relative importance of the *s*- and *r*-process, where the pure *r*-process ratio is [Ba/Eu] ~ -0.8 and the pure *s*-process ratio is ~ 1 dex (Arlandini et al. 1999). Figure 12 shows a clear rapid rise in [Ba/Eu] going from [Fe/H] $= -2$ to -1. Sculptor stars tend towards [Ba/Eu] $= +0.5$ at [Fe/H] $= -1$, which is 0.8 dex above the corresponding MW value. The dG chemical evolution model predicts a similar trend, owing to greater retention of lower-mass stellar ejecta as compared with SN ejecta.

6.   The Future: Linking the High- and Low-Redshift Systems

Structure formation, the birth of stars, and feedback from SNe are intimately connected with the chemical evolution of galaxies and the IGM. The IGM contains the material that seeds galaxy formation, while star formation within these galaxies affects the IGM through chemical, radiative and mechanical feedback. The details of these interactions are not fully understood, however, as most of this activity occurred at high redshifts. Piecing together the early evolution of galaxies and the connections between different types of galaxies, starbursts, absorption systems and the IGM remains a primary challenge in astrophysics.

Acknowledgments.   I am grateful to the organizing committee of the Frank N. Bash Symposium 2005 and the Astronomy Department at the University of Texas.

References

Afflerbach, A., Churchwell, E., & Werner, M. W. 1997, ApJ, 478, 190
Argast, D., Samland, M., Gerhard, O. E., & Thielemann, F.-K. 2000, A&A, 356, 873
Argast, D., Samland, M., Thielemann, F.-K., & Gerhard, O. E. 2002, A&A, 388, 842
Arlandini, C., et al. 1999, ApJ, 525, 886
Arnone, E., Ryan, S. G., Argast, D., Norris, J. E., & Beers, T. C. 2005, A&A, 430, 507
Barklem, P. S., et al. 2005, A&A, 439, 129
Bensby, T., Feltzing, S., Lundström, I., & Ilyin, I. 2005, A&A, 433, 185
Boissier, S., & Prantzos, N. 2000, MNRAS, 312, 398
Busso, M., Gallino, R., & Wasserburg, G. J. 1999, ARA&A, 37, 239
Chiappini, C., Matteucci, F., & Romano, D. 2001, ApJ, 554, 1044
Dessauges-Zavadsky, M., Prochaska, J. X., D'Odorico, S., Calura, F., & Matteucci, F. 2006, A&A, 445, 93
Dolphin, A. E., Weisz, D. R., Skillman, E. D., & Holtzman, J. A. 2005, preprint (astro-ph/0506430)
Elmegreen, B. G. 2002, ApJ, 577, 206
Fenner, Y., & Gibson, B. K. 2003, Pub. of the Astronomical Soc. of Australia, 20, 189
Fenner, Y., Gibson, B. K., Lee, H.-c., Karakas, A. I., Lattanzio, J. C., Chieffi, A., Limongi, M., & Yong, D. 2003, Pub. of the Astronomical Soc. of Australia, 20, 340
Fragile, P. C., Murray, S. D., & Lin, D. N. C. 2004, ApJ, 617, 1077
Fujita, A., Mac Low, M.-M., Ferrara, A., & Meiksin, A. 2004, ApJ, 613, 159
Gay, P. L., & Lambert, D. L. 2000, ApJ, 533, 260
Geisler, D., Smith, V. V., Wallerstein, G., Gonzalez, G., & Charbonnel, C. 2005, AJ, 129, 1428
Goswami, A., & Prantzos, N. 2000, A&A, 359, 191
Grebel, E. K. 2005, in AIP Conf. Proc. 752: Stellar Astrophysics with the World's Largest Telescopes, ed. Joanna Mikolajewska & A. Olech, 161
Hansen, B. M. S., et al. 2002, ApJ, 574, L155
Hou, J., Chang, R., & Fu, C. 1998, in ASP Conf. Ser. 138, Pacific Rim Conf. on Stellar Astrophysics, ed. K. L. Chan, K. Cheng, & H. Singh, (San Francisco: ASP), 143
Iwamoto, K., et al. 1999, ApJS, 125, 439
Karakas, A. I. & Lattanzio, J. C. 2003, Publ. Astron. Soc. Aust., 20, 279
Kotoneva, E., Flynn, C., Chiappini, C., & Matteucci, F. 2002, MNRAS, 336, 879
Kroupa, P., Tout, C. A., & Gilmore, G. 1993, MNRAS, 262, 545
Marcolini, A., Brighenti, F., & D'Ercole, A. 2004, MNRAS, 352, 363
Marlowe, A. T., Heckman, T. M., Wyse, R. F. G., & Schommer, R. 1995, ApJ, 438, 563
Martin, C. L. 1998, ApJ, 506, 222
Martin, C. L., Kobulnicky, H. A., & Heckman, T. M. 2002, ApJ, 574, 663
Murakami, I., & Babul, A. 1999, MNRAS, 309, 161
Ryan, S. G., Norris, J. E., & Beers, T. C. 1996, ApJ, 471, 254
Schaye, J. 2001, ApJ, 559, L1
Shaver, P. A., McGee, R. X., Newton, L. M., Danks, A. C., & Pottasch, S. R. 1983, MNRAS, 204, 53
Shetrone, M., Venn, K. A., Tolstoy, E., Primas, F., Hill, V., & Kaufer, A. 2003, AJ, 125, 684
Sneden, C., & Cowan, J. J. 2003, Science, 299, 70
Tinsley, B. M. 1980, Fundamentals of Cosmic Physics, 5, 287
Tosi, M. 2003, Ap&SS, 284, 651
Venn, K. A., Irwin, M., Shetrone, M. D., Tout, C. A., Hill, V., & Tolstoy, E. 2004, AJ, 128, 1177
Wada, K., Spaans, M., & Kim, S. 2000, ApJ, 540, 797
Woosley, S. E. & Weaver, T. A. 1995, ApJS, 101, 181
Yong, D., Lambert, D. L., & Ivans, I. I. 2003, ApJ, 599, 1357

Victor Debattista responds to a question about the dynamics of galactic bars.

Frank N. Bash Symposium 2005: New Horizons in Astronomy
ASP Conference Series, Vol. 352, 2006
S. J. Kannappan, S. Redfield, J. E. Kessler-Silacci, M. Landriau, and N. Drory

The N-Body Approach to Disk Galaxy Evolution

Victor P. Debattista[1]

Astronomy Department, University of Washington, WA, USA

Abstract.   I review recent progress from N-body simulations in our understanding of the secular evolution of isolated disk galaxies. I describe some of the recent controversies in the field which have been commonly attributed to numerics. The numerical methods used are widely used in computational astronomy and the problems encountered are therefore of wider interest.

1. Introduction

The fragility of disk galaxies suggests that a significant fraction of their history was spent experiencing mild external perturbations. Thus their evolution is likely to have been driven partly by internal processes. Dissipational gas physics probably plays a prominent role in this evolution but the effects of collisionless evolution very near equilibrium are also significant. Collisionless secular evolution is driven largely by non-axisymmetric structures, especially by bars, which represent large departures from axisymmetry, and are thus efficient agents of mass, energy and angular momentum redistribution. Moreover bars occur in over 50% of disk galaxies (Knapen 1999; Eskridge et al. 2000) and simulations starting from the 1970s (Miller et al. 1970; Hohl 1971) have shown that they form readily via global instabilities (Kalnajs 1972; Toomre 1981). Here I review recent progress in understanding the secular evolution of isolated disk galaxies.

2. Numerical Methods

At present there is considerable debate on the role numerics play in the evolution of N-body simulations. This review therefore treats collisionless isolated galaxy simulations as the necessary prerequisites for confidence in any other N-body study of galaxy formation and evolution. The physics of isolated collisionless systems consist only of Newton's law of gravity and Newtonian mechanics, both of which are well understood. Other than initial conditions, the main difficulty lies in computing the gravitational field of a given mass distribution accurately and efficiently. Any gravodynamical code must include at least this much, including those used for studying planet and cosmological structure formation.

Many algorithms have been devised for calculating gravitational fields of isolated galaxies. The simplest is direct particle-particle interactions (e.g., Aarseth 1963), but this scales as $\mathcal{O}(N^2)$, N being the number of particles, making it

---

[1]Brooks Fellow.

impractical for other than special applications. Particle-Mesh (PM/grid) codes (Miller & Prendergast 1968; Miller 1976) solve the potential on a grid by binning the mass distribution. Several geometries are possible, including cartesian, cylindrical and spherical. Grid codes are very efficient, scaling largely with the number of grid cells, N_g, as $\sim \mathcal{O}(N_g)$ (Sellwood 1997). Their main drawbacks are their inflexible geometry and their trade-off between volume and spatial resolution, but hybrid codes can alleviate many of these problems (e.g., Fux 1999; Sellwood 2003). Adaptive Mesh Refinement (AMR) codes (e.g., Bryan & Norman 1995; Kravtsov et al. 1997), on the other hand, let the grid evolve dynamically to increase the resolution in dense regions. Tree codes (e.g., Barnes & Hut 1986; Hernquist 1987) instead group particles by location, computing direct forces for nearby particles and a few multipoles for the more distant ones. These codes can be vectorized efficiently and scale as $\mathcal{O}(N \log N)$ (Dubinski 1996; Stadel 2001; Springel et al. 2001) while Dehnen (2000) presented an $\mathcal{O}(N)$ extension. Tree codes have been widely implemented and several are publicly available. Self-Consistent Field (SCF) codes (e.g., Clutton-Brock 1972; Hernquist & Ostriker 1992; Earn & Sellwood 1995) expand the density in a set of orthogonal basis functions, from which the potential can then be computed. SCF codes scale as $\mathcal{O}(N)$ and do not require softening because truncating the expansion is enough to suppress small scale noise. However they require that the basis set be chosen with care or that a large number of basis functions be included (which re-introduces small scale noise). Weinberg (1996) solves this problem by making the basis set adaptive.

2.1. Code Testing

The first requirement of any N-body code used to study galaxy evolution is that it be collisionless. Relaxation rates of conserved quantities are useful diagnostics for this purpose (e.g., Hohl 1973; Hernquist & Barnes 1990; Hernquist & Ostriker 1992; Valenzuela & Klypin 2003).

Testing the gravodynamical part of a code requires more effort. Comparisons with the limited number of exact analytic results known provide stringent code tests; such systems (in 2-D) include the Kalnajs disk (Kalnajs 1972), the Mestel disk (Zang 1976), the isochrone disk (Kalnajs 1978) and the power-law disks (Evans & Read 1998a,b). Several code tests using these predictions have been carried out. Earn & Sellwood (1995) compared the eigenfrequencies of instabilities in isochrone disks with simulations using an SCF and a PM method. They were able to reproduce the predicted values to within 5% with just $15K$ particles using the SCF method whereas softening caused the PM code to never quite converge to the predicted value. Meanwhile, Sellwood & Evans (2001) presented N-body examples of the power-law disks including reproducing a challenging case of a perfectly stable disk. They reported that aliasing rendered SCF codes ill-suited to these disks. Instabilities of spherical systems have also been used for code testing. For example Weinberg (1996) tested his adaptive SCF code on the instabilities of spherical generalized polytropes investigated by Barnes et al. (1986) and found generally good agreement. Analytic solutions of non-equilibrium evolution, which include the solution of 1-D plane wave collapse (Zel'Dovich 1970) and of spherical infall in an expanding universe (Fillmore &

Goldreich 1984; Bertschinger 1985) have also been used to test N-body codes (Kravtsov et al. 1997; Davé et al. 1997).

Different codes can be tested also by direct comparison. Inagaki et al. (1984) compared bar formation between an N-body model and a direct numerical integration of the collisionless Boltzmann equation (CBE) and Poisson equation (Nishida et al. 1981; Watanabe et al. 1981). The two simulations matched each other to better than 2% in bar amplitude well into the non-linear regime and eventual discrepancies were due to the inability of the CBE code to handle large gradients in the distribution function (DF).

3. Cusp Evolution

A prediction of cold dark matter (CDM) cosmology is that dark matter halos are cusped, with densities $\rho \sim r^{-\beta}$ at small radii and $1 \leq \beta \leq 1.5$ (Navarro et al. 1997; Moore et al. 1998; Jing & Suto 2000; Power et al. 2003). Several arguments have been advanced against the presence of such cusps in real galaxies, including detailed rotation curve fits of dwarf and low surface brightness galaxies (Blais-Ouellette et al. 2001; de Blok et al. 2001; Matthews & Gallagher 2002), bar gas flows (Weiner et al. 2001; Pérez et al. 2004) and pattern speeds (Debattista & Sellwood 1998, 2000, see § 4 below), and the microlensing optical depth and gas dynamics in the Milky Way (Binney et al. 2000; Bissantz et al. 2003).

Several ways to erase cusps have been considered including new dark matter physics (Spergel & Steinhardt 2000; Peebles 2000; Goodman 2000; Kaplinghat et al. 2000; Cen 2001), feedback from star formation (Navarro et al. 1996; Gnedin & Zhao 2002), or an initial power spectrum with decreased power on small scales (Hogan & Dalcanton 2000; Colín et al. 2000; Bode et al. 2001). Binney et al. (2001) suggested that bars in young galaxies were able to torque up cusps and expel them, while Hernquist & Weinberg (1992) had found that such torques can reduce the density of a spheroid by a factor of $\sim$100 out to $\sim$0.3a_B (where a_B is the bar's semi-major axis). Weinberg & Katz (2002, hereafter WK02), following Hernquist & Weinberg (1992), presented perturbation theory calculations and simulations of imposed rigidly-rotating non-slowing (IRRNS) bars to argue that cusps are erased in a few bar rotations ($\sim$10^8 years). Their bars were large ($\sim$10 kpc), but they argued for a scenario in which large primordial bars destroy cusps while the current generation of smaller bars formed later. Large reductions of dark halo densities were not seen in self-consistent cuspy halo simulations (e.g., O'Neill & Dubinski 2003; Valenzuela & Klypin 2003); WK02 suggested that this was because the resonant dynamics responsible for cusp removal are very sensitive to numerical noise. They needed $N > 10^6$ for their SCF code and argued that even larger N would be needed for grid, tree or direct codes.

The scenario of WK02 makes three main claims: (1) that bars cause a decrease in cusp density and can destroy cusps if sufficiently large and strong (2) that such bars formed via interactions at high redshift and (3) that N needs to be large in order that the relevant phase space is adequately covered and that orbit diffusion does not destroy the resonant dynamics causing cusp destruction.

3.1. When Do Bars Destroy Cusps?

Sellwood (2003, hereafter S03) presented simulations of the same IRRNS bar as was used by WK02 and found cusp destruction occurred in a runaway process.

The inclusion of the $l = 1$ spherical harmonic terms in the potential had a large effect on the evolution; in their absence, cusp removal took $\sim$5–6 times longer than when they were present. For the same IRRNS bar, McMillan & Dehnen (2005, hereafter MD05) found that the cusp moves off-center by as much as 30% of a_B. This centering instability gives the appearance that the cusp has been erased when halo density is measured by spherically averaging about the origin which, they argued, caused WK02 to over-estimate a bar's ability to erase a cusp. Evidence for lopsidedness in the simulation of WK02 can be inferred from its asymmetric bar-induced halo wake (their figure 2). When MD05 suppressed this purely numerical instability they still found that the cusp is destroyed, although after a much longer time. The possibility that an offset cusp is being confused for cusp destruction was investigated by Holley-Bockelmann et al. (2005, hereafter HB05). Their self-consistent simulations formed bars in a disk of particles, thus damping any centering instability. While confirming that some of the evolution seen by WK02 was due to the centering instability, they found that when odd l terms were excluded in their simulations that the damage to the cusp was not significantly diminished.

Thus these idealized bars at least are able to destroy cusps. What about more realistic bars? S03 showed that when, instead of assuming a fixed pattern speed, Ω_p, he allowed it to decrease such that total angular momentum is conserved assuming a constant moment of inertia (an IRRS bar), that the cusp was damaged significantly less. This happens because an IRRNS bar must transfer more angular momentum to the halo than it can plausibly have in order to destroy the cusp. Shorter, more realistically sized bars were also ineffective bar destroyers. Similarly MD05 found that an IRRS bar is unable to destroy a cusp in a Hubble time. S03 also presented self-consistent simulations in which a disk of particles was grown inside a halo. In these cases, besides forming smaller bars which slow, cusp erasure was also inhibited by the increase in halo density as the disk grew and again as the central density of the disk increased because of angular momentum transport outwards. HB05 instead found that the large relative angular momentum gained by the inner halo in their self-consistent simulations flattened the cusp, but only out to 900 pc scaled to the Milky Way.

3.2.   Do Bars Get as Large as Needed?

Thus all self-consistent simulations find that bars of sizes typical of those observed are unable to remove cusps on scales of several kpc. Do bars ever get to be large enough to do so? Bars that form via disk instabilities (Toomre 1981) generally extend to roughly the radius of the rotation curve turnover. However, externally triggered bars may be substantially larger. HB05 showed an example of such a bar; scaled to the Milky Way, external triggering produced a bar of $a_B = 12$ kpc, as opposed to $\sim$4.5 kpc via the bar instability. Whether the required triggering actually occurs is a question for hierarchical simulations while direct observations at high redshift should establish whether bars get to be as large as 10 kpc.

3.3.   Is the Evolution in Simulations Compromised by Too Small N?

Weinberg & Katz (2005) provided several reasons for the need of large N. They focused especially on the inner Lindblad resonance (ILR) which always extends down to the cusp. One possible problem they identified is two-body scattering which leads to particles executing a random walk and therefore lingering in resonance for significantly less time than they would otherwise. In the terminology of Tremaine & Weinberg (1984), scattering causes particles to traverse the ILR in the fast rather than the slow regime. Since resonant torques scale as m_P^2 in the fast regime and as $\sqrt{m_P}$ in the slow regime (where m_P is the fractional mass of the perturber) two-body scattering causes the friction to be substantially reduced. For a typical bar, they argued that $N > 10^8$ within the virial radius is required. A second limiting factor they identified is phase space coverage. Whether a resonant particle gains or loses angular momentum depends on its phase. If the phase space of the resonance is not sufficiently sampled by particles then the correct ensemble average is not attained.

HB05 found no difference between their tidally perturbed simulations with $N = 5.5M$ and $N = 11M$, while $N = 1.1M$ resulted in a weaker bar and less friction than in the other two cases. This, they argued, proved that the $N = 1.1M$ simulation suffered from too much orbital diffusion to follow resonances correctly. However, Sellwood (2005) argued plausibly that this N dependence was due to the bar being weaker, and not because friction depends on N. The weaker bar in the $N = 1.1M$ simulation was caused by the unavoidably larger $m = 2$ *seed* amplitude in the initial disk which, once swing-amplified, interfered destructively with the distortion induced by the externally applied tidal field.

The simulations of S03 (using both a PM and an SCF code) and MD05 (using a tree code) are instructive because they studied the same physical system: a Hernquist (1990) halo with an IRRNS bar of $a_B = 0.7r_s$ (r_s being the halo scale radius). Both studies found that the onset of the runaway cusp destruction depended on N and occurs earlier with *decreasing* N. S03 found good agreement between evolution using the PM code and the SCF code; additionally he showed that using only $l = 0$ and 2 terms was sufficient for the evolution, with little change when larger even values of l were included. On the other hand, MD05 forced symmetry about the origin to suppress the centering instability. Thus the two studies were very similar; but while their results are in qualitative agreement, there are surprising quantitative differences. S03 found that the onset of the runaway appears to be converging by $N = 3M$ at ~ 130 bar rotations (his figure 2b) but for the same N MD05 saw no evidence for convergence with runaway at ~ 210 bar rotations (their figure 5b). Yet both agreed that the radius containing 1% of the total mass is driven from $\sim 0.1r_s$ to $\sim 0.5r_s$. Unless the difference is due to the initial conditions in some worrisome way, noise would seem to responsible for these differences. A more careful comparison between these different simulations seems particularly worthwhile.

Sellwood & Debattista (2006) report an entirely different test of whether scattering overwhelms bar evolution. They presented simulations (using a PM code) in which they perturbed the system into a metastable state with nearly constant Ω_p (more on this in § 4). Their metastable state resulted because Ω_p was driven up such that the principal resonances are trapped in *shallow* local minima of the DF. Systems persisted in this metastable state for ~ 5 Gyr, which would not have been possible if orbit scattering had been strong.

Unlike two-body scattering, phase space coverage can be investigated by means of test particle simulations, where the self-gravitating halo response is not included. Using such experiments, Sellwood (2005) found that $N \sim 1M$

was sufficient for the $\Omega_{\rm p}$ evolution of a low mass bar $M_{\rm bar} = 0.005 M_{\rm halo}$ to converge, and a factor of $\sim$100 fewer particles were needed for $M_{\rm bar} = 0.02 M_{\rm halo}$. When self-gravity was introduced (thus adding scattering), $1M$ particles were needed for the $M_{\rm bar} = 0.02 M_{\rm halo}$ case. As in S03, he argued that WK02 needed large N because of the extremely difficult nature of their experiments with a fixed-amplitude bar rotating at a fixed $\Omega_{\rm p}$. When $\Omega_{\rm p}$ is a function of time, the resonances are broadened to orbits near resonance, which makes phase space coverage a less stringent constraint on N. Weak bars in particular may be prone to such difficulties, but it is strong bars that are most interesting.

In summary, there may be some evidence that noise is compromising some if not all of the evolution in some cases, in ways not yet fully understood. But it is clear that only quite large bars can erase cusps to scales larger than 1 kpc.

4.   Evolution of Pattern Speeds

The pattern speed of a bar is usually parametrized by the ratio $\mathcal{R} \equiv D_L / a_B$, where D_L is the corotation radius, at which the gravitational and centrifugal forces cancel. A self-consistent bar must have $\mathcal{R} \geq 1$ (Contopoulos 1980). A bar is termed fast when $1.0 \leq \mathcal{R} \leq 1.4$ and slow otherwise. This definition does not distinguish fast from slow in terms of $\Omega_{\rm p}$ alone: a bar in a galaxy with rotation velocity of 200 km s^{-1} is slow at $\Omega_{\rm p} = 100$ km s^{-1} kpc^{-1} if $a_B = 1$ kpc, but fast at $\Omega_{\rm p} = 20$ km s^{-1} kpc^{-1} if $a_B = 10$ kpc. Observational evidence points to fast bars, both in early-type barred galaxies (Merrifield & Kuijken 1995; Gerssen et al. 1999; Debattista et al. 2002; Aguerri et al. 2003; Gerssen et al. 2003; Debattista & Williams 2004) and in late types (Lindblad et al. 1996; Lindblad & Kristen 1996; Weiner et al. 2001; Pérez et al. 2004).

Tremaine & Weinberg (1984) developed the perturbation theory of dynamical friction for perturbers in spheroidal systems, showing that friction arises near resonances, when $m\Omega_{\rm p} = k\Omega_r + l\Omega_\phi$, where Ω_r and Ω_ϕ are the radial and angular frequencies, respectively (see Weinberg 2004, for a time-dependent treatment). Weinberg (1985) applied this theory to a bar rotating in a dark halo and found that the bar is braked such that $D_L \gg a_B$ unless (1) angular momentum is added to the bar, (2) the bar is weak or (3) the halo has low mass.

The transfer of angular momentum from disk to spheroid (bulge or halo) was reported in several early simulations (Sellwood 1980; Little & Carlberg 1991; Hernquist & Weinberg 1992). Fully self-consistent, 3-D simulations of bar-unstable disks embedded in dark halos were presented by Debattista & Sellwood (1998, 2000, hereafter DS00) who found that disks needed to be the dominant mass component if the bars that formed were to remain at $\mathcal{R} < 1.4$. Bar slow-down has also been found in other simulations since then (O'Neill & Dubinski 2003; Holley-Bockelmann et al. 2005; Berentzen et al. 2005; Martinez-Valpuesta et al. 2005). O'Neill & Dubinski (2003) included a comparison with DS00 and found generally good agreement.

The constraint of DS00 severely limits CDM cusps to be present in dark halos. Valenzuela & Klypin (2003, hereafter VK03) presented simulations of a CDM Milky Way, with halo concentration $c \simeq 15$. These simulations were evolved on an AMR code with a maximum refinement corresponding to a resolution of 20–40 pc. They found that the bar that formed remained fast for $\sim$4–5

Gyr, thereafter slowing to $\mathcal{R} = 1.7$. They concluded that the slow bars in the simulations of DS00 were an artifact of low resolution. They also argued that their bars remained modest in length (1–2 scale-lengths) but became too long in lower resolution simulations, which they concluded again proved the limitations of low resolution simulations. Sellwood & Debattista (2006) repeated one of the simulations of VK03 using the same initial conditions but evolved on a hybrid grid code (Sellwood 2003) with a fixed softening comparable to that of VK03. They found instead that the bar reached $\mathcal{R} = 2$ within 4 Gyr. They accounted for the result of VK03 by noting that as the bar formed, the central density increased (see § 5). Thus in an AMR code there is a tendency for the spatial resolution to increase; the disk in the model of VK03 being rather thin, this led to enhanced forces and a corresponding artificial increase in Ω_p. Once this happened, the bar found itself with its principal resonances at local phase space minima previously generated by the forming bar. Although angular momentum is exchanged at resonances, the sign of the torque depends on the phase space gradient of the DF at the resonances; in the absence of a gradient no friction is possible. Thus the numerics induced a metastable state. This state proved fragile to realistic perturbations and is not likely to last long in nature, but in the quieter environment of an isolated N-body system it can persist for many Gyrs, enough to fully account for the behavior found by VK03.

In a series of papers, Athanassoula presented several arguments against the conclusion of DS00. In Athanassoula (2003) she compared the evolution of two halo-dominated systems, MQ2 and MHH2, with nearly identical disk and halo rotation curves. However, their velocity dispersions were different, being larger in MHH2 because its halo extended to larger radii. As a result Ω_p decreased significantly in MQ2 but hardly at all in MHH2. She concluded that this "argues against a link between relative halo content and bar slowdown..." This claim however is contradicted by her own simulations. The confusion arises from her relying on the change in Ω_p to constrain the halo. Not only is this unobservable, but if we compute $\mathcal{R}$ for her models, we find $1.4 < \mathcal{R} \leq 1.7$ for run MQ2 and $\mathcal{R} \simeq 3$ for MHH2, *i.e.*, both these halo-dominated systems are slower than observed. Far from being in disagreement with the results of DS00, her simulations support them. She also showed that weak bars are less able to drive angular momentum exchanges and may remain fast; since observational measurements have only been obtained for strong bars, this is not worrisome.

Athanassoula & Misiriotis (2002) argued that bar lengths are difficult to measure in simulations, making a comparison with corotation difficult. Bar lengths are certainly not always easy to measure, but it is still possible to define values which straddle the real value; for example, DS00 had cases in which the uncertainty in a_B was as large as $\Delta a_B / \overline{a_B} \sim 0.3$, not substantially different than in observations (Debattista 2003). But these are not Gaussian errors and the probability of a_B falling outside the given range is practically zero. Moreover, the same problem afflicts observations; therefore measurement uncertainties in a_B are a nuisance but not a repudiation of the constraint.

Athanassoula (2002) showed that loss of angular momentum from the bar leads to a growing bar, as first suggested by DS00. VK03 pointed out that bars become excessively long in the presence of strong friction. Even though angular momentum redistribution leads to larger disk scale-lengths, bars extending $\gtrsim 10$ kpc are not common (Erwin 2005). Strong bar growth does not seem to have occurred through the history of the current generation of bars and presumably, neither has strong friction.

Thus to date no well-resolved simulation has provided a valid counter-example to the claim by Debattista & Sellwood (1998) that dense halos cannot support fast strong bars. If anything, recent simulations have lent support to it.

5.   The Secular Evolution of Disk Densities

The excellent recent review by Kormendy & Kennicutt (2004) presents the observational evidence for pseudobulge formation via secular evolution and discusses some of the older N-body results in that field. Here I review recent developments not covered by those authors.

5.1.   Disk Profile Evolution

Bar formation is accompanied by a rearrangement of disk material as first shown by Hohl (1971). Generally a nearly double-exponential profile develops, with a smaller central scale-length than the initial and a larger one further out. In this respect these profiles resemble bulge+disk profiles and comparisons with observations show that these profiles are reasonable approximations to observed profiles (Debattista et al. 2004; Avila-Reese et al. 2005). Debattista et al. (2006) showed that the degree by which the profile changes depends on the initial disk temperature Q. In hot disks ($Q \sim 2$) little angular momentum needs to be shed and the azimuthally-averaged density profile is practically unchanged. Thus the distribution of disk scale-lengths depends not only on the initial angular momentum of the baryons (and presumably that of the dark halo) but also the disk temperature. Angular momentum redistribution continues also after the bar forms, especially to the halo. This leads to a further increase in the central density of the disk even when the evolution is collisionless. This may be sufficient to render an initially halo-dominated system into one dominated by baryons in the inner parts (Debattista & Sellwood 2000; Valenzuela & Klypin 2003).

The angular momentum lost by the bar as it forms may be transported out to large radii via a resonant coupling between the bar and spirals (Debattista et al. 2006) of the kind found by Masset & Tagger (1997) and Rautiainen & Salo (1999). Debattista et al. (2006) show that this transport leads to breaks in the density distribution which, when viewed edge-on, are indistinguishable from those observed in real galaxies (Pohlen 2002).

5.2.   Vertical Evolution: Peanut-Shaped Bulges

N-body simulations of the vertical evolution of disk galaxies have concentrated on box- or peanut- (B/P-) shaped bulges, which are present in some 45% of edge-on galaxies (Lütticke et al. 2000). Simulations have shown that these form via the secular evolution of bars (Combes & Sanders 1981), either through resonant scattering or through bending (a.k.a. buckling) instabilities (Pfenniger 1984; Combes et al. 1990; Pfenniger & Friedli 1991; Raha et al. 1991). Although it is often thought that a peanut requires that a bulge was built by secular processes, simulations show that peanuts can also form when the initial conditions include a bulge, as would happen if bulges form through mergers at high redshift (Athanassoula & Misiriotis 2002; Debattista et al. 2005).

Observations seeking to establish the connection between B/P-shaped bulges and bars (Kuijken & Merrifield 1995; Merrifield & Kuijken 1999; Bureau & Freeman 1999; Chung & Bureau 2004), by looking for evidence of bars in edge-on B/P-bulged systems, have benefited from comparisons with the edge-on stellar velocity distributions of N-body bars (Bureau & Athanassoula 1999, 2005). Bureau & Athanassoula (2005) characterized the signature of an edge-on bar as having (1) a Freeman type II profile, (2) a rotation curve with a local maximum interior to its flat part (3) a broad velocity dispersion profile with a plateau at intermediate radii (4) a correlation between velocity and the third-order Gauss-Hermite moment h_3 (Gerhard 1993; van der Marel & Franx 1993). A diagnostic of B/P-shaped bulges in face-on galaxies was developed, and tested on high mass and force resolution simulations, by Debattista et al. (2005). Vertical velocity dispersions constitute a poor diagnostic because they depend on the local surface density. Instead, their diagnostic is based on the fact that peanut shapes are associated with a flat density distribution in the vertical direction. The kinematic signature corresponding to such a distribution is a minimum in the fourth-order Gauss-Hermite moment h_4.

The buckling instability itself has also been studied with simulations. Debattista et al. (2006) showed that an otherwise vertically-stable bar is destabilized when it slows and grows. Moreover, the instability can occur more than once for a given bar (Martinez-Valpuesta et al. 2005). Finally, after Raha et al. (1991) showed that buckling weakens bars, it has often been assumed that bars are destroyed by buckling. Debattista et al. (2006) presented a series of high force and mass resolution simulations demonstrating that this is not the case.

5.3. Spirals

There is broad agreement that spirals constitute density waves (Lin & Shu 1964). At least three different dynamical mechanisms have been proposed for exciting them: swing-amplification (Toomre 1981), long-lived modes (Bertin et al. 1989a,b) and recurrent instabilities seeded by features in the angular momentum distribution (Sellwood & Lin 1989; Sellwood & Kahn 1991; Sellwood 2000). All three are still viable and not many new N-body results have been obtained in recent years, but Sellwood & Binney (2002) used simulations to demonstrate that spirals cause a considerable radial shuffling of mass at nearly fixed angular momentum distribution. This happens at corotation and is not accompanied by substantial heating — stars on nearly circular orbits can be scattered onto other nearly circular orbits. For example, they estimate that a star born at the solar radius can be scattered nearly uniformly within $\Delta R = \pm 4$ kpc. Thus the idea of a Galactic habitable zone becomes somewhat suspect, as does the need for infall to maintain the metallicity distribution observed at the solar circle.

6. Bar Destruction by CMCs

Central massive concentrations (CMCs), whether supermassive black holes (hard CMCs) or gas condensations several hundred parsecs in size (soft CMCs), could destroy bars. Early studies of this phenomenon were inspired by the similar work for slowly-rotating triaxial elliptical galaxies (e.g., Gerhard & Binney 1985), where the loss of triaxiality results from the destruction of box orbits by scattering off the CMC. Such scattering may not be efficient in bars, since the main bar-supporting orbits are loops which avoid the center. Hasan & Norman (1990) and Hasan et al. (1993) argued that, when a CMC in a barred galaxy grows sufficiently massive, it quickly destroys bar-supporting orbits by driving an ILR, around which orbits are unstable, to large radii. The more centrally concentrated the CMC grew, the further out was the ILR and therefore the more disruptive it was. How massive the CMC needed to be required N-body simulations to establish. Friedli & Benz (1993) modeled gas and stars with PM+SPH simulations and found that gas inflows destroyed bars when 2% of the baryonic mass ended up in a hard CMC. Norman et al. (1996) pursued collisionless 2-D and 3-D simulations, with CMCs grown by slowly contracting a massive component. They needed a CMC of mass 5% of disk mass, M_d, to destroy bars. Then bar destruction was rapid and led to a bulge-like spheroid.

Shen & Sellwood (2004, hereafter SS04) presented a series of high-quality simulations including high mass, force and time resolution and found that bars are more robust than previously thought. They varied the growth rates, compactness and mass of the CMCs and considered both weak and strong bars within which they grew CMCs at fixed compactness. A rigid halo with a large core radius was also included. They obtained a fast decay while a CMC was growing followed by a more gradual decay once the CMC reached its full mass. The time over which the bar was grown proved unimportant, with only the final mass and compactness mattering. Hard CMCs cause more damage than soft ones, needing 4-5%M_d and $> 10\%M_d$, respectively, to destroy bars. Their tests showed that time steps need to be as small as 10^{-4} of a dynamical time in order that more rapid but incorrect bar destruction is avoided. They interpreted the two-phase bar destruction as scattering of low energy bar-supporting x_1 orbits during the CMC growth phase and continued gradual global structural adjustment, which further destroyed high energy x_1 orbits, thereafter. They predicted that massive halos, which lead to bar growth (Debattista & Sellwood 2000; Athanassoula 2003) render bars even more difficult to destroy; this prediction, as well as the two-phase bar weakening, was confirmed in live-halo simulations by Athanassoula et al. (2005).

Bournaud & Combes (2002) and Bournaud et al. (2005) have argued for a radically different picture. They simulated gas accretion and noted that bars were destroyed and reformed 3 or 4 times over a Hubble time. The amount of gas required to destroy bars is not, however, wholly consistent in these simulations: a system with $\sim 7\%$ *total* gas mass fraction lost its bar within 2 Gyr in Bournaud et al. (2005) whereas a system with three times more gas maintained its bar in Bournaud & Combes (2002). Possibly star-formation, included in the later simulations, somehow quenched the infall onto the center. SS04 hinted that the timestep used by Bournaud & Combes (2002) was too large but Bournaud et al. (2005) reported that using a timestep 0.125× their standard value (and close to that advocated by SS04) still led to recurrent bar destruction. Bournaud et al. (2005) argued that their results are correct and that other studies had erred in mimicking gas accretion by simply growing a massive object because this neglected the important effects of angular momentum transport from gas to the bar. This is perhaps consistent with the simulations of Friedli & Benz (1993) and of Berentzen et al. (1998); the latter were able to destroy bars by gas inflow leading to a CMC of mass fraction just 1.6%. On the other hand, the fully live simulations of Debattista et al. (2006) only destroyed bars when soft CMCs reached $\sim$20% M_d, in good agreement with SS04. One possibly important difference is that the simulations of Bournaud et al. (2005) included a rigid halo, which prevents it from accepting angular momentum from the bar, while Debattista et al. (2006) had live halos.

Hozumi & Hernquist (2005), using a 2-D SCF code, also concluded that hard CMCs can destroy bars with smaller masses, 0.5%M_d. SS04 (see also Sellwood 2002) speculated that low order SCF expansions may not be able to simultaneously maintain a system axisymmetric near the center and non-axisymmetric further out.

7. Multiple Patterns

An emerging field in the past few years has been galaxies with multiple patterns. These are challenging to study because traditional tools such as surfaces-of-section are no longer viable. N-body simulations, therefore, are vital for studying systems such as bars in triaxial halos and bars within bars.

7.1. Bars in Triaxial Halos

CDM predicts that dark matter halos are triaxial (Barnes & Efstathiou 1987; Frenk et al. 1988; Dubinski & Carlberg 1991; Jing & Suto 2002). The condensation of baryons inside triaxial halos drives them to rounder shapes, but systems do not generally become wholly axisymmetric (Dubinski 1994; Kazantzidis et al. 2004). Using rigidly-rotating bars, El-Zant & Shlosman (2002) computed the Liapunov exponents of orbits and showed that chaos quickly dominates the evolution when halos are triaxial and cuspy. N-body simulations by Berentzen et al. (2005) indeed show that bars are destroyed in cases where the triaxiality in the potential is as small as $c/a \sim 0.9$ and the halo is cuspy. One way in which this fate can be avoided is for the bar to alter the shape of the inner halo. At present it is not clear which systems can accomplish this and which cannot. A better understanding of when bars are destroyed in triaxial halos could possibly lead to an important new constraint on the shapes and profiles of dark matter halos.

7.2. Bars Within Bars

While observations of largely gas-free early-type galaxies find an abundance of nuclear bars within large scale bars (Erwin & Sparke 2002 found them in $\sim$30% of barred S0–Sa galaxies), simulating them has proved difficult. Moreover, it is only recently that direct observational evidence for kinematically decoupled primary and nuclear bars in one system has been obtained (Corsini et al. 2003). Therefore their dynamics have been poorly understood, in spite of the fact that they have been postulated to be an important mechanism for driving gas to small radii to feed supermassive black holes (Shlosman et al. 1989).

Most numerical studies have required gas to form secondary bars (Friedli & Martinet 1993; Shlosman & Heller 2002; Englmaier & Shlosman 2004), but their presence in gas-poor early-type galaxies suggests that gas is not the main ingredient for forming secondary bars. Stellar counter-rotation can lead to counter-rotating bars (Sellwood & Merritt 1994; Friedli 1996). Rautiainen et al. (2002) were the first to succeed in producing collisionless N-body simulations with both bars rotating in the same sense. Their secondary bars, which were vaguely spiral-like and possibly hollow, rotated faster than the primary bars and survived for several Gyrs. Debattista & Shen (2006, in preparation, see also Shen & Debattista in these proceedings) present further examples and explore the mutual evolution of the two bars. Now that N-body simulations can achieve the high force and mass resolution needed to form self-consistent nested bar systems it is hoped that progress in understanding these systems will be more rapid than in the 30 years since their discovery (de Vaucouleurs 1975).

8.   Desiderata for the Future

The collisionless simulation of isolated galaxies is an endeavor over thirty years old. The subject is still rich, with several open problems, and continues to be very active. Algorithmically, if not conceptually, it is the simplest problem that can be considered. Various gravity solvers for its study are available (which are used also in other areas of astronomy). Despite much progress, the degree of disagreement in the field, as described above, is surprising. The way to progress from this point is to compare directly the results of different codes with each other, as has been done in other fields (e.g., Kang et al. 1994; Frenk et al. 1999). Unfortunately the number of such tests in galaxy evolution has been small. Therefore a detailed comparison between many different codes would be very valuable at this time. Ideally this would involve several implementations of the same code type to establish behaviors in the different types. The actual tests to be performed should include systems in which an analytic result is known (useful to establish values of numerical parameters which are optimal) as well as systems for which the result is not known in advance. Furthermore, the N-body tests of the type recently proposed by Weinberg & Katz (2005) provide a challenging and useful basis for assessing the effect of noise on simulations. Such a comparison will give the community greater confidence that we are able to model correctly the most basic level of galaxy evolution.

Acknowledgments.   V. P. D. is supported by a Brooks Prize Fellowship in Astrophysics at the University of Washington. I thank Tom Quinn, Jerry Sellwood, Juntai Shen, Martin Weinberg, Peter Erwin and Kelly Holley-Bockelmann for useful discussions, and the organizers of Bash '05 for inviting me to this symposium.

References

Aarseth, S. J. 1963, MNRAS, 126, 223
Aguerri, J. A. L., Debattista, V. P., & Corsini, E. M. 2003, MNRAS, 338, 465
Athanassoula, E. 2002, ApJ, 569, L83
Athanassoula, E. 2003, MNRAS, 341, 1179
Athanassoula, E., Lambert, J. C., & Dehnen, W. 2005, MNRAS, 363, 496
Athanassoula, E., & Misiriotis, A. 2002, MNRAS, 330, 35
Avila-Reese, V., Carrillo, A., Valenzuela, O., & Klypin, A. 2005, MNRAS, 361, 997
Barnes, J., & Efstathiou, G. 1987, ApJ, 319, 575
Barnes, J., & Hut, P. 1986, Nat, 324, 446
Barnes, J., Hut, P., & Goodman, J. 1986, ApJ, 300, 112
Berentzen, I., Heller, C. H., Shlosman, I., & Fricke, K. J. 1998, MNRAS, 300, 49

Berentzen, I., Shlosman, I., & Jogee, S. 2006, ApJ, 637, 582
Bertin, G., Lin, C. C., Lowe, S. A., & Thurstans, R. P. 1989a, ApJ, 338, 78
Bertin, G., Lin, C. C., Lowe, S. A., & Thurstans, R. P. 1989b, ApJ, 338, 104
Bertschinger, E. 1985, ApJS, 58, 39
Binney, J., Bissantz, N., & Gerhard, O. 2000, ApJ, 537, L99
Binney, J., Gerhard, O., & Silk, J. 2001, MNRAS, 321, 471
Bissantz, N., Englmaier, P., & Gerhard, O. 2003, MNRAS, 340, 949
Blais-Ouellette, S., Amram, P., & Carignan, C. 2001, AJ, 121, 1952
Bode, P., Ostriker, J. P., & Turok, N. 2001, ApJ, 556, 93
Bournaud, F., & Combes, F. 2002, A&A, 392, 83
Bournaud, F., Combes, F., & Semelin, B. 2005, MNRAS, 364, L18
Bryan, G. L., & Norman, M. L. 1995, Bulletin of the American Astronomical Society,
    27, 1421
Bureau, M., & Athanassoula, E. 1999, ApJ, 522, 686
Bureau, M., & Athanassoula, E. 2005, ApJ, 626, 159
Bureau, M., & Freeman, K. C. 1999, AJ, 118, 126
Cen, R. 2001, ApJ, 546, L77
Chung, A., & Bureau, M. 2004, AJ, 127, 3192
Clutton-Brock, M. 1972, Ap&SS, 16, 101
Colín, P., Avila-Reese, V., & Valenzuela, O. 2000, ApJ, 542, 622
Combes, F., Debbasch, F., Friedli, D., & Pfenniger, D. 1990, A&A, 233, 82
Combes, F., & Sanders, R. H. 1981, A&A, 96, 164
Contopoulos, G. 1980, A&A, 81, 198
Corsini, E. M., Debattista, V. P., & Aguerri, J. A. L. 2003, ApJ, 599, L29
Davé, R., Dubinski, J., & Hernquist, L. 1997, New Astronomy, 2, 277
de Blok, W. J. G., McGaugh, S. S., & Rubin, V. C. 2001, AJ, 122, 2396
de Vaucouleurs, G. 1975, ApJS, 29, 193
Debattista, V. P. 2003, MNRAS, 342, 1194
Debattista, V. P., Carollo, C. M., Mayer, L., & Moore, B. 2004, ApJ, 604, L93
Debattista, V. P., Carollo, C. M., Mayer, L., & Moore, B. 2005, ApJ, 628, 678
Debattista, V. P., Corsini, E. M., & Aguerri, J. A. L. 2002, MNRAS, 332, 65
Debattista, V. P., Mayer, L., Carollo, C. M., Moore, B., Wadsley, J., & Quinn, T. 2006,
    ApJ, in press (astro-ph/0509310)
Debattista, V. P., & Sellwood, J. A. 1998, ApJ, 493, L5
Debattista, V. P., & Sellwood, J. A. 2000, ApJ, 543, 704
Debattista, V. P., & Williams, T. B. 2004, ApJ, 605, 714
Dehnen, W. 2000, ApJ, 536, L39
Dubinski, J. 1994, ApJ, 431, 617
Dubinski, J. 1996, New Astronomy, 1, 133
Dubinski, J., & Carlberg, R. G. 1991, ApJ, 378, 496
Earn, D. J. D., & Sellwood, J. A. 1995, ApJ, 451, 533
El-Zant, A., & Shlosman, I. 2002, ApJ, 577, 626
Englmaier, P., & Shlosman, I. 2004, ApJ, 617, L115
Erwin, P. 2005, MNRAS, 364, 283
Erwin, P., & Sparke, L. S. 2002, AJ, 124, 65
Eskridge, P. B., et al. 2000, AJ, 119, 536
Evans, N. W., & Read, J. C. A. 1998a, MNRAS, 300, 83
Evans, N. W., & Read, J. C. A. 1998b, MNRAS, 300, 106
Fillmore, J. A., & Goldreich, P. 1984, ApJ, 281, 1
Frenk, C. S., et al. 1999, ApJ, 525, 554
Frenk, C. S., White, S. D. M., Davis, M., & Efstathiou, G. 1988, ApJ, 327, 507
Friedli, D. 1996, A&A, 312, 761
Friedli, D., & Benz, W. 1993, A&A, 268, 65
Friedli, D., & Martinet, L. 1993, A&A, 277, 27
Fux, R. 1999, A&A, 345, 787

Gerhard, O. E. 1993, MNRAS, 265, 213
Gerhard, O. E., & Binney, J. 1985, MNRAS, 216, 467
Gerssen, J., Kuijken, K., & Merrifield, M. R. 1999, MNRAS, 306, 926
Gerssen, J., Kuijken, K., & Merrifield, M. R. 2003, MNRAS, 345, 261
Gnedin, O. Y., & Zhao, H. 2002, MNRAS, 333, 299
Goodman, J. 2000, New Astronomy, 5, 103
Hasan, H., & Norman, C. 1990, ApJ, 361, 69
Hasan, H., Pfenniger, D., & Norman, C. 1993, ApJ, 409, 91
Hernquist, L. 1987, ApJS, 64, 715
Hernquist, L. 1990, ApJ, 356, 359
Hernquist, L., & Barnes, J. E. 1990, ApJ, 349, 562
Hernquist, L., & Ostriker, J. P. 1992, ApJ, 386, 375
Hernquist, L., & Weinberg, M. D. 1992, ApJ, 400, 80
Hogan, C. J., & Dalcanton, J. J. 2000, Phys.Rev.D, 62, 063511
Hohl, F. 1971, ApJ, 168, 343
Hohl, F. 1973, ApJ, 184, 353
Holley-Bockelmann, K., Weinberg, M., & Katz, N. 2005, MNRAS, 363, 991
Hozumi, S., & Hernquist, L. 2005, PASJ, 57, 719
Inagaki, S., Nishida, M. T., & Sellwood, J. A. 1984, MNRAS, 210, 589
Jing, Y. P., & Suto, Y. 2000, ApJ, 529, L69
Jing, Y. P., & Suto, Y. 2002, ApJ, 574, 538
Kalnajs, A. J. 1972, ApJ, 175, 63
Kalnajs, A. J. 1978, in IAU Symp. 77: Structure and Properties of Nearby Galaxies,
    ed. E. M. Berkhuijsen & R. Wielebinski, 113
Kang, H., Ostriker, J. P., Cen, R., Ryu, D., Hernquist, L., Evrard, A. E., Bryan, G. L.,
    & Norman, M. L. 1994, ApJ, 430, 83
Kaplinghat, M., Knox, L., & Turner, M. S. 2000, Physical Review Letters, 85, 3335
Kazantzidis, S., Kravtsov, A. V., Zentner, A. R., Allgood, B., Nagai, D., & Moore, B.
    2004, ApJ, 611, L73
Knapen, J. H. 1999, in ASP Conf. Ser. 187: The Evolution of Galaxies on Cosmological
    Timescales, ed. J. E. Beckman & T. J. Mahoney, 72
Kormendy, J., & Kennicutt, R. C. 2004, ARA&A, 42, 603
Kravtsov, A. V., Klypin, A. A., & Khokhlov, A. M. 1997, ApJS, 111, 73
Kuijken, K., & Merrifield, M. R. 1995, ApJ, 443, L13
Lin, C. C., & Shu, F. H. 1964, ApJ, 140, 646
Lindblad, P. A. B., & Kristen, H. 1996, A&A, 313, 733
Lindblad, P. A. B., Lindblad, P. O., & Athanassoula, E. 1996, A&A, 313, 65
Little, B., & Carlberg, R. G. 1991, MNRAS, 250, 161
Lütticke, R., Dettmar, R.-J., & Pohlen, M. 2000, A&AS, 145, 405
Martinez-Valpuesta, I., Shlosman, I., & Heller, C. 2006, ApJ, 637, 214
Masset, F., & Tagger, M. 1997, A&A, 322, 442
Matthews, L. D., & Gallagher, J. S. 2002, ApJS, 141, 429
McMillan, P. J., & Dehnen, W. 2005, MNRAS, 363, 1205
Merrifield, M. R., & Kuijken, K. 1995, MNRAS, 274, 933
Merrifield, M. R., & Kuijken, K. 1999, A&A, 345, L47
Miller, R. H. 1976, Journal of Computational Physics, 21, 400
Miller, R. H., & Prendergast, K. H. 1968, ApJ, 151, 699
Miller, R. H., Prendergast, K. H., & Quirk, W. J. 1970, ApJ, 161, 903
Moore, B., Governato, F., Quinn, T., Stadel, J., & Lake, G. 1998, ApJ, 499, L5
Navarro, J. F., Eke, V. R., & Frenk, C. S. 1996, MNRAS, 283, L72
Navarro, J. F., Frenk, C. S., & White, S. D. M. 1997, ApJ, 490, 493
Nishida, M. T., Yoshizawa, M., Watanabe, Y., Inagaki, S., & Kato, S. 1981, PASJ, 33,
    567
Norman, C. A., Sellwood, J. A., & Hasan, H. 1996, ApJ, 462, 114
O'Neill, J. K., & Dubinski, J. 2003, MNRAS, 346, 251

Peebles, P. J. E. 2000, ApJ, 534, L127
Pérez, I., Fux, R., & Freeman, K. 2004, A&A, 424, 799
Pfenniger, D. 1984, A&A, 134, 373
Pfenniger, D., & Friedli, D. 1991, A&A, 252, 75
Pohlen, M. 2002, Ph.D. Thesis, Ruhr-Universitat
Power, C., Navarro, J. F., Jenkins, A., Frenk, C. S., White, S. D. M., Springel, V., Stadel, J., & Quinn, T. 2003, MNRAS, 338, 14
Raha, N., Sellwood, J. A., James, R. A., & Kahn, F. D. 1991, Nat, 352, 411
Rautiainen, P., & Salo, H. 1999, A&A, 348, 737
Rautiainen, P., Salo, H., & Laurikainen, E. 2002, MNRAS, 337, 1233
Sellwood, J. A. 1980, A&A, 89, 296
Sellwood, J. A. 1997, in ASP Conf. Ser. 123: Computational Astrophysics; 12th Kingston Meeting on Theoretical Astrophysics, ed. D. A. Clarke & M. J. West, 215
Sellwood, J. A. 2000, Ap&SS, 272, 31
Sellwood, J. A. 2002, in The Shapes of Galaxies and Their Dark Halos, ed. P. Natarajan (Singapore: World Scientific), 123
Sellwood, J. A. 2003, ApJ, 587, 638
Sellwood, J. A. 2005, ApJ, 637, 567
Sellwood, J. A., & Binney, J. J. 2002, MNRAS, 336, 785
Sellwood, J. A., & Debattista, V. P. 2006, ApJ, 639, 868
Sellwood, J. A., & Evans, N. W. 2001, ApJ, 546, 176
Sellwood, J. A., & Kahn, F. D. 1991, MNRAS, 250, 278
Sellwood, J. A., & Lin, D. N. C. 1989, MNRAS, 240, 991
Sellwood, J. A., & Merritt, D. 1994, ApJ, 425, 530
Shen, J., & Sellwood, J. A. 2004, ApJ, 604, 614
Shlosman, I., Frank, J., & Begelman, M. C. 1989, Nat, 338, 45
Shlosman, I., & Heller, C. H. 2002, ApJ, 565, 921
Spergel, D. N., & Steinhardt, P. J. 2000, Physical Review Letters, 84, 3760
Springel, V., Yoshida, N., & White, S. D. M. 2001, New Astronomy, 6, 79
Stadel, J. G. 2001, Ph.D. Thesis, University of Washington
Toomre, A. 1981, in Structure and Evolution of Normal Galaxies, ed. S. M., Fall & D. Lynden-Bell (Cambridge: Cambridge University Press), 111
Tremaine, S., & Weinberg, M. D. 1984, ApJ, 282, L5
Valenzuela, O., & Klypin, A. 2003, MNRAS, 345, 406
van der Marel, R. P., & Franx, M. 1993, ApJ, 407, 525
Watanabe, Y., Inagaki, S., Nishida, M. T., Tanaka, Y. D., & Kato, S. 1981, PASJ, 33, 541
Weinberg, M. D. 1985, MNRAS, 213, 451
Weinberg, M. D. 1996, ApJ, 470, 715
Weinberg, M. D. 2005, preprint (astro-ph/0404169)
Weinberg, M. D., & Katz, N. 2002, ApJ, 580, 627
Weinberg, M. D., & Katz, N. 2005, preprint (astro-ph/0508166)
Weiner, B. J., Sellwood, J. A., & Williams, T. B. 2001, ApJ, 546, 931
Zang, T. A. 1976, Ph.D. Thesis, Massachusetts Institute of Technology
Zel'Dovich, Y. B. 1970, A&A, 5, 84

Jon Miller and Eric Gawiser enjoy the opening reception.

Frank N. Bash Symposium 2005: New Horizons in Astronomy
ASP Conference Series, Vol. 352, 2006
S. J. Kannappan, S. Redfield, J. E. Kessler-Silacci, M. Landriau, and N. Drory

Galaxy Formation

Eric Gawiser[1]

*Yale Astronomy Department and Yale Center for Astronomy &
Astrophysics, New Haven, CT, USA*

Abstract.   I summarize current knowledge of galaxy formation with emphasis
on the initial conditions provided by the ΛCDM cosmology, integral constraints
from cosmological quantities, and the demographics of high-redshift protogalaxies. Tables are provided summarizing the number density, star formation rate
and stellar mass per object, cosmic star formation rate and stellar mass densities, clustering length and typical dark matter halo masses for Lyman break
galaxies, Lyman alpha emitting galaxies, distant red galaxies, sub-millimeter
galaxies, and damped Lyman α absorption systems. I also discuss five key unsolved problems in galaxy formation and prognosticate advances that the near
future will bring.

1.   Boundary Conditions for Galaxy Formation

1.1.   Initial Conditions: ΛCDM Cosmology

The initial conditions for the formation of galaxies are provided by the now-
standard Λ Cold Dark Matter (CDM) cosmological model. The combined results of the *WMAP* satellite study of Cosmic Microwave Background (CMB)
anisotropies, large-scale structure, and Type Ia supernova observations yield
best-fit values for the cosmological parameters of roughly $\Omega_\Lambda = 0.7$, $\Omega_m = 0.3$,
$\Omega_b = 0.04$, and $H_0 = 70h_{70}$ km s^{-1} Mpc^{-1} (Bennett et al. 2003).[2] The original model of galaxy formation was Monolithic Collapse (Eggen, Lynden-Bell,
& Sandage 1962), where gravitational collapse of a cloud of primordial gas very
early in the lifetime of the Universe formed all parts of each galaxy at the same
time. Modern evidence rules out this model on two fronts; the widely varying ages of different components of the Galaxy provide a counter-example, and
the ΛCDM cosmology predicts "bottom-up" i.e. hierarchical rather than "top-
down" structure formation.

Hierarchical structure formation is a generic feature of CDM models. Small
overdensities are able to overcome the cosmological expansion and collapse first,
and the resulting dark matter "halos" merge together to form larger halos which
serve as sites of galaxy formation. This process continues until the present
day, making galaxy formation an ongoing process. The nearly-scale-invariant
primordial power spectrum inferred from combining *WMAP* with large-scale

---

[1]NSF Astronomy & Astrophysics Postdoctoral Fellow (AAPF)

[2]We include h_{70}, analogous to the traditional parameter $h \equiv h_{100}$, even though its value appears
quite close to 1.

structure observations provides power on all scales in the distribution of CDM. The baryons fall into the CDM potential wells after decoupling, leaving only trace evidence of their previous acoustic oscillations as a series of low-amplitude peaks in the matter power spectrum. The non-linear collapse of dark matter overdensities occurs on larger and larger scales, so the typical collapsed halo mass grows with time, but no preferred scale is introduced. ΛCDM therefore provides a distribution of halos where galaxies can form, with the details of the process up to baryonic physics.

Despite the lack of preferred galaxy scales in the distribution of dark matter halos, baryonic physics causes galaxies to have minimum and maximum masses. The maximum mass is that of cD galaxies in cluster centers with baryonic masses $\sim 10^{12}$ $M_\odot$ and virial masses $\sim 10^{13}$ $M_\odot$; there are $\sim 10^{14}$ $M_\odot$ of baryons available in a rich cluster but virialization of galaxies and heating of gas to the high virial temperature prevent most of this mass from finding its way to the central galaxy. The minimum mass observed today is that of dwarf galaxies, $\sim 10^8$ $M_\odot$, but galaxies may initially have formed as small as 10^6 $M_\odot$ (the baryonic Jeans mass after recombination, i.e., the minimum mass for which gravity overwhelmed pressure support). Explaining the lack of observed galaxies with circular velocities below 30 km/s is a major goal; it is suspected that feedback from supernova explosions may have quenched star formation in such low-mass objects immediately after a single burst of star formation (Dekel & Silk 1986).

The growth of cosmological structure and collapse of dark matter halos is a feature of the matter-dominated epoch. During radiation domination, perturbations on scales smaller than the sound horizon were unable to grow due to acoustic oscillations in the photon-baryon fluid that gave rise to the famous peaks in the CMB angular power spectrum and the lower-amplitude peaks in the matter power spectrum. Now that we have entered a phase of dark energy domination, structure growth is slowing and will cease entirely as the universe enters a new phase of inflation. This cosmological "freeze-out" in structure formation is recent, since equality between the dark energy and matter densities occurred at $z_{eq} = 0.4$. The slowing of structure formation occurs gradually, so the growth of cosmological structure continued nearly unabated until z_{eq}, even though we see strong observational evidence for "downsizing" at $z < 1$ where high-mass galaxies grow far more slowly than lower-mass galaxies (e.g., Treu et al. 2005; Smith 2005). Another term being used by some is "anti-hierarchical," which is basically a synonym for "downsizing" but seems to imply inconsistency with hierarchical cosmology. However, the observed freeze-out in galaxy (and possibly supermassive black hole) formation in massive galaxies is not inconsistent with CDM models; rather, it appears to be caused by baryonic feedback which is not well understood at present (see § 6.). The slowing of cosmological structure growth since $z \simeq 0.4$ may, however, play a role in the recent decline of the cosmic star formation rate density discussed by Bell et al. (2005).

1.2.   Final Conditions: Low-Redshift Galaxies

The study of galaxy formation is made easier by having full boundary conditions. The final conditions are the Hubble sequence of mature galaxies we see in the nearby universe at redshift zero. Indeed, much has been learned about galaxy formation from "archaeological" evidence in the ages and chemical abundances of various Galactic stellar populations, and expanding these studies to the rest of the Local Group and beyond is quite useful. Nonetheless, there are great advantages to observing galaxies in the act of formation, which motivates the study of high-redshift galaxies. At $z > 2$, galaxy-mass halos are rare so the majority of galaxies we observe reside in dark matter halos that have only recently collapsed, i.e., at high redshift most galaxies are young. In this sense, $z > 2$ can be considered the epoch of galaxy formation.

2. Integral Constraints: Cosmological Quantities

Instead of studying galaxies as discrete objects residing in dark matter halos, one can track the cosmological quantities that comprise the baryon budget. Galaxy formation and evolution plays the fundamental role in the processing of baryons from neutral hydrogen to molecular gas to stars to metals. Star formation is inextricably linked with galaxy formation; whether you choose to define a galaxy as a large conglomeration of stars or an overdensity of baryons inside a collapsed dark matter halo, the galaxies in our universe form great numbers of stars. The cosmological quantities of interest provide integral constraints on star formation. The cosmic star formation rate density (SFRD) is an integral constraint averaged over the volume of the universe observable at a given redshift. The cosmic density of neutral gas, Ω_{gas}, the cosmic density of metals, Ω_Z, and the cosmic stellar mass density all provide integral constraints on the SFRD over time, as will be discussed below. The sum of the cosmic infrared background (CIB) and cosmic far-infrared background (FIRB) radiation provides an integral constraint on the SFRD from the Big Bang all the way to $z = 0$ by tracing the energy generated by nuclear reactions in stars.

2.1. Cosmic Density of Neutral Gas

The Damped Lyman α Absorption systems (DLAs, Wolfe et al. 1986) are quasar absorption line systems with H I column densities $\geq 2 \times 10^{20}$ cm^{-2}, sufficient to self-shield against the high-redshift ionizing background. Studying quasar absorption line spectra provides a (nearly) unbiased sample of lines-of-sight through the cosmos ideal for measuring cosmological quantities. The DLAs have been found to contain the majority of neutral hydrogen atoms at high redshift (see the recent review by Wolfe, Gawiser, & Prochaska 2005). Moreover, DLAs contain the vast majority of neutral gas, by which we mean neutral hydrogen and helium in regions that are sufficiently neutral to cool and participate in star formation, as lower column density systems are predominantly ionized. Hence the DLAs provide the reservoir of neutral gas that is available for star formation. In a simple closed box model, $d\rho_{gas}/dt = -d\rho_*/dt$, and the net decrease in the cosmic density of neutral gas from $z = 3$ to $z = 0$ is assumed to have all been turned into stars (see Fig. 5 of Wolfe et al. 2005). In that case, the DLAs appear to have formed about half of the stars seen in galaxies today. The truth is more complicated in hierarchical cosmology, where an open box model must be used;

$$\frac{d\rho_{gas}}{dt} = -\frac{d\rho_*}{dt} + \text{infall} + \text{merging} - \text{winds}. \quad (1)$$

Cosmological models for infall of gas from the intergalactic medium (IGM), merging of lower column density systems, and gas loss due to galactic winds are still quite uncertain, but the star formation rates actually measured for DLAs (Wolfe, Gawiser, & Prochaska 2003a; Wolfe, Prochaska, & Gawiser 2003b, Wolfe et al. 2004) imply that DLAs could have formed all present-day stars. Unfortunately, large uncertainties in the source and sink terms prevent us from using changes in the cosmic density of neutral gas as an integral constraint on the cosmic SFRD at the present time.

2.2.   Star Formation Rate Density

The cosmic star formation rate density has now been measured out to $z \simeq 6$ (Giavalisco et al. 2004). The high-redshift points are taken from only the Lyman break galaxies, and it is unclear how severe the resulting incompleteness is since we are not sure if all star-forming galaxy populations at these redshifts are known. The plot is traditionally shown in misleading units of $M_\odot Mpc^{-3} yr^{-1}$ versus redshift; in order to integrate-by-eye, one should plot this quantity versus time, and this has the effect of greatly increasing the apparent amount of star formation at low redshifts. Despite significant uncertainties in the SFRD at $z > 3$ due to incompleteness and large dust corrections, it appears that most stars in the present-day universe formed at $z < 2$ (see Fig. 33 of Pettini 2004).

2.3.   Stellar Mass Density

The cosmic stellar mass density provides an integral constraint on the SFRD, $\rho_*(t) = \int_0^t d\rho_*/dt$. See Dickinson et al. (2003) for a recent compilation. Note that the stellar masses of galaxies are not direct observables but are inferred from rest-frame optical (and near-infrared) photometry by modelling each object's star formation history using an assumed initial mass function (IMF).

2.4.   Cosmic Metal Enrichment History

The cosmic metal density is really a history of cosmic metal enrichment due to star formation, $\rho_{metals}(t) = 1/42 \int_0^t d\rho_*/dt$ (Pettini 2004). Wolfe et al. (2005, see their Fig. 7) show that the cosmic metallicity traced by DLAs rises gradually from a mean value of $[M/H] = -1.5$ at $z \simeq 4$ to a mean value of -0.7 at $z \simeq 1$. The range of observed DLA metallicities is somewhat higher than that of halo stars but overlaps, and is somewhat lower than that of thick disk stars and far lower than the near-solar values seen for thin disk stars in the Milky Way. The DLAs uniformly show greater metal enrichment than the Lyman α forest but less than values inferred for Lyman break galaxies or quasars at the same epoch (see Figs. 8, 32 of Pettini 2004, and see Leitherer 2005 for a review). The values given above are the cosmic mean metallicity of the neutral gas traced by DLAs, but they do not represent a full census of metals, which can also be found in heavily star-forming regions that have already used up their neutral gas or can be expelled by galactic winds into the IGM, which is predominantly ionized. It is therefore useful to compare the observed DLA metallicities with those expected from the DLA star formation rates; this leads to a factor of ten deficit in the observed metallicities called the "Missing Metals Problem" (Wolfe et al. 2005; Hopkins et al. 2005; Pettini 1999). The most likely explanation is that the star-forming regions of the galaxies seen as DLAs have superwinds sufficiently strong to move most of the metals produced into the IGM.

3. Theoretical Advances

Theoretical efforts to understand and model galaxy formation are mostly beyond the analytical realm, where they divide into semi-analytic models and cosmological simulations. These two approaches have been converging in recent years, as the practitioners of cosmological hydrodynamic simulations are using more detailed "recipes" for star formation, supernova feedback, and winds and in some cases have claimed grandiose results from purely N-body simulations with many semi-analytic recipes added (e.g. Springel et al. 2005). For examples of state-of-the-art cosmological hydrodynamic simulations of high-redshift galaxies and AGN, see Nagamine, Springel, & Hernquist (2004) and Di Matteo, Springel, & Hernquist (2005).

Semi-analytic models reproduce observations moderately well but have yet to demonstrate much success in predicting future observations, making them more of a tool for interpreting results than theoretical models in the classic sense. Somerville, Primack, & Faber (2001) tuned their models to reproduce galaxy properties at $z = 0$ and found one of their models to be in good agreement with the dust-corrected points at $z > 2$ in the cosmic SFRD diagram. However, as mentioned above, semi-analytic models for infall, merging, and winds are highly uncertain and it is not clear if observations of the cosmic density of neutral gas and the missing metals problem are consistent with the predictions. Similar scatter is seen in theoretical predictions of the cosmic stellar mass density.

4. Protogalaxy Demographics

DLAs dominate the neutral gas, making DLA-based studies appropriate for determining its cosmic density. However, other cosmological quantities should be summed over all high-redshift objects rather than just DLAs or just Lyman break galaxies, which trace the bright end of the high-redshift rest-UV galaxy luminosity function. Another motivation for studying all types of objects is the search for the progenitors of typical spiral galaxies like the Milky Way, which have not yet been pinpointed amongst the zoo of high-redshift galaxies. In designing the Multiwavelength Survey by Yale-Chile (MUSYC, Gawiser et al. 2005a, http://www.astro.yale.edu/MUSYC), it was decided to focus on selecting all known populations of galaxies at $z \simeq 3$, where most objects are young and several selection techniques overlap (see review by Stern & Spinrad 1999). The various populations at this epoch are labelled by three-letter acronyms (TLAs). We discuss each below.

4.1. Lyman Break Galaxies (LBGs)

The Lyman Break Galaxies (LBGs) are selected via the Lyman break at 912Å in the rest-frame. Higher-energy photons are unable to escape the galaxies or travel far in the IGM due to the large cross-section for absorption of ionizing photons by neutral hydrogen (for an illustration of the technique first successfully applied by Steidel & Hamilton 1992, see Fig. 19 of Pettini 2004). At $z \simeq 3$, the Lyman break generates a very red color in $U - V$, which could also be observed for an intrinsically red object such as an M dwarf or elliptical galaxy, leading to the additional requirement of a blue continuum color in e.g. $V - R$, consistent with the expected starburst nature of young galaxies. This makes the LBG technique insensitive to heavily dust-reddened or evolved stellar populations.

The selected population of galaxies is described in detail by Giavalisco (2002) and Steidel et al. (2003). Star formation rates range from 10–1000 $M_\odot$ yr^{-1} with a median value of $\sim$50 $M_\odot$ yr^{-1} after correction for reddening values ranging over $0 \lesssim E(B - V) \lesssim 0.4$ (Pettini 2004). Inferred stellar masses range over 6×10^8 $M_\odot \lesssim M_* \lesssim 10^{11}$ $M_\odot$ with median value 2×10^{10} $M_\odot$. Implied stellar ages range over 1 Myr $\lesssim t_* \lesssim 2$ Gyr with median age 500 Myr (Shapley et al. 2005). Observed qualities of LBGs are summarized in Tables 1 and 2 below, giving values for the space density, clustering length and dark matter halo masses from Adelberger et al. (2005), the SFR and stellar mass per object and stellar mass density from Shapley et al. (2001) and the cosmic SFRD from Steidel et al. (1999).

4.2.   Lyman Alpha Emitters (LAEs)

Starbursting galaxies can emit most of their ultraviolet luminosity in the Lyman α line. Because Lyman α photons are resonantly scattered in neutral hydrogen, even a small amount of dust will quench this emission. Hence, selecting objects with strong Lyman α emission lines is expected to reveal a set of objects in the early phases of rapid star formation. These could either be young objects in their first burst of star formation or evolved galaxies undergoing a starburst due to a recent merger. Selecting galaxies with strong emission lines also allows us to probe the high-redshift luminosity function dimmer than the "spectroscopic" continuum limit of magnitude $R = 25.5$ that is used to select the Steidel et al. LBG samples, since continuum detection is not necessary for spectroscopic confirmation using the emission line.

Observed qualities of the Lyman Alpha Emitting galaxies (LAEs) are summarized in Tables 1 and 2 below, giving values for the SFR per object from Hu, Cowie, & McMahon (1998) and the space density, SFRD, clustering length and dark matter halo masses from MUSYC (Gawiser et al. 2005b).

4.3.   Distant Red Galaxies (DRGs)

The inability of the Lyman break selection technique to find intrinsically red objects can be overcome by using observed NIR imaging to select high-redshift galaxies via their rest-frame Balmer/4000Å break. Looking for a continuum break in $J - K$ selects objects at $2 < z < 4$, labelled Distant Red Galaxies (DRGs) (Franx et al. 2003; van Dokkum et al. 2003). Reddy et al. (2005) offer a comparison of the redshift distributions of objects selected by LBG/star-forming colors, DRGs selected in $J - K$, and the passive evolution and star-forming samples selected through their BzK colors by Daddi et al. (2004). Note that this comparison is somewhat biased as the spectroscopic follow-up was performed on a sample originally selected only by the LBG/star-forming criteria. van Dokkum et al. (2005) report MUSYC results for an analogous comparison derived from a K-selected sample with inferred stellar masses $>10^{11}$ $M_\odot$.

Observed qualities of DRGs are summarized in Tables 1 and 2 below, giving values for the SFR and stellar mass per object from van Dokkum et al. (2004) and for the space density, SFRD, stellar mass density, clustering length and dark matter halo masses from MUSYC (Gawiser et al. 2006).

4.4. Sub-Millimeter Galaxies (SMGs)

The Sub-Millimeter Galaxies (SMGs) are selected using sub-millimeter bolometer arrays, e.g. SCUBA or MAMBO, which have poor spatial resolution, $\sim 15''$. Complementary high-resolution radio imaging is needed to obtain positions accurate enough to find optical counterparts or perform spectroscopy. This means that the SMGs with redshifts are really jointly selected in both sub-mm and radio. Observed qualities of SMGs are summarized in Tables 1 and 2 below, giving values for the space density from Chapman et al. (2003), the SFR per object and SFR density from Chapman et al. (2005), the clustering length from Webb et al. (2003) and the dark matter halo masses from MUSYC (Gawiser et al. 2006).

4.5. Damped Lyman α Absorption Systems (DLAs)

The Damped Lyman α Absorption systems (DLAs) were introduced above in § 2.1. Observed qualities of DLAs are summarized in Tables 1 and 2 below, giving the range of SFR per object for the two DLAs for which this quantity has been determined (Møller et al. 2002; Bunker 2004, see Wolfe et al. 2005 for a review). Also shown are the SFR density from Wolfe et al. (2003a) and the clustering length and dark matter halo masses determined by Cooke et al. (2006).

5. Clustering of Protogalaxies

It seems appropriate to provide a brief summary of the method used to generate the clustering lengths and inferred dark matter halo masses given in the Tables. The spatial correlation function $\xi(r) = (r/r_0)^{-\gamma}$ is inferred by fitting a power-law to either the observed spatial or angular correlation function of the sample. If only angular positions are observed, the redshift distribution $N(z)$ must be measured spectroscopically and used to invert the Limber equation as described in Giavalisco et al. (1998). The Landy-Szalay estimator is typically used to estimate the angular or spatial correlation function of the datapoints and to correct for the so-called "integral constraint" caused by measuring the mean density of the population from the observed survey volume (Landy & Szalay 1993). The LBG, LAE, and DRG samples are large enough to use the correlation length r_0 measured from the auto-correlation function to determine the bias factor, e.g., b_{LBG}, following

$$\xi_{LBG-LBG}(r) = (r/r_0)^{-\gamma} = b_{LBG}^2 \xi_{DM}(r), \tag{2}$$

where $\xi_{DM}(r)$ is the dark matter autocorrelation function predicted by the ΛCDM cosmology. The SMG and DLA samples are small, so their cross-correlation with the more numerous LBGs is used to determine their bias factor, e.g.

$$\xi_{DLA-LBG}(r) = (r/r_0)^{-\gamma} = b_{DLA} b_{LBG} \xi_{DM}(r). \tag{3}$$

The bias factor of each family of protogalaxies determines its typical dark matter halo mass following the method of Mo & White (1996), whose application to the cross-correlation function was first suggested by Gawiser et al. (2001). This method also allows one to predict the number abundance of dark matter halos with mass above the given threshold mass and to compare this with the observed number density of the population to infer the average halo occupation number.

Table 1.   The $z = 3$ universe. References for entries are given in the text, with a few entries still to be determined from MUSYC, ALMA, and JWST. Typical systematic uncertainties are a factor of two.

TLA	Space density $[h_{70}^3 \ \mathrm{Mpc}^{-3}]$	SFR per object $[h_{70}^{-2} \ \mathrm{M_\odot yr}^{-1}]$	Stellar mass per object $[h_{70}^{-2} \ \mathrm{M_\odot}]$	Clustering length (r_0) $[h_{70}^{-1} \ \mathrm{Mpc}]$
LBG	2×10^{-3}	30	10^{10}	6 ± 1
LAE	3×10^{-4}	6	MUSYC	4 ± 1
DRG	3×10^{-4}	200	2×10^{11}	9 ± 2
SMG	2×10^{-6}	1000	MUSYC	16 ± 7
DLA	ALMA	1–50	JWST	4 ± 2

Table 2.   Cosmological quantities. References for entries are given in the text, with a few entries still to be determined from MUSYC and JWST. Typical systematic uncertainties are a factor of three.

TLA	SFR density $[\mathrm{M_\odot yr}^{-1} h_{70} \ \mathrm{Mpc}^{-3}]$	Stellar mass density $[\mathrm{M_\odot} h_{70} \ \mathrm{Mpc}^{-3}]$	Dark matter halo mass $[\mathrm{M_\odot}]$
LBG	0.1	10^7	3×10^{11}
LAE	0.002	MUSYC	10^{11}
DRG	0.06	6×10^7	3×10^{12}
SMG	0.02	MUSYC	10^{13}
DLA	0.03	JWST	10^{11}

6.   Five Unsolved Problems in Galaxy Formation

1.   *What does a protogalaxy look like?* The term protogalaxy has been used loosely here and in the literature to describe young galaxies at high redshift. Part of the difficulty is that once an object has sufficient stars to be observed in rest-frame UV or optical radiation, we consider it a galaxy. But before this time it is either unobservable or only observable in absorption (e.g., DLAs), X-ray emission from a supermassive black hole (quasars/AGN), or in far-infrared radiation from dust which could be enshrouding either a powerful AGN or rapid star formation. If dark matter halo collapse, initial star formation and supermassive

black hole formation all occur simultaneously, the formation epoch of the galaxy is well-defined, and the picture is simple. But it is possible that many collapsed halos remain quiescent clouds of neutral gas until star formation is triggered by later mergers; these objects could comprise the half of DLAs that fail to show significant cooling in the [CII] 158 micron line (Wolfe et al. 2004). The distribution of lag times between dark matter halo collapse, supermassive black hole formation, and rapid star formation remains uncertain.

2. *When/how did each component of the Galaxy form?* Observations indicate that the thin disk formed at $z \simeq 2$, but simulations have trouble creating disk galaxies. One area now receiving attention is the manner in which angular momentum coupling between dark matter and baryons affects bar/disk formation and the cuspiness of bulges. It is still not clear if the globular clusters should be considered Galactic components or were all formed earlier and captured, despite evidence that some globular clusters are captured dwarf galaxies. Could globular clusters have formed in the same low-mass halos that met the Jeans threshold for collapse after recombination and hosted the Population III stars?

3. *When/how did galaxy sequences evolve?* HST observations of morphologies of galaxies at $z > 2$ imply that the Hubble sequence was not yet present. This is somewhat subtle, as cosmological surface brightness dimming would make a face-on spiral appear very different at high redshift, but most objects display irregular morphology and even the most promising edge-on disk candidates show spectroscopic kinematics inconsistent with the presence of disks (Erb et al. 2004). However, in the low-redshift ($z < 1$) universe we see a clear bimodality in the distribution of galaxy properties, the so-called red and blue sequences (e.g. Bell et al. 2004; Kannappan 2004). Such bimodalities are unlikely to arise from cosmological structure formation but are presumably caused by baryonic physics and appear directly linked to the "downsizing" behavior discussed in § 1.1.

4. *What role did feedback play?* The non-linear baryonic physics of star formation leads to highly energetic processes (ultraviolet radiation, stellar winds, supernova explosions) that can ionize or expel neutral gas that would otherwise participate in further star formation. It is now clear that the processes of galaxy and supermassive black hole formation are intimately connected, as evidenced by striking correlations between the masses of black holes and the velocity dispersions (or masses) of bulges in which they are embedded (Gebhardt et al. 2000; Ferrarese & Merritt 2000; Kormendy & Richstone 1995; Magorrian et al. 1998). Possible explanations include simultaneous hierarchical growth of galaxies and their central black holes through mergers (Haehnelt & Kauffmann 2000; Di Matteo et al. 2005), a strong coupling between black hole accretion and star formation in proto-disks at high redshift (e.g., Burkert & Silk 2001), and the effects of AGN feedback on the surrounding intergalactic medium (Scannapieco, Silk, & Bouwens 2005). One way or another, it appears that feedback from AGN, supernovae, and galactic winds must regulate the joint formation of the bulge and central black hole. Feedback may also play a role in determining the cuspiness of the dark matter halos, which does not appear consistent with profiles predicted from N-body simulations (Silk 2004). The galactic winds play a critical role in metal enrichment of the intergalactic medium and probably play a lesser role in ejecting neutral gas from the galaxies. As mentioned above, supernova feedback may explain the apparent minimum galaxy mass.

5. *When/how was the universe reionized?* A major area of ongoing investigation is the reionization epoch when the intergalactic medium was ionized. Slightly inconsistent results have been reported for the reionization redshift from *WMAP* observations of the temperature-E-mode cross-power-spectrum ($z_r = 20 \pm 9$, Bennett et al. 2003) and from the apparent end of reionization where the neutral hydrogen fraction dropped to 0.01 as seen in SDSS quasar spectra at $z \simeq 6.3$ (Fan et al. 2002). It seems premature to hypothesize bimodal models of reionization where separate families of sources produce the "early" ionization seen by *WMAP* and the completion of reionization seen by SDSS. Nonetheless, it is unclear at present which sources reionized the universe, and the leading candidates are the first generation of zero-metallicity stars (Population III) and starbursting galaxies including LBGs and LAEs. The quasars have very hard, ionizing spectra but were not numerous enough to reionize the universe at $z > 6$; they appear to dominate HeII reionization at $z \sim 3$. Significant uncertainties exist regarding the nature of the Population III stars: did they form in 10^6 M$_\odot$ dark matter halos that collapsed after recombination, or in larger galaxies later on? A top-heavy initial mass function is presumed for Population III, but what was the exact mass range and nature of stellar death? Did multiple stars occur per halo, or did the death of the first very massive star prevent further star formation or cause sufficient metal enrichment to generate Population II stars?

7.   Conclusions: Coming Attractions

The speakers have been asked to discuss major advances expected in the coming decade. For galaxy formation, I will go on record with three promising predictions and one slightly facetious warning.

The coming years will see the unification of galaxy formation and evolution. Until very recently, galaxy formation was studied at $z > 2.5$ and galaxy evolution was studied at $z < 1$ and the period $1 < z < 2.5$ was referred to as the "redshift desert." But technological advances in NIR imaging and spectroscopy have made the rest-frame Balmer/4000Å break and nearby emission lines available for study in distant galaxies. Development of these "needle-in-a-haystack" techniques now allows us to successfully find evolved galaxies at $z > 2$ even though these objects may be rare at those epochs. Hence we are beginning to study objects at $z \sim 3$ that formed at $z > 6$ which may turn out to be much easier than observing $z > 6$ galaxies directly. Imaging with the Spitzer satellite is enabling the first studies of rest-frame near-infrared emission from $z > 2$ galaxies, breaking degeneracies between age and dust. Deep imaging and slitless spectroscopy with the GALEX satellite are revealing the analogs of Lyman break galaxies at low redshift (Burgarella et al. 2005). These combined studies may make it possible to piece together a rough evolutionary sequence, e.g. DLA→LAE→LBG→SMG→DRG, that would form part of a *grand unified* model of high-redshift galaxies and AGN.

We will be able to study the interstellar medium in emission at high redshift. ALMA will enable studies of molecular gas in young galaxies through high-order CO lines. The [CII] 158 micron line, which dominates the cooling of the cold neutral medium phase at both low and high redshift, should be detectable for galaxies with large gas mass or rapid cooling equilibrating the heating due to starbursts. The current set of Early Universe Molecular Emission Line Galaxies consist mostly of quasars and are reviewed (and assigned the questionable TLA "EMG") by Solomon & Vanden Bout (2005). Both CO and [CII] have now been detected in $z > 6$ SDSS quasars, where they provide the best direct probes of the quasar host galaxies (Bertoldi et al. 2003; Walter et al. 2004; Maiolino et al. 2005). Detecting these lines and the sub-millimeter dust continuum from protogalaxies with ALMA will allow us to probe a multivariate mass function of gas mass, molecular mass, dust mass, and stellar mass. Even ALMA sensitivity may only allow detections of the tip of the gas-mass function, but this will provide a complementary set of objects to the tip of the rest-frame-UV and rest-frame-optical luminosity functions currently studied at high redshift, and much can be learned from the intersection and union of these samples.

High-redshift galaxies will be used to constrain dark energy properties. It has recently been shown (Seo & Eisenstein 2003; Linder 2003; Blake & Glazebrook 2003) that the scale of baryon acoustic oscillations provides a "standard rod" that can be measured in the clustering of high-redshift galaxies. The measurement will constrain the dark energy equation-of-state as a function of redshift, $w(z)$, via its influence on the expansion history of the universe. The measurement can be performed at any redshift where the line-of-sight starting at $z = 0$ is sufficiently influenced by the dark energy, making both $z = 1$ and $z = 3$ acceptable. Of order a million redshifts are needed, and the most likely surveys to accomplish this ambitious goal appear to be HETDEX using the VIRUS instrument under construction for HET and the wide-field multi-fiber spectrograph KAOS proposed for Gemini.

The rapidly increasing sophistication of studies of the high-redshift universe will generate even more jargon. We are already debating proper nomenclature for special categories of DLAs at lower column density (sub-DLAs) and those found in gamma-ray burst afterglows (burst-DLAs or bDLAs). Four-letter object acronyms (FLOAs?) are going to be part of the future.

Acknowledgments. In terms of organization, camaraderie, talks, and facilities, the 2005 Bash Symposium was a 5σ event, which is inconsistent with gaussian random initial conditions, thereby proving "intelligent design" by the organizing committees. I thank the organizers for inviting me to speak on my favorite topic and the editors for their hard work assembling this volume. I acknowledge valuable conversations with Pieter van Dokkum, Priya Natarajan, Jason Tumlinson and Meg Urry while outlining this talk. I thank the MUSYC Collaboration for allowing me to show results in preparation. This material is based upon work supported by the National Science Foundation under Grant No. AST-0201667, an NSF Astronomy and Astrophysics Postdoctoral Fellowship awarded to E.G.

References

Adelberger, K. L., Steidel, C. C., Pettini, M., Shapley, A. E., Reddy, N. A., & Erb, D. K. 2005, ApJ, 619, 697
Bell, E. F., Wolf, C., Meisenheimer, K., Rix, H.-W., Borch, A., Dye, S., Kleinheinrich, M., Wisotzki, L., & McIntosh, D. H. 2004, ApJ, 608, 752
Bell, E. F. et al. 2005, ApJ, 625, 23

Bennett, C. L. et al. 2003, ApJS, 148, 1

Bertoldi, F. et al. 2003, A&A, 409, L47

Blake, C. & Glazebrook, K. 2003, ApJ, 594, 665

Bunker, A. 2004, in Astrophysics and Space Science Library Volume 301, Multiwavelength Cosmology, ed. M. Plionis (Dordrecht, The Netherlands: Kluwer Academic Publishers), 67

Burgarella, D., Perez-Gonzalez, P., Buat, V., Takeuchi, T. T., Lauger, S., Rieke, G., & Ilbert, O. 2005, in The Fabulous Destiny of Galaxies: Bridging Past and Present, in press (astro-ph/0509388)

Burkert, A. & Silk, J. 2001, ApJ, 554, L151

Chapman, S. C., Blain, A. W., Ivison, R. J., & Smail, I. R. 2003, Nat, 422, 695

Chapman, S. C., Blain, A. W., Smail, I., & Ivison, R. J. 2005, ApJ, 622, 772

Cooke, J., Wolfe, A. M., Gawiser, E., & Prochaska, J. X. 2006, ApJ, 636, L9

Daddi, E., Cimatti, A., Renzini, A., Fontana, A., Mignoli, M., Pozzetti, L., Tozzi, P., & Zamorani, G. 2004, ApJ, 617, 746

Dekel, A. & Silk, J. 1986, ApJ, 303, 39

Di Matteo, T., Springel, V., & Hernquist, L. 2005, Nat, 433, 604

Dickinson, M., Papovich, C., Ferguson, H. C., & Budavári, T. 2003, ApJ, 587, 25

Eggen, O. J., Lynden-Bell, D., & Sandage, A. R. 1962, ApJ, 136, 748

Erb, D. K., Steidel, C. C., Shapley, A. E., Pettini, M., & Adelberger, K. L. 2004, ApJ, 612, 122

Fan, X., Narayanan, V. K., Strauss, M. A., White, R. L., Becker, R. H., Pentericci, L., & Rix, H.-W. 2002, AJ, 123, 1247

Ferrarese, L. & Merritt, D. 2000, ApJ, 539, L9

Franx, M. et al. 2003, ApJ, 587, L79

Gawiser, E., Wolfe, A. M., Prochaska, J. X., Lanzetta, K. M., Yahata, N., & Quirrenbach, A. 2001, ApJ, 562, 628

Gawiser, E. et al. 2005a, ApJS, in press, astro-ph/0509202

—. 2005b, submitted to ApJLetters

—. 2006, in preparation

Gebhardt, K. et al. 2000, ApJ, 539, L13

Giavalisco, M. 2002, ARA&A, 40, 579

Giavalisco, M., Steidel, C. C., Adelberger, K. L., Dickinson, M. E., Pettini, M., & Kellogg, M. 1998, ApJ, 503, 543

Giavalisco, M. et al. 2004, ApJ, 600, L93

Haehnelt, M. G. & Kauffmann, G. 2000, MNRAS, 318, L35

Hopkins, A. M., Rao, S. M., & Turnshek, D. A. 2005, ApJ, 630, 108

Hu, E. M., Cowie, L. L., & McMahon, R. G. 1998, ApJ, 502, L99

Kannappan, S. J. 2004, ApJ, 611, L89

Kormendy, J. & Richstone, D. 1995, ARA&A, 33, 581

Landy, S. D. & Szalay, A. S. 1993, ApJ, 412, 64

Leitherer, C. 2005, in IAU Symp. 228, From Lithium to Uranium: Elemental Tracers of Early Cosmic Evolution, ed. V. Cill, P. Francois & F. Primas (Cambridge: Cambridge University Press), 551

Linder, E. V. 2003, Phys.Rev.D, 68, 083504

Magorrian, J. et al. 1998, AJ, 115, 2285

Maiolino, R. et al. 2005, A&A, 440, L51

Mo, H. J. & White, S. D. M. 1996, MNRAS, 282, 347

Møller, P., Warren, S. J., Fall, S. M., Fynbo, J. U., & Jakobsen, P. 2002, ApJ, 574, 51

Nagamine, K., Springel, V., & Hernquist, L. 2004, MNRAS, 348, 421

Pettini, M. 1999, in Chemical Evolution from Zero to High Redshift, 233–+

Pettini, M. 2004, in Cosmochemistry: The Melting Pot of the Elements, XIII Canary Islands Winter School of Astrophysics, ed. C. Esteban, R. J. Garci Lopez, A. Herrero, & F. Sanchez (Cambridge: Cambridge University Press), 257

Reddy, N. A., Erb, D. K., Steidel, C. C., Shapley, A. E., Adelberger, K. L., & Pettini, M. 2005, ApJ, 633, 748

Scannapieco, E., Silk, J., & Bouwens, R. 2005, ApJ, 635, L13

Seo, H.-J. & Eisenstein, D. J. 2003, ApJ, 598, 720

Shapley, A. E., Steidel, C. C., Adelberger, K. L., Dickinson, M., Giavalisco, M., & Pettini, M. 2001, ApJ, 562, 95

Shapley, A. E., Steidel, C. C., Erb, D. K., Reddy, N. A., Adelberger, K. L., Pettini, M., Barmby, P., & Huang, J. 2005, ApJ, 626, 698

Silk, J. 2004, in AIP Conf. Proc. 743, The New Cosmology: Conference on Strings and Cosmology, ed. R. E. Allen, D. V. Nanopoulos, & C. N. Pope (New York: American Institute of Physics), 33

Smith, R. J. 2005, MNRAS, 359, 975

Solomon, P. M. & Vanden Bout, P. A. 2005, ARA&A, 43, 677

Somerville, R. S., Primack, J. R., & Faber, S. M. 2001, MNRAS, 320, 504

Springel, V., White, S. D. M., Jenkins, A., Frenk, C. S., Yoshida, N., Gao, L., Navarro, J., Thacker, R., Croton, D., Helly, J., Peacock, J. A., Cole, S., Thomas, P., Couchman, H., Evrard, A., Colberg, J., & Pearce, F. 2005, Nat, 435, 629

Steidel, C. C., Adelberger, K. L., Giavalisco, M., Dickinson, M., & Pettini, M. 1999, ApJ, 519, 1

Steidel, C. C., Adelberger, K. L., Shapley, A. E., Pettini, M., Dickinson, M., & Giavalisco, M. 2003, ApJ, 592, 728

Steidel, C. C. & Hamilton, D. 1992, AJ, 104, 941

Stern, D. & Spinrad, H. 1999, PASP, 111, 1475

Treu, T., Ellis, R. S., Liao, T. X., & van Dokkum, P. G. 2005, ApJ, 622, L5

van Dokkum, P. G. et al. 2003, ApJ, 587, L83

—. 2004, ApJ, 611, 703

—. 2005, submitted to ApJ

Walter, F., Carilli, C., Bertoldi, F., Menten, K., Cox, P., Lo, K. Y., Fan, X., & Strauss, M. A. 2004, ApJ, 615, L17

Webb, T. M. et al. 2003, ApJ, 582, 6

Wolfe, A. M., Gawiser, E., & Prochaska, J. X. 2003a, ApJ, 593, 235

—. 2005, ARA&A, 43, 861

Wolfe, A. M., Howk, J. C., Gawiser, E., Prochaska, J. X., & Lopez, S. 2004, ApJ, 615, 625

Wolfe, A. M., Prochaska, J. X., & Gawiser, E. 2003b, ApJ, 593, 215

Wolfe, A. M., Turnshek, D. A., Smith, H. E., & Cohen, R. D. 1986, ApJS, 61, 249

Matteo Viel presents a lively review of cosmology and the Lyman-α forest.

Frank N. Bash Symposium 2005: New Horizons in Astronomy
ASP Conference Series, Vol. 352, 2006
S. J. Kannappan, S. Redfield, J. E. Kessler-Silacci, M. Landriau, and N. Drory

The Lyman-α Forest As a Cosmological Probe

Matteo Viel[1]

Institute of Astronomy, University of Cambridge, Cambridge, UK

Abstract. The Lyman-α forest consists of the absorptions produced by intervening neutral hydrogen along lines of sight to distant quasars. About 80% percent of the baryons at $z > 2$ are believed to reside in the filamentary structures probed by the absorptions. These intergalactic structures trace the underlying dark matter density field at scales and redshifts which cannot be probed by any other observable. After describing the essential physical aspects and a brief historical introduction, I will describe the first analytical models of the Lyman-α forest in the framework of cold dark matter scenarios. Then, I will focus on possible ways of extracting cosmological parameters from a set of observed quasar spectra, by running a grid of cosmological hydro-dynamical simulations. In particular, I will recover the linear dark matter power spectrum at $z > 2$ and at scales of ∼1–40 comoving Mpc. I will address the significance of the results obtained especially when combined with the larger scales measurement of the power spectrum made by WMAP, giving constraints on the power spectrum amplitude, spectral index and its running. I will critically compare all the results obtained with those of the SDSS collaboration, based on a set of more than 3000 quasars at low resolution. Several physical aspects, which could affect the constraints on cosmological parameters, will be briefly discussed: feedback effects in the form of galactic winds, metal enrichment, the thermal state of the intergalactic medium and the amplitude and nature of the ultraviolet background. Finally, I will address further improvements that could be achieved in the next few years in this field both under the observational and the theoretical sides. In particular, the perspectives of measuring the cosmic expansion using absorption lines.

1. Brief Historical Introduction and Essential Physical Aspects

In this Section, I will discuss the physics of the absorption deriving a relation between the observed flux and some astrophysical and cosmological parameters.

The Lyman-α forest is produced by the absorption of the continuum of a distant source (typically a quasar) caused by the neutral hydrogen present along the line-of-sight that connects the observer to the source. It is observed in the optical and UV bands in quasar (QSO) spectra and in the redshift interval from $z = 1.5$ up to $z \sim 6.3$. The absorption is caused by the redshifted Lyman-α resonant line at an observed wavelength $\lambda = \lambda_0(1 + z)$, where $\lambda_0 = 1215.67$Å. Even if the amount of neutral hydrogen present in the universe after reionization is very small (one part in 10^5) the probability of the transition is high and this absorption mechanism is the most sensitive probe of baryons at $z > 2$. In fact,

---

[1]Particle Physics Astronomy Research Council (PPARC) Fellow.

recent estimates based on simple semi-analytical models that I will discuss below predict that 60–80% of the baryons in the universe at $z = 3$ are in the Lyman-α forest (Rauch 1998; Weinberg et al. 1999).

The observational discovery of the Lyman-α forest can be traced back to Lynds in 1971 (Lynds 1971) who identified several absorption lines in a QSO spectra blueward of Lyman-α emission. However, the existence of an intergalactic medium (IGM) with a high degree of ionization was postulated in 1965 by Bahcall & Salpeter (1965), Gunn & Peterson (1965) and Scheuer (1965) who calculated upper limits on the amount of neutral hydrogen in the universe. In particular, in Bahcall & Salpeter (1965) the gas responsible for the absorption lines was thought to be associated to groups of galaxies: this model was afterwards quickly abandoned since it failed to reproduce the correct number of absorbers. The first generation of models were based on a sharp distinction between the medium that produces the absorption, the IGM or ICM (inter-cloud medium) that confined the clouds either with pressure effect (Sargent 1982) or gravitationally with dark matter mini-haloes (Rees 1986). Only from the early 90s some theories that aimed at describing the physics of Lyman-α clouds in the context of hierarchical models of structure formation (cold dark matter (CDM) models) succeeded in reproducing many observed properties of the Lyman-α forest. These models did not assume any more a two-phase medium but instead were based on the dynamical evolution of a *unique medium*, the IGM, with small fluctuations around the mean (e.g., Oort 1971).

The basic idea is very simple: at high redshift, gas and dark matter evolve together under the influence of gravity. The situation changes at later times with the formation of bound structures: the dark matter, not collisional, can collapse on any scale, while the gas, having pressure, produces shocks and heat converting the kinetic energy in thermal energy. Both dynamical processes, that describe the growth of structure, and thermal ones, that describe more closely the physics of absorption are thereby important in modelling the Lyman-α forest.

The main ingredients of these models (Bi 1993; Bi & Davidsen 1997) can be summarized as follows.

Dynamic evolution. The growth of cosmic structures is specified by the cosmological model chosen and in particular by the power spectrum of matter density fluctuations $P(k)$ and by the linear theory of density perturbations. More precisely one needs: *i)* Models of hierarchical structure formation based on CDM and linear Jeans theory of growth of density perturbations for baryons and DM. These models should take into account the effect of baryon pressure below a given scale that is related to the Jeans length $\lambda_J = c_s \sqrt{(\pi/G\rho)}$; *ii)* Non-linear evolution of density perturbations to better describe the dark matter and baryon density field at $z \sim 3$ using different models such as the lognormal model of Bi & Davidsen (1997) or methods based on N-body or hydro-simulations (Viel et al. 2002a,b).

Thermal and chemical evolution. Different thermal and chemical processes are at work in the IGM (e.g., Hui & Gnedin 1997): *i)* adiabatic cooling due to the expansion of the universe, the overall expansion drives the a temperature fall-off as $(1+z)^2$ as z decreases; *ii)* photoionization heating due to the UV background; *iii)* Compton cooling which consists of the scattering of free electrons with the (colder) Cosmic Microwave Background (CMB) photons (this process is important at high redshifts $z > 9$); *iv)* recombination cooling due to the re-

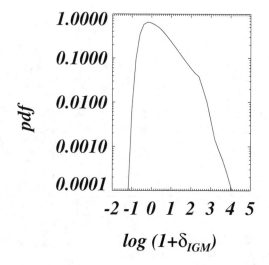

Figure 1.   Probability distribution function (mass-weighted) of the gas overdensity at $z = 3$ as extracted from a hydro-dynamical simulation for a "standard" ΛCDM model.

combination of protons and electrons. The basic assumption is that the IGM
is in photoionization equilibrium with the UV background. This determines a
power-law equation of state for the temperature-density relation which reads:
$T(z) = T_0(z)(1 + \delta_{IGM})^{\gamma(z)-1}$.

In Figure 1, I plot the probability distribution function (pdf) of the gas distribution as extracted from an hydro-dynamical simulation at $z = 3$. One can see
that most of the gas is at densities around the mean density, as postulated by
the models of Oort (1971) and by the lognormal approximation of Bi & Davidsen
(1997) and Viel et al. (2002a,b). In fact, the pdf is very close to being lognormal. As shown in Bi & Davidsen (1997) the evolution of the lognormal pdf with
redshift shows that at $z \sim 3$ about 30% of the mass of the IGM is in regions that
are underdense and that fill about 70% of the volume of the universe, [1] while at
$z = 0$ due to structure formation this volume goes up to 90% (10% of the total
mass).

In Figure 2, I reproduce the gas density-temperature relation of the IGM
at $z \sim 3$. In particular, we can see the three phases of the IGM. The phase
responsible for Lyman-α absorption is described by a power-law relation set by
the balance between adiabatic cooling and photoheating; for temperature colder
than this the cooling time is shorter than the Hubble time. Then, we have a
phase in which the gas becomes hotter due to shocks: in this region the Hubble
time is shorter than the cooling time and the gas cannot cool. Finally, there is a

---

[1]At this z clouds at the mean density typically produce absorptions of $10^{13.5}$ cm^{-2}, which can
be seen as an average Lyman-α absorption. While clouds with $\rho = 10 < \rho >$ produce $10^{14.3}$
cm^{-2} absorptions (see Schaye 2001).

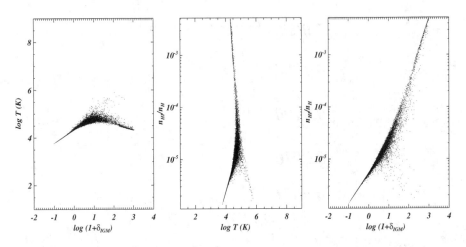

Figure 2.   *Left:* gas temperature-density relation at $z = 3$ from a hydro-dynamical simulation. One can see the three phases: the tight relation of the gas responsible for Lyman-α absorption at low densities; the shocked gas at $T > 10^5$ K; the colder phase of gas which is falling in the potential wells of dark matter haloes. *Middle:* Ionization fraction (n_{HI}/n_H) vs. temperature; *Right:* Ionization fraction vs. gas (IGM) overdensity. Only a random 10% subset of the particle distribution is plotted here. The original simulation is for a standard ΛCDM model with 2×200^3 gas+DM particles.

tail at high densities that is produced by the shock-heated gas that is cold and dense enough to cool down in the potential wells of dark matter haloes (mainly due to line cooling and Bremsstrahlung).

The ionization fraction as a function of temperature and density is shown in the middle and right panel of Figure 2, respectively. We note that denser regions usually contain more neutral hydrogen, in fact this quantity is $\propto \rho_{IGM}^2$, but if these regions are hot enough then the ionization fraction drops (middle panel). This is due to the fact that the recombination coefficient $\alpha_{rec} \propto T^{-0.7}$ (see Hui & Gnedin 1997).

Having outlined the above dynamical and thermal/chemical features of the IGM we can now, in a relatively easy way, sketch an expression for the observed flux. The neutral hydrogen density at z and in $\mathbf{x}$ reads:

$$n_{HI}(\mathbf{x}, z) \sim 10^{-5}\bar{n}_{IGM}(z)\frac{\Omega_b h^2}{0.019}\frac{0.5}{\Gamma}\left(\frac{T(\mathbf{x}, z)}{10^4 K}\right)^{-0.7}\left(\frac{1+z}{4}\right)^3 (1 + \delta_{IGM}(\mathbf{x}, z))^2 \tag{1}$$

with Γ the hydrogen ionization rate; while the HI optical depth at velocity v is:

$$\tau_{HI}(v) = \frac{\sigma_{0\alpha}c}{H(z)}\int dy \, n_{HI}(y)V(v_{pec}, b_{param}), \tag{2}$$

with $\sigma_{0\alpha} = 4.45\text{e-}18$ cm^2 the Lyman-α cross-section, c the speed of light and $H(z)$ the Hubble parameter at z, b_{param} the Doppler parameter and V is the intrinsic shape of the line, the Voigt profile, which depends on the peculiar velocity as well (e.g. Bechtold (2002)). Note that in the above expression the integral is made in velocity space. Finally, the transmitted flux is $F = \exp\left(-\tau_{HI}\right)$.

The idea that is behind the use of the Lyman-α forest as a cosmological probe is to recover with suitable methods the dark matter density field from the observed flux. It is important to "deconvolve" the thermal information from the dynamical one in order to have insights on the structure formation mechanism. Statistical properties of the flux, such as the flux power spectrum, will then be closely related to the matter power spectrum of density fluctuations.

The results obtained from the semi-analytical models described in this section were afterwards confirmed by the more refined hydro-dynamical simulations that showed in a very convincing way how the Lyman-α absorptions were better described by continuous temperature and density fields rather than discrete absorption lines. With the results of Cen et al. (1994); Hernquist et al. (1996); Theuns et al. (1998), the forest became a fundamental cosmological tool: the clouds were not considered any more as isolated objects with dimensions comparable to the Jeans length but were better described by a continuous density field that traced the underlying dark matter distribution in a simple manner.

2. Cosmological Significance of the Lyman-α Forest

In this section I will outline some of the most important aspects that characterize the use of the forest as a cosmological probe.

- **Physics of reionization** The forest is sensitive to the presence of neutral hydrogen, thereby the study of absorptions at high redshift places tight limits on the degree of ionization of the IGM. QSOs at $z \sim 6$ show almost no transmitted flux blueward of Lyman-α emission probably indicating that we are approaching the epoch of reionization. These results are however still controversial and are maybe in contrast with the high redshift of reionization as inferred from WMAP (Spergel et al. 2003; Fan et al. 2004; Furlanetto & Oh 2005; Songaila 2004).

- **Thermal state of the IGM** The study of absorption lines and of their profiles allows us to measure the parameters T_0 and γ that describe the thermal state of the IGM. Given the relationships between gas density and gas temperature and the definition of Doppler parameter it is easy to see that one can recover a relation in the $b - N$ plane (N is the column density) whose envelope constrains the thermal state. This analysis has been carried by Ricotti et al. (2000); Schaye et al. (2000); McDonald et al. (2001).

- **Metal enrichment of the IGM** The forest is contaminated by metals even in underdense regions and presents and average metallicity of $Z \sim 10^{-3} Z_\odot$. Moreover, there are systems whose column density is very high. The analysis of strong absorption systems, produced by galaxies along the line-of-sight, and weaker ones in the more general IGM can allow us to discriminate between enrichment processes that take place at high or low redshift (Schaye et al. 2003; Adelberger et al. 2005).

- **Evolution of the UV background** As we saw in the previous section the neutral hydrogen abundance depends on the metagalactic hydrogen ionization rate Γ. It is thus possible to constrain this value and get insights on the nature and abundance of the sources that contribute to the background. In order to do that it is necessary to calibrate the mean flux using the observed spectra and to use hydro-dynamical simulations with different thermal histories (e.g., Bolton et al. 2005).

- **Cosmological parameters** The forest offers the possibility of constraining the matter power spectrum at scales and redshifts that are not probed by other observables. I will expand this last point below.

3. Pre-WMAP Lyman-α Cosmology

If we consider the flux as a continuous field (as we showed above this seems to be the most correct interpretation of the forest), the idea is to determine the biasing function that relates flux and matter:

$$P_F(k) = b^2(k) P(k). \tag{3}$$

The method was pioneered by Croft et al. (1998, 2002) and is very simple. We rely on a grid of simulations of the forest exploring, as widely as possible, the astrophysical and cosmological parameter space. Then, we find out which is the simulation that best fits the observed flux power spectrum and we compute the function $b^2(k)$ for this simulation. This can be done since we know the matter power spectrum for any cosmological model. At this point we simply divide the observed flux power by this function to get an estimate of the *observed matter power spectrum*. This method was critically discussed by Gnedin & Hamilton (2002) who found it to be robust at least in the cases for which the error bars on the observed 3D flux power were of the order of 20% (as in the Croft et al. 1998, 2002, analysis). The main uncertainty in this framework is in the accuracy of the simulations that are used to model the flux power. Croft et al. (1998) used DM simulations only, Croft et al. (2002) used a few small box-size hydro-dynamical ones, Gnedin & Hamilton (2002) instead analysed Croft et al. results with a grid of HPM (Hydro-Particle-Mesh) simulations: an approximate and faster way to model the IGM distribution (see Gnedin & Hui (1998); Viel, Haehnelt & Springel (2005)). All the physical effects that have an impact on the flux power, such as peculiar velocities, thermal histories, mean flux, UV background evolution, impact of strong absorption systems etc., are implicitly taken into account in the function $b(k)$.

With this method one can constrain the amplitude, shape and curvature of the matter power spectrum. For example σ_8 (the r.m.s. fluctuation amplitude on a scale of 8 Mpc/h) is a measure of its amplitude; the spectral index n, that parametrises the power spectrum $P(k) = AT^2(k)k^n$ (with A normalization factor and $T(k)$ the transfer function), is a measure of the shape; and the running of a spectral index $n_{run} = dn/d\ln k$ calculated at $k_0 = 0.05$ Mpc^{-1} quantifies the curvature. The information contained in the forest is complementary to that of the CMB: it is on smaller scales and at smaller redshifts. When the two estimates are combined together the range of wavenumber used increases significantly and the error bars on the recovered cosmological parameters halve.

The results obtained by Croft et al. (2002) suggested in a very convincing way that the power spectrum of the density fluctuations derived from the Lyman-α forest was very well fitted by a model with $\sigma_8 = 0.7$ and a large "tilt" in the power spectrum, i.e. a value $n = 0.93$.

When the results of the WMAP team came out (Spergel et al. 2003; Verde et al. 2003), it was evident that when combining the Lyman-α forest data as inferred by Croft et al. (2002) some discrepancies from a "standard" ΛCDM model started to appear. In particular, there was an evidence for a spectral index smaller than one ($n = 0.93 \pm 0.03$) and a 2σ evidence for a running spectral index $n_{run} = -0.031 \pm 0.016$ and a value of $\sigma_8 = 0.80 \pm 0.05$, somewhat smaller than that inferred by the CMB alone ($\sigma_8 = 0.9 \pm 0.1$). These results have had fundamental implications for the standard scenario of structure formation. For example, with these "new" values there is a suppression of power on small scales that delays the formation of dark matter haloes (Yoshida et al. 2003). Moreover, very tight limits on inflationary parameters/models can be set (Peiris et al. 2003) and on the amount of gravitational waves produced in the primordial universe (Spergel et al. 2003).

4. Post-WMAP Lyman-α Cosmology

Two groups in the last two years have re-analysed these results using very different data sets and theoretical models.

- **Data:** The first group (Kim et al. 2004) based their analysis on 27 high resolution ($R \sim 50000$) high signal-to-noise (~ 50) obtained with the UVES spectrograph at a median redshift $z \sim 2.1$. This sample is the LUQAS (Large Uves Quasar Absorption Spectra) sample. The second group (McDonald et al. 2005a) used ~ 3000 SDSS (Sloan Digital Sky Survey) spectra which have $R \sim 1000$ and signal-to-noise ~ 5. However, this latter data set is extremely useful to probe the growth of cosmic structures since it spans the redshift range $z = 2.4 - 4.2$.

- **Numerical simulations:** The first group (Viel, Haehnelt & Springel 2004) used a grid of high-resolution large box sizes fully hydrodynamical simulations run with the GADGET-II code (Springel 2005) trying to model the biasing function to use the method of Croft et al. (2002). The advantage of this approach consists in the reliability of the hydro code used but the drawbacks are severe since it is very time consuming to explore the parameter space widely with these kind of simulations. The second group (McDonald et al. 2005b) used the HPM method to explore the parameter space (~ 34 astrophysical and cosmological parameters), since HPM simulations are significantly faster than fully hydro ones. This method is intrinsically less precise and the simulated flux power needs to be calibrated with that of full hydro-dynamical runs. Another difference between the two groups is in the fact that LUQAS decided to use the Croft et al. method which is based on the 3D flux power spectrum and allows the recovery of the matter power spectrum as a function of the wavenumber. The SDSS instead uses the 1D flux power and does a forward modelling of it without trying to invert the flux-matter relation: this prevents a re-

	LUQAS	SDSS
DATA	27 high res z~2.1 QSO spectra	3000 QSO spectra z>2.2
THEORY: SIMULATIONS	full hydro simulations	HPM `calibrated' sims
THEORY: METHOD	3D flux Inversion	1D flux fwd modelling
ADVANTAGES	band power	large redsh.range
DRAWBACKS	only z=2.1 (and 2.72) and larger error bars	34 parameters modelling
RESULTS	no running, scale invariant, `high' power spectrum amplitude	

Figure 3.   Schematic comparison between LUQAS (Viel, Haehnelt & Springel 2004; Kim et al. 2004) and SDSS (McDonald et al. 2005a; Seljak et al. 2005).

covery of the power spectrum in bandpowers but gives smaller error bars, since the 1D flux power has smaller error bars then the 3D one (which is obtained by differentiating the 1D).

- **Results:** Even if the methods and data used are very different the two groups came to very similar conclusions: *i)* "high" values of the power spectrum amplitude: $\sigma_8 = 0.94 \pm 0.08$; *ii)* scale-invariant power spectrum $n = 0.99 \pm 0.03$; *iii)* no or little running of the spectral index $n_{run} = -0.033 \pm 0.025$. The above results are those found by Viel, Weller & Haehnelt (2004) that agree perfectly with those found subsequently by Seljak et al. (2005) (these latter have smaller error bars given the fact that for the final estimates many other observables have been combined and the actual errors on the flux power are also smaller).

Both these two groups showed that the earlier results obtained by the WMAP team when they had combined the CMB data and the forest data were underestimating the systematic errors and in particular the mean flux calibration as we will see below (Seljak et al. 2003; Viel, Haehnelt & Springel 2004).

In Figure 3, I give a schematic representation of the main differences between the two groups.

In Figure 4, I give a sample of the 2D likelihood contours that I get from combining the WMAP data with the LUQAS and Croft et al. (2002) samples. The grayscale contours represent the mean likelihoods while continuous lines are for the marginalized ones. The error bars that can be obtained by adding the forest approximately decreases by a factor 2 compared from the CMB alone. Some degeneracies in the planes are clearly visible such as the one in the $n - \sigma_8$ plane and those concerning the optical depth/reionization redshift.

4.1. Observational and Theoretical Systematics

The **mean flux** is a fundamental observed quantity that is used in order to calibrate the simulated spectra. The physical quantity which is better related to the growth of structure is not the flux F itself but $F/ < F >$. Thereby, it is crucial to get an accurate estimate of $< F >$ as a function of redshift. A more common way to present the above quantity is with the effective optical depth $\tau_{eff} = -\ln < F >$. In Figure 5 I plot several determinations of this quantity as found in the literature from Kim et al. (2002); Schaye et al. (2003); Press et al. (1993) and the LUQAS estimates. There is a systematic difference between the values inferred from low resolution spectra, such as those of Press et al. and the high resolution ones. It is now well established that the correct values are those inferred from high resolution spectra. Even if the difference between the two estimates is tiny (of the order of 10% or so) this translates into a big difference in terms of the inferred matter power spectrum. The WMAP team took the power spectrum inferred by Croft et al. (2002) for a value of the effective optical depth given by the low resolution spectra. If they had taken the power as inferred from the high resolution values of the optical depth they would have inferred a higher power spectrum amplitude ($\sigma_8 \sim 0.9$ instead of $\sigma_8 \sim 0.7$) and no evidence for a tilt or a running of the spectral index. The calibration of the mean flux level is thus the main systematic error.

Other sources of systematic errors are the uncertainties in the poorly known **thermal history of the IGM**. Since the error bars on the SDSS flux power are of the order of the percent, the thermal history — at least for interpreting this data set — should be modelled at this level. Measurements obtained from different methods discussed in the previous section give the following constraints on the amplitude and slope of the temperature-density relation: at $z = 3$, $\gamma = 1.2 \pm 0.3$ and $T_0 = 24000 \pm 8000$ K (Ricotti et al. 2000). These error are still large and different 1D flux power spectra that fit the above typically differ between each other by a quantity between a few percent up to 10% in the wavenumber range of interest (Viel & Haehnelt 2006). For example, if we take a look at the bias function $b(k)$ the uncertainties due to different values of T_0 translate into a 6% error on the power spectrum amplitude, while the error on γ gives an 8% contribution. These effects are of course degenerate with the mean flux and a study of the correlations of these parameters is fundamental.

Another potentially severe source of systematic errors is the different scalings given by **different simulations**. This was investigated in Viel, Haehnelt & Springel (2004); McDonald et al. (2005b). The best approach is of course to investigate the properties of the IGM by using full hydro-dynamical simulations. However, at present, this is not feasible due to limitations in the CPU time and in the data disk storage. In Viel, Haehnelt & Springel (2004), it was found that the bias function strongly depends on the kind of simulation used and part of the discrepancy with the earlier Croft et al. (2002) results was due to the different scalings obtained. However, in the final estimates the LUQAS group

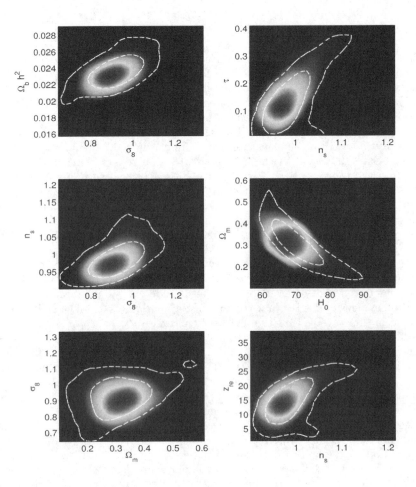

Figure 4.   2D likelihoods contours obtained from a combined analysis of WMAP data (Spergel et al. 2003) and the power spectrum inferred from the Lyman-α forest by Viel, Haehnelt & Springel (2004). The grayscale contours refer to the mean likelihood, while the lines refer to the marginalized likelihood. z_{re} is the reionization redshift, τ the optical depth, Ω_b the baryon fraction (other parameters are defined in the text).

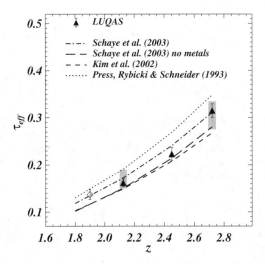

Figure 5.   Different values of the effective optical depth from the literature. Estimates obtained from high and low resolution spectra can significantly bias the recovered cosmological parameters.

deliberately chose a conservative approach and allow for all the different scalings of the literature to contribute to the final error bar.

In the rest of this subsection I would like to briefly address two (numerical) aspects that will need some improvements in the near future. First, as mentioned above, the **HPM method** used by the SDSS is far from being accurate. In this method, it is assumed that *all* the gas that gives rise to Lyman-α absorption obeys the temperature-density relation $T = T_0(1 + \delta)^{\gamma-1}$. Unfortunately, while this approximation is good at high redshifts it clearly breaks down at later stages when non linear evolution of the cosmic structures increases the amount of shock-heated gas. As shown in Viel, Haehnelt & Springel (2005) the discrepancies in terms of flux power at $z = 2$ could be of the order of 20%. Even if the HPM simulations are calibrated against full hydro ones, the box sizes and resolution of the latter should be appropriate since the amount of shock-heated gas is very likely to depend on these two quantities. Secondly, the impact of **galactic winds** on the forest at high redshift should be addressed with better simulations. This has been done recently by many groups (e.g., Theuns et al. 2002; Kollmeier et al. 2005) and the impact on the simulated flux power is likely to be very small. However, as we will discuss in the next section, so far no simulation can successfully reproduce the observed metallicity pattern in the forest suggesting that at least some physical aspects are poorly understood.

5.   Future Perspectives

In this section I will briefly address what could be the most important achieve-
ments on the theoretical and observational sides in the Lyman-α forest community in the next few years.

5.1.   Observations

At the present day the amount of data available to the community is huge.
Not only there are the 3000 spectra of the SDSS that sample a large interval in
redshift but the scientific community has about 200 spectra at high and medium
resolution (see also Tytler et al. 2004). Since the data sets are so different it will
be very important to quantify some systematic discrepancies in the raw data. A
re-observation at low resolution of high resolution spectra will be fundamental
in improving the study of the forest. Another area in which we will have some
improvement will be the understanding of the quasar continuum that can allow
the recovery of the power spectrum at larger scales than those that are currently
probed.

The accurate estimate of the mean flux level will be crucial in reducing the
most important systematic error. Presently the error bar is of the order of 7–8%
but I believe that in a couple of years this error can be lowered below 2%.

Another fundamental problem is the precise measurement of the level of metal
contamination in the forest and the impact on the flux power. At smaller scales
the effect is very big (see Kim et al. 2004) and better measurements could allow
us to extend the region used to recover cosmological parameters at large k getting
better constraints even on astrophysical parameters.

The poorly known thermal history of the IGM should be better constrained
in the near future maybe by using the bispectrum statistics (e.g., Viel et al.
2004) or some other clever way to measure the temperature from absorption
lines. This will improve the estimates on the cosmological parameters.

5.2.   Simulations

I think that the most significant results in the next few years will be achieved
on the theoretical side. At the present time it is very difficult to model the
flux power spectrum to the percent level taking into account all the possible
effects that can have an impact on this quantity. The conservative approach
that is used most of the time is to simulate some physical effects with approximate methods. In the next few years it will be possible to implement "new"
physical processes into the hydro-dynamical simulations and to run large grids
of simulations able to explore the parameter space. In particular, I think that
the major progresses will be the following: *i)* simulations of metal enrichment
with galactic winds and precise estimates of the level of metal contamination in
the Lyman-α forest can be made; *ii)* radiative transfer effects which can cause
UV fluctuations and temperature fluctuations in the IGM able to influence the
flux power; *iii)* accurate simulations of "non-standard" cosmological models in
which the dark matter is made of warm particles such as gravitinos or sterile
neutrinos or has a contribution of neutrinos as addressed, for example, by Viel
et al. (2005); Seljak et al. (2005). All three points will be crucial in exploring the
range of allowed wavenumbers k especially at smaller scales (< 2 Mpc) where the

uncertainties due to the thermal state and to the metal contamination prevent a reliable recovery of the matter power spectrum.

5.3. Measuring the Cosmic Expansion with QSO Absorption Lines

Absorption lines can be used to probe the expansion of the universe directly. The measurements consists in quantifying the change of cosmological redshifts with time (Sandage 1962; Loeb 1998). It is easy to see that the derivative of redshift w.r.t. time is:

$$\frac{dz}{dt} = (1+z)H_0 - H(t_e) \qquad (4)$$

with H_0 the value of the Hubble constant at the present time and $H(t_e)$ the value at the time the photon was emitted. For a Friedmann-Robertson-Walker universe with no curvature and neglecting the radiation energy density we get:

$$\Delta v \sim 31h(1 - E(z))\frac{\Delta t}{10yr}\text{cm/s} \qquad (5)$$

with $h = H_0/(100\text{km/s/Mpc})$ and $E(z) = (1+z)^{-1} \cdot [\Omega_{matter}(1+z)^3 + \Omega_{DE}(1+z)^{(3+3w)}]^{0.5}$. The wavelength shift corresponds to a Doppler shift of about 1–10 cm/s over a period of 10 yrs. Measuring the wavelength shift requires about a factor 10–100 times improved wavelength calibration compared to that currently achieved with an instrument like HARPS at the ESO 3.6m. For an Einstein-de-Sitter universe the redshift of an object at fixed coordinate distance decreases with time for any z of the object. However, for flat models with a contribution of dark energy to the total energy density the redshift increases for objects at low redshift while it decreases for objects at high redshifts. This feature is the most direct proof of the accelerating effect of the dark energy. The experiment CODEX (CosmicDynamicsEXperiment) aims at detecting this spectral feature change using Lyman-α forest absorption lines by observing bright QSOs in the sky with the new (60m?) ESO telescope OWL. A concept study for this project has been carried out recently to find that the accuracy (r.m.s. values) that needs to be achieved in order to measure the velocity shift is the following:

$$\sigma_v = 2\left(\frac{S/N}{2350}\right)^{-1}\left(\frac{N_{QSO}}{30}\right)^{-1/2}\left(\frac{1+z_{QSO}}{5}\right)^{-1.8}\text{cm/s}. \qquad (6)$$

The scaling relations are easy to understand. The higher the S/N the better the measurements, the number of QSO as well contributes to the final signal, while the dependency on the redshift is due to the fact that distant sources will in general produce bigger shifts up to a given redshift above which the QSO spectra will be almost everywhere saturated. The resolution R at 5000Å needs to be higher than 30000 but above this value the σ_v saturates and will not benefit more from an even higher resolution spectrograph (above this level the spectral features are well resolved and this is the thing that matters). Peculiar velocities (acceleration) could be the most severe source of systematic errors. Typical peculiar velocities of the gas responsible for the Lyman-α forest with optical depth of order unity are ~ 100 km/s. The characteristic timescale on which these peculiar velocities change is of the order of the dynamical timescale of the absorbing gas which is usually larger than 10^9 yr. Thus, a simple calculation shows that the effect is more than one order of magnitude smaller than the signal we are interested in.

6.   Conclusion

In this talk I discussed the cosmological significance of the Lyman-α forest. I focussed on the current interpretation of the forest: a continuous field of absorption features that traces in a relatively simple way the underlying dark matter distribution. I described the essential physical aspects of the absorption mechanism, stressing the fact that, especially at large scales, the flux power spectrum can allow the measurement of the dark matter power spectrum. In particular, the amplitude, shape and curvature of the power spectrum can be constrained at scales of the order of ~ 10 Mpc that cannot be probed by other observables. This measurement is done by using accurate hydro or pseudo-hydro dynamical simulations. When the recovered data are combined with the larger scales estimates of the CMB data, the constraints on the parameters become tighter. I showed that in the last two years two groups analysed Lyman-α forest spectra using two very different analysis techniques and very different data sets. Nevertheless, the conclusions in terms of cosmological parameters are very similar: the model that best fits the data has a relatively high power spectrum amplitude, $\sigma_8 \sim 0.93$; a spectral index close to be scale invariant $n \sim 1$; no or little running of the spectral index $|n_{run}| < 0.025$. I discussed some systematic errors that can affect the recovery of the power spectrum and I focussed in particular on the mean flux level that is the most important one. Finally, I briefly addressed some improvements that, in my opinion, will be made in the next few years both on the theoretical and observational side. An experiment to measure the cosmic expansion using absorption lines has been presented.

Acknowledgments.   I would like to thank all my many collaborators, the CODEX team, PPARC for financial support, the COSMOS UK-CCC facility. Finally, I thank Frank Bash, the organizers, the attendees and the supporters of this wonderful meeting.

References

Adelberger, K. L. 2005, ApJ, 629, 636
Bolton, J. et al. 2005, MNRAS, 357, 1178
Bahcall, J. N. & Salpeter, E. E. 1965. ApJ, 142, 1677
Bechtold, J. 2002, in Galaxies at High Redshift, ed. I. Pérez-Fournon, M. Balcells, F. Moreno-Insertis & F. Sánchez, (Cambridge: Cambridge University Press), chapter 4
Bi, H. 1993, ApJ, 405, 479
Bi, H., & Davidsen, A. F. 1997, ApJ, 479, 523
Cen, R. et al. 1994, ApJ, 437, L9
Croft, R. A. C., Weinberg, D. H., Katz, N. & Hernquist, L. 1998, ApJ, 495, 44
Croft, R. A. C., Weinberg, D. H., Bolte, M., Burles, S., Hernquist, L., Katz, N., Kirkman, D. & Tytler, D. 2002, ApJ, 581, 20
Furlanetto, S. & Oh, S. P. 2005, MNRAS, 363, 1031
Fan, X., et al. 2004, AJ, 128, 515
Gnedin, N. Y. & Hamilton, A. J. S. 2002, MNRAS, 334, 107
Gnedin, N. Y. & Hui, L. 1998, MNRAS, 296, 44

Gunn, J. E. & Peterson, B. A. 1965, ApJ, 142, 1633
Hernquist, L. et al. 1996, ApJ, 457, 51
Hui, L., Burles, S., Seljak, U., Rutledge, R. E., Magnier, E. & Tytler, D. 2001, ApJ, 552, 15
Hui, L. & Gnedin, N. 1997, MNRAS, 292, 27
Kim, T.-S., Carswell, R. F., Cristiani, S., D'Odorico, S. & Giallongo, E. 2002, MNRAS, 335, 555
Kim, T.-S., Viel, M., Haehnelt, M. G., Carswell, R. F. & Cristiani, S. 2004, MNRAS, 347, 355
Kollmeier, J., Miralda-Escude, J., Cen, R., & Ostriker, J. P. 2006, ApJ, 638, 52
Lewis, A. & Bridle, S. 2002, Phys.Rev.D, 66, 103511
Lynds, R. 1971, ApJ, 164, 73
Loeb, A. 1998, ApJ, 499, L111
McDonald, P. et al. 2001, ApJ, 562, 52
McDonald, P. et al. 2005a, ApJ, 635, 761
McDonald, P. et al., 2005b, MNRAS, 360, 1471
Oort, H. J. 1981, A&A, 94, 359
Peiris, H. et al. 2003, ApJS, 148, 213
Press, W. H., Rybicki, G. B., Schneider, D. P. 1993, ApJ, 414, 64
Rauch, M. 1998, ARA&A, 36, 267
Rees, M. J. 1986, MNRAS, 218, 25
Ricotti, M., Gnedin, N. & Shull, M. 2000, ApJ, 534, 1
Sandage, A. 1962, ApJ, 136, 319
Sargent, W. L. W., Young, P. J. & Schneider, D. P. 1982, ApJ, 256, 374
Schaye, J. 2001, ApJ, 559, 507
Schaye, J., 2000, MNRAS, 318, 817
Schaye, J., ApJ, 596, 768
Scheuer, P. A. G. 1965. Nat, 207, 963
Seljak, U., McDonald, P. & Makarov A. 2003, MNRAS, 342, L79
Seljak, U. et al. 2005, Phys.Rev.D, 71, 103515
Songaila, A. 2004, AJ, 127, 2598
Spergel, D. N. et al. 2003 ApJS, 148, 175
Springel, V. 2005, MNRAS, 364, 1105
Theuns, T. et al. 1998, MNRAS, 301, 502
Theuns, T. et al., 2002, ApJ, 578, L5
Tytler, D. et al. 2004, ApJ, 617, 1
Verde, L. et al. 2003, ApJS, 148, 195
Viel, M., Matarrese, S., Mo, H. J., Theuns, T. & Haehnelt, M. G. 2002, MNRAS, 336, 685
Viel M., Matarrese, S., Mo, H. J., Haehnelt, M. G. & Theuns, T. 2002, MNRAS, 329, 848
Viel, M. et al. 2004, MNRAS, 347, L26
Viel, M., Haehnelt, M. G. & Springel, V. 2004, MNRAS, 354, 684
Viel, M., Weller, J. & Haehnelt, M. G. 2004, MNRAS, 355, L23
Viel, M. & Haehnelt, M. G. 2006, MNRAS, 365, 231
Viel, M., Haehnelt, M. G. & Springel, V., 2005, preprint (astro-ph/0504641)
Viel, M. et al. 2005, Phys.Rev.D, 71, 063534
Weinberg, D. et al. 1999, in 'Evolution of Large Scale Structure: From Recombination to Garching, ed. A. J. Banday, R. K. Sheth, & L. N. da Costa (Enschede: PrintPartners, Ipskamp), 346
Yoshida, N. et al. 2003, ApJ, 598, 73

Mike Siegel moderates questions for Naomi McClure-Griffiths after her talk.

Frank Bash meets Matteo Viel.

Scott Sheppard responds to a question on the solar system during a panel discussion.

Carlos Allende-Prieto and Yeshe Fenner engage with another panelist on the topic of Galactic chemical abundance patterns.

Sera Markoff and Craig Wheeler visualize the geometry of accretion and jets.

Jillian Bornak and Pey-Lian Lim meet Sheila Kannappan in the poster hall.

Don Winget and Jason Kalirai analyze the physics of white dwarf stars.

Eva Noyola explains her poster to Victor Debattista and Juntai Shen.

Professors Neal Evans and Don Winget reminisce about their postdoc years.

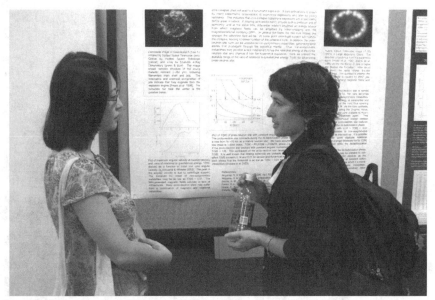

Shizuka Akiyama and Sera Markoff compare theoretical modeling techniques.

Marcelo Alvarez and Mark Krumholz brainstorm at a whiteboard.

Asteroid expert Judit Györgyey Ries learns about cosmology from Matteo Viel.

Niv Drory and Frank Bash listen to McDonald Observatory Director David
Lambert's closing remarks at the end of the symposium.

Part II

Frank N. Bash Symposium 2005: New Horizons in Astronomy
ASP Conference Series, Vol. 352, 2006
S. J. Kannappan, S. Redfield, J. E. Kessler-Silacci, M. Landriau, and N. Drory

Non-Axisymmetric Instabilities in Core-Collapse Supernovae

Shizuka Akiyama and J. Craig Wheeler

Department of Astronomy, University of Texas, Austin, TX, USA

Abstract. Observations indicate that core-collapse supernova explosions are axisymmetric and bipolar in nature. A rotating core would easily give such a preferred axis of symmetry; at the same time subsequent differential rotation becomes an energy source from which magnetic fields can be amplified. A rapidly and differentially rotating core can also be unstable to non-axisymmetric instabilities, generating MHD waves that propagate through the accreting mantle. We discuss the possible effects of the rotational instabilities in the core on the mantle and the core itself.

1. Introduction

The observational association of bi-polar explosions of supernovae with their gaseous remnants and relativistic jets from gamma-ray bursts motivates the jet-induced explosion mechanism as an alternative theory of core-collapse explosions. Since pulsars are rotating and magnetized neutron stars, there is no doubt that rotation and magnetic fields are inherent to the exploding engine. The question is whether and how rotation and magnetic fields generate magnetohydrodynamic (MHD) jets that can cause successful supernova explosions and how they evolve to what are observed in pulsars today. In general, when a rotating and weakly magnetized iron core collapses, it results in a rapidly and differentially rotating proto-neutron star (Akiyama et al. 2003; Akiyama & Wheeler 2005). Differential rotation with a negative angular velocity gradient can feed the magnetorotational instability (MRI: Balbus & Hawley 1991, 1998), and the initially weak magnetic fields may be amplified to the order of 10^{15-17} G in the core-collapse environment (Akiyama et al. 2003). Meanwhile, differential rotation may drive the proto-neutron star to oscillate due to rotational non-axisymmetric instabilities. We discuss possible effects due to the non-axisymmetric instabilities of the proto-neutron star on supernova dynamics and on proto-neutron star evolution.

2. Non-Axisymmetric Instabilities in Core-Collapse

It is well-known that MacLaurin spheroids are unstable to non-axisymmetric instabilities when the ratio of rotational to gravitational energy, $T/|W|$, exceeds 0.14 and 0.27 for secular and dynamical instability, respectively. Recent studies show that the critical value for some dynamically unstable modes is decreased for high differential rotation to as low as $T/|W| \sim 0.01$ (Shibata et al. 2003). Core collapse will produce a strongly differentially rotating environment, and these instabilities will be relevant to a wide range of rotating core-collapse models, as our previous work suggests (Fig. 1). In addition, as the proto-neutron star spins up during the de-leptonization phase, T/|W| increases and more cores would cross the threshold of non-axisymmetric instabilities, affecting the final pulsar rotation. If non-axisymmetric instabilities are efficient in transporting rotational energy from the proto-neutron star core to the mantle, they could alter the explosion dynamics, and angular momentum transport affects the spin rate of the proto-neutron star. The caveat is that the core-collapse environment offers non-axisymmetric instabilities not in an isolated neutron star in equilibrium, but rather in the dynamical environment of a hot proto-neutron star surrounded by an accreting mantle and growing magnetic fields.

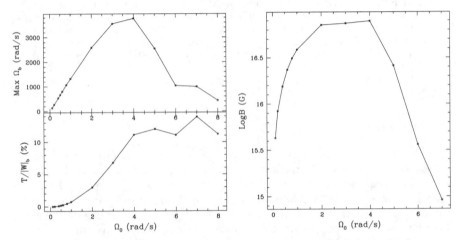

Figure 1.   Plot of angular velocity of proto-neutron star and the ratio of rotational to gravitational energy, T/|W|, as a function of initial angular velocity in iron core (left). The rotating iron core collapses while conserving angular momentum, thus increasing the proto-neutron star rotation monotonically until bounce occurs at sub-nuclear density due to the strong centrifugal force. This creates non-monotonic behavior in the proto-neutron star rotation. The model at the peak in the angular velocity was found with T/|W| = 11%. The non-monotonic behavior in angular velocity is reflected in the magnetic field strength (right).

2.1.   Effects on Explosion

The magnetic field saturation timescale due to the MRI (τ_{MRI}) is tens of msec after bounce. The competition between the growth timescales for the magnetic field and non-axisymmetric instabilities (τ_{naxi}) determines whether magnetic fields exist before the onset of non-axisymmetric mode formation. In typical cases, we expect $\tau_{MRI} < \tau_{naxi}$, and the proto-neutron star core that becomes unstable to non-axisymmetric instabilities is surrounded by strong MRI-generated fields. Because the dense mantle is efficient at transmitting compressional waves, the oscillation of the proto-neutron star is able to drive acoustic flux, thereby exciting magnetosonic waves (Akiyama & Wheeler 2006). The estimated power carried by the magnetosonic waves generated by non-axisymmetric perturbations is given by (Akiyama & Wheeler 2006)

$$L_{mhd} \simeq 6r^4 \rho v_\phi \Omega^2 \left(\frac{\delta\rho}{\rho}\right)^{2/3} \sim 6 \times 10^{51} \mathrm{erg\, s^{-1}}, \tag{1}$$

where r is the radius, ρ the density and v_ϕ the magnetosonic velocity,

$$v_\phi^2 = \frac{1}{2}\left[v_a^2 + c_s^2 + \sqrt{(v_a^2 + c_s^2)^2 - 4v_a^2 c_s^2 cos^2\theta}\right], \tag{2}$$

v_a is Alfvén velocity, c_s is sound velocity, and $cos\theta = \vec{k} \cdot \vec{v_a}/kv_a$, where $\vec{k}$ is the wavevector. For the case of sub-Keplerian rotation, saturation fields are relatively weak, and hence $v_a < c_s$ and $v_\phi \sim v_a$. In this limit the power radiated into fast magnetosonic waves and that into ordinary hydrodynamic waves is virtually the same.

The timescale for dissipation of the rotational kinetic energy by this MHD sound flux is:

$$\tau_{mhd} = 0.16\frac{M}{\rho v_\phi r^2 \left(\frac{\delta\rho}{\rho}\right)^{2/3}} \sim 160\mathrm{ms}. \tag{3}$$

This result is somewhat sensitive to specific parameters, especially the ambient density which will vary with position and time, but indicates that for a proto-neutron star of radius 50 km the timescale could be rather short even for small density perturbations. The MHD power generated by the proto-neutron star oscillation may propagate through the mantle anisotropically due to the density structure or magnetic field. Recently Burrows et al. (2005) found an acoustic flux driven explosion in their non-rotating 2-D simulation, which was monopolar in nature. If the magnetic field is strong enough, the MHD power may be channeled along the field lines.

2.2. Effects on Pulsar Evolution

Soon after the formation of a proto-neutron star, it extends to about 50 km in radius. Initially hot proto-neutron stars de-leptonize, cool, and contract over a timescale of seconds to become cold neutron stars with radii ~ 10 km. The contraction of the proto-neutron star spins up the star if angular momentum is conserved. Effectively, the gravitational potential energy is converted to differential rotational energy, thus providing a constant supply of energy for the non-axisymmetric instabilities. For example, a proto-neutron star with an initial period of 15 msec has T/|W| ~ 0.08 (Fig. 2). This is below the threshold for secular instability, but above the low T/|W| dynamical instability. If the timescale of contraction is shorter than the instability, the proto-neutron star evolves along the constant angular momentum curve. When it contracts to about 30 km, T/|W| ~ 0.14. Further contraction may provide enough extra rotational energy for the star to be unstable to non-axisymmetric instabilities and the dissipation lowers T/|W| to a stable value. Even after the neutron star has contracted to 10 km, the star may still be unstable to non-axisymmetric instabilities until T/|W| drops below the lowest threshold, 0.01. In this scenario, there will be episodes of rotational energy dissipation over long timescales. This may regulate the spin rate of neutron stars and affect the dynamos that are responsible for their magnetic fields.

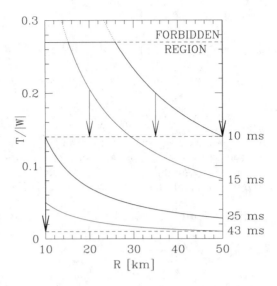

Figure 2.   Schematic diagram illustrating the possible $T/|W|$ evolution of a contracting proto-neutron star. As the proto-neutron star contracts, the radius changes from 50 km to 10 km, resulting in an increase of $T/|W|$ if angular momentum is constant. The three horizontal lines are at $T/|W| = 0.27$, 0.14, and 0.01, the thresholds for dynamical, secular, and low $T/|W|$ dynamical instabilities, respectively. The curved lines trace the evolution when angular momentum is conserved. The period of the newly formed proto-neutron star is indicated for each curve. The vertical arrows indicate the possibility of angular momentum dissipation due to non-axisymmetric instabilities.

3.   Conclusion

The aspherical explosion model is required by observations and favored by theory. It is an open issue whether rotation and magnetic fields play major roles in the explosion mechanism. Non-axisymmetric instabilities may provide a new mechanism to dissipate rotational energy into the supernova, resulting in the spinning down of the proto-neutron star.

Acknowledgments.   This work was supported in part by NASA Grant NAG59302 and NSF Grant AST-0098644.

References

Akiyama, S., Wheeler, J. C., Meier, D. L., & Lichtensdat, I. 2003, ApJ, 584, 954
Akiyama, S., & Wheeler, J. C. 2005, ApJ, 629, 414
Akiyama, S., & Wheeler, J. C. 2006, in preparation
Balbus, S. A. & Hawley, J. F. 1991, ApJ, 376, 214
Balbus, S. A. & Hawley, J. F. 1998, Review of Modern Physics, 70, 1
Burrows, A., Livne, E., Dessart, L., Ott, C., & Murphy, J. 2005, preprint (astro-ph/0510687)
Shibata, M., Karino, S., & Eriguchi, Y. 2003, MNRAS, 343, 919

Frank N. Bash Symposium 2005: New Horizons in Astronomy
ASP Conference Series, Vol. 352, 2006
S. J. Kannappan, S. Redfield, J. E. Kessler-Silacci, M. Landriau, and N. Drory

Studying Distant Dwarf Galaxies with GEMS and SDSS

Fabio D. Barazza,[1] Shardha Jogee,[1] Hans-Walter Rix,[2] Marco Barden,[2] Eric F. Bell,[2] John A. R. Caldwell,[3] Daniel H. McIntosh,[4] Klaus Meisenheimer,[2] Chien Y. Peng,[5] and Christian Wolf[6]

[1] *Department of Astronomy, University of Texas, Austin, TX, USA*
[2] *Max-Planck Institute for Astronomy, Heidelberg, Germany*
[3] *McDonald Observatory, University of Texas, Fort Davis, TX, USA*
[4] *Department of Astronomy, University of Massachusetts, Amherst, MA, USA*
[5] *Space Telescope Science Institute, Baltimore, MD, USA*
[6] *Department of Physics, University of Oxford, Oxford, UK*

Abstract. We study the colors, structural properties, and star formation histories of a sample of ~1600 dwarfs over look-back times of ~3 Gyr ($z = 0.002 - 0.25$). The sample consists of 401 distant dwarfs drawn from the Galaxy Evolution from Morphologies and SEDs (GEMS) survey, which provides high resolution *Hubble Space Telescope* Advanced Camera for Surveys images and accurate redshifts, and of 1291 dwarfs at 10–90 Mpc compiled from the Sloan Digitized Sky Survey (SDSS). We find that the GEMS dwarfs are bluer than the SDSS dwarfs, which is consistent with star formation histories involving starbursts and periods of continuous star formation. The full range of colors cannot be reproduced by single starbursts or constant star formation alone. We derive the star formation rates of the GEMS dwarfs and estimate the mechanical luminosities needed for a complete removal of their gas. We find that a large fraction of luminous dwarfs are likely to retain their gas, whereas fainter dwarfs are susceptible to significant gas loss, *if* they experience single starbursts.

1. Introduction

The evolution of dwarfs is a complex problem, where evolutionary paths may depend on a variety of external and internal factors. Our knowledge of the local volume (<8 Mpc) has deepened, in particular due to strong efforts in determining distances to many nearby galaxies (Karachentsev et al. 2003, and references therein). However, it is still unclear what governs the evolution of dwarfs in low density regions and how the different morphological types form.

Here, we present a study of the properties of dwarf galaxies over the last 3 Gyr ($z = 0.002 - 0.25$)[7] drawn from GEMS (Rix et al. 2004) and SDSS (Abazajian et al. 2004).

---

[7] We assume a flat cosmology with $\Omega_M = 1 - \Omega_\Lambda = 0.3$ and H_0=70 km s^{-1} Mpc^{-1}.

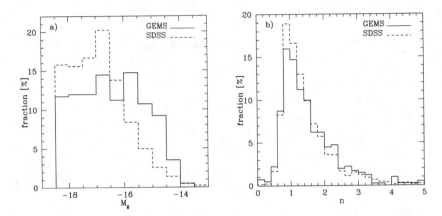

Figure 1.   a) The g-band luminosity distribution of the GEMS and SDSS samples. The median value for the GEMS dwarf sample is -16.51 mag and for SDSS -16.95 mag. b) Histograms of the Sersic index n. The median values are 1.32 and 1.25 for GEMS and SDSS, respectively.

2.   Basic Properties of the Sample

Our starting sample consists of 988 dwarfs from the GEMS survey in the redshift range $z \sim 0.09 - 0.25$ (corresponding to look-back times of 1 to 3 Gyr), and a comparison local sample of 2847 dwarfs with $z < 0.02$ from the NYU Value-Added low-redshift Galaxy Catalog (NYU-VAGC, Blanton et al. 2005) of the SDSS, which have been identified and extracted by applying an absolute magnitude cut of $M_g > -18.5$ mag. The surface brightness profiles of the dwarfs in this sample have subsequently been fitted with a Sersic model using GALFIT (Peng et al. 2002). Finally, we limited the sample to objects with an effective surface brightness brighter than 22 mag arcsec^{-2} in the z band, which corresponds to the completeness limit of the SDSS sample (Blanton et al. 2005). The final sample consists of 401 dwarfs from GEMS and 1291 dwarfs from SDSS. Figure 1 shows the distributions of the luminosities and Sersic indices.

3.   Global Colors and Star Formation Histories

The comparison of the global colors of the two samples shows that the GEMS dwarfs are significantly bluer than the SDSS dwarfs, which is apparent in the histograms shown in Figure 2a. A KS-test yields a probability of $\sim 2 \times 10^{-41}$ that the two color distributions stem from the same parent distribution. In order to examine the origin of this color difference, we compare the colors of our sample dwarfs to star formation (SF) models from *Starburst 99* (Leitherer et al. 1999) in Figure 2b. The general color difference between the two samples is consistent with the color evolution of models combining a starburst (SB) with continuous SF on a low level. The model tracks show rather long periods of time with roughly constant $U - B$ colors, while the $B - V$ colors are becoming significantly redder. This reddening occurs over time spans comparable to the

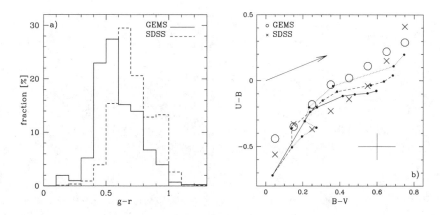

Figure 2. a) The $g - r$ color distributions for both samples. The median colors are 0.57 and 0.70 for GEMS and SDSS, respectively. b) Color-color plot for the two samples. The distribution of the galaxies is represented by the mean $U - B$ colors in 0.1 mag $B - V$ color bins. The lines represent models, where continuous SF with a constant rate of 0.03 $M_\odot$ yr^{-1} and a metallicity of $Z = 0.004$ has been combined with various single SBs starting at different times. For all models a Kroupa IMF has been used. Models start at 0.1 Gyr (left) and end at 15 Gyr (right) *Solid line:* A single SB with a mass of 3×10^8 $M_\odot$ and $Z = 0.0004$ starts at 0.1 Gyr. *Dashed line:* A single SB with a mass of 3×10^8 $M_\odot$ and $Z = 0.004$ starts at 0.9 Gyr. *Dotted line:* A single SB with a mass of 5×10^8 $M_\odot$ and $Z = 0.02$ starts at 3.9 Gyr. The error bars represent the errors of the single color measurements. The arrow indicates the effect of dust on the colors (Schlegel et al. 1998).

average look-back time of the GEMS sample and the amount of reddening is in good agreement with the color difference between the two samples.

4. Star Formation Rates and Feedback

Using the rest-frame luminosity in a synthetic UV band centered on the 2800Å line, we estimate the star formation rate (SFR) of the dwarfs in the GEMS sample, assuming continuous SF over the last 10^8 years, which is likely the case for a majority of our dwarfs. In Figure 3a we plot the *normalized* SFR versus M_B. In a next step, we estimate the mechanical luminosities (MLs) needed for the complete removal of the gas from the dwarfs. The estimate is based on the blowaway model by Mac Low & Ferrara (1999). In this model, the MLs depend only on the mass, which we derive from the V-band luminosities, and on the ellipticity. In Figure 3b we compare these MLs with the ones expected for the measured SFRs. We find that for their derived SF histories, the luminous ($M_B = -18$ to -16 mag) dwarfs are likely to retain their gas and avoid blowaway. However, a fair number of low luminosity dwarfs ($M_B = -14$ to -16) are susceptible to a complete blowaway of gas, *if* one of them were to experience a SB. However, in practice, only a small fraction of these low luminosity dwarfs *may be actually undergoing* SBs. Even though we do not have any clear evidence that some

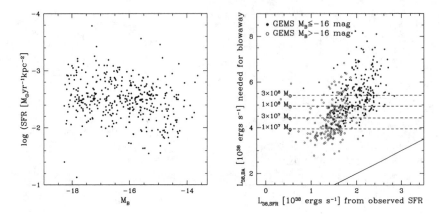

Figure 3.   a) Normalized SFR versus M_B. SFRs have been estimated from 2800Å continuum fluxes (L_{2800}) using $SFR[M_\odot \mathrm{yr}^{-1}] = 3.66 \times 10^{-40} L_{2800}$ [ergs s^{-1} λ^{-1}] from Kennicutt (1998). These SFRs have then been divided by the isophotal area provided by SExtractor. b) Plot of the MLs needed for complete blowaway (BA) of the gas in dwarfs versus the MLs inferred from the SFRs. The four dashed lines mark the peak MLs reached by SBs with the indicated masses. The solid line corresponds to $L_{38,SFR} = L_{38,BA}$.

dwarfs in our sample are experiencing SBs at the time of observation, we are also not able to rule this out. The derived MLs stem from the SFRs, which have been determined assuming that the dwarfs had a constant SFR over the last 10^8 years. In addition, we used the near-UV luminosities, which could be affected by dust. In view of these uncertainties, the derived SFRs have to be considered as lower limits.

Acknowledgments.   F.D.B. and S.J. acknowledge support from the National Aeronautics and Space Administration (NASA) LTSA grant NAG5-13063 and from HST-GO-10395 and HST-GO-10428. E.F.B. was supported by the European Community's Human Potential Program under contract HPRN-CT-2002-00316 (SISCO). C.W. was supported by a PPARC Advanced Fellowship. D.H.M acknowledges support from the NASA LTSA Grant NAG5-13102. Support for GEMS was provided by NASA through number GO-9500 from STScI, which is operated AURA, Inc., for NASA, under NAS5-26555.

References

Abazajian, K., et al. 2004, AJ, 128, 502
Blanton, M. R., et al. 2005, AJ, 129, 2562
Karachentsev, I. D., et al. 2003, A&A, 398, 479
Kennicutt, R. C. 1998, ARA&A, 36, 189
Leitherer, C., et al. 1999, ApJS, 123, 3
Mac Low, M.-M., & Ferrara, A. 1999, ApJ, 513, 142
Peng, C. Y., Ho, L. C., Impey, C. D., & Rix, H. 2002, AJ, 124, 266
Rix, H., et al. 2004, ApJS, 152, 163
Schlegel, D. J., Finkbeiner, D. P., & Davis, M. 1998, ApJ, 500, 525

Frank N. Bash Symposium 2005: New Horizons in Astronomy
ASP Conference Series, Vol. 352, 2006
S. J. Kannappan, S. Redfield, J. E. Kessler-Silacci, M. Landriau, and N. Drory

TEXES Observations of Molecular Hydrogen Emission from AB Aurigae

Martin A. Bitner,[1] Matthew J. Richter,[2] John H. Lacy,[1]
Daniel T. Jaffe,[1] Jacqueline E. Kessler-Silacci,[1]
Thomas K. Greathouse,[3] and Geoffrey A. Blake[4]

[1] McDonald Observatory and Department of Astronomy, University of
Texas, Austin, TX, USA
[2] Department of Physics, University of California, Davis, CA, USA
[3] Lunar and Planetary Institute, Houston, TX, USA
[4] Department of Geological and Planetary Sciences, California Institute
of Technology, Pasadena, CA, USA

Abstract.
    We have used TEXES, the Texas Echelon-cross-Echelle Spectrograph on the
NASA Infrared Telescope Facility (IRTF) to search for pure rotational H_2 emission from young stars with disks. From the ground, three pure rotational tran-
sitions of molecular hydrogen are accessible: $J = 6\text{--}4$ ($\lambda = 8.025$ μm), $J = 4\text{--}2$
($\lambda = 12.27$ μm), and $J = 3\text{--}1$ ($\lambda = 17.035$ μm). Thus far, studies of gas in
protoplanetary disks have focused on either small radii using near-infrared CO
observations as a probe or on large radii with observations in the millimeter
wavelength range. Observations of molecular emission in the mid-infrared offer
the potential to study gas in disks at intermediate radii (1-10 AU). Molecular
hydrogen can be a useful probe since it is the dominant constituent in disks and
therefore allows for mass determinations which avoid CO/H_2 conversion factors.
When coupled with knowledge of the stellar mass and inclination, high resolu-
tion observations ($R \approx 60{,}000$ for $J = 3\text{--}1$ and $R \approx 80{,}000$ for $J = 4\text{--}2$ and
$J = 6\text{--}4$) may allow us to study line profiles and determine the radial location
of the emission. In the case of AB Aurigae, we have detected $J = 4\text{--}2$ emission
with FWHM of 7 km s^{-1}. Assuming an inclination of 20° and a stellar mass of
2.5 $M_\odot$ centers the emission at $\approx$16 AU in the disk.

1. Introduction

Disks around young stars are a natural part of the star formation process and,
as the likely site of planet formation, have generated significant interest. Much
has been learned about the structure of these disks through study of dust emission and detailed modeling of the spectral energy distributions (SED) of these
stars. However, if we assume the standard gas-to-dust ratio for the interstellar
medium, dust makes up only 1% of the mass of the disk. Thus there is considerable motivation to develop tracers of the gas in these disks. Studies of gas
in disks have so far focused mainly on gas at either large radii ($>$50 AU) using
observations at submillimeter wavelengths (Semenov et al. 2005) or small radii
($<$0.5 AU) with observations of CO near-infrared (NIR) lines (Najita, Carr, &
Mathieu 2003).

Observations of pure rotational mid-infrared H_2 lines may be useful probes for studying intermediate radii of disks (1-10 AU). Three pure rotational H_2 lines are accessible from the ground: $J = 3\text{--}1$ ($\lambda \approx 17\,\mu\text{m}$), $J = 4\text{--}2$ ($\lambda \approx 12\,\mu\text{m}$), and $J = 6\text{--}4$ ($\lambda \approx 8\,\mu\text{m}$). These transitions probe temperatures in the range of 200-1000 K.

2.   TEXES and the TEXES/IRTF H_2 in Disks Survey

We have used TEXES, the Texas Echelon-Cross-Echelle Spectrograph (Lacy et al. 2002), to look for H_2 emission from young stars with disks. The high resolution of TEXES is advantageous for this search for several reasons. There are telluric features near the spectral regions containing the H_2 lines. The high spectral resolution of TEXES allows for the avoidance of these atmospheric features. In addition, TEXES also offers the opportunity to study resolved line profiles in detail and to determine the radial location of the emission.

On the Infrared Telescope Facility (IRTF), TEXES has proven to be sensitive to small amounts of warm gas. In one hour of observing time on a source, we can set 3σ upper limits of 1×10^{-14} erg s^{-1} cm^{-2} on a 5 km s^{-1} $J = 3\text{--}1$ line and 2×10^{-14} erg s^{-1} cm^{-2} on a 10 km s^{-1} full width at half-maximum (FWHM) $J = 4\text{--}2$ line. Assuming temperatures between 150-450 K and a distance of 140 pc, we can convert these line fluxes to a mass of H_2. Based on these assumptions and our demonstrated line flux sensitivities, we find that we are sensitive to H_2 masses as small as $2.5\,M_\oplus$ if there is no extinction and the gas is dust-free. In the case of AB Aur, see Figure 1, we have detected $J = 4\text{--}2$ emission with a flux of 1.8×10^{-14} erg s^{-1} cm^{-2} and FWHM of 7 km s^{-1}.

3.   AB Aurigae and Modeling

AB Aurigae is one of the nearest and brightest of the Herbig Ae stars. This naturally leads to its being among the best studied of its class. It is located at a distance of 144 pc based on Hipparcos measurements, has a spectral type of A0-A1 (Hernández et al. 2004) and is surrounded by a $r \approx 450$ AU disk (Mannings & Sargent 1997). Using the *Hubble Space Telescope* (*HST*) Space Telescope Imaging Spectrograph (STIS), Grady et al. (1999) observed an extended circumstellar nebulosity with radius $\approx$1300 AU. Observations at 11.7 and 18.7 μm by Chen & Jura (2003) are well fit by models including an optically thick disk encased by an optically thin spherical envelope. The mass of the star is found to be $2.4\,M_\odot$ by van den Ancker, de Winter, & Tjin A Djie (1998). There have been a wide range of inclination estimates for AB Aur. Mannings & Sargent (1997) derive an inclination of 76° based on the aspect ratio of the major and minor semiaxes of the disk seen in the ^{13}CO(1-0) line. Whereas, using near-infrared interferometric observations at the 1 AU scale, Millan-Gabet, Schloerb, & Traub (2001) observed a nearly spherical image implying a nearly face-on inclination (0°). Eisner et al. (2003) derived an inclination between 27°-35° using the Palomar Testbed Interferometer. Fukagawa et al. (2004) found an inclination of 30° using NIR coronographic imaging on the Subaru telescope. Semenov et al. (2005) derived an inclination angle of 17° using millimeter observations.

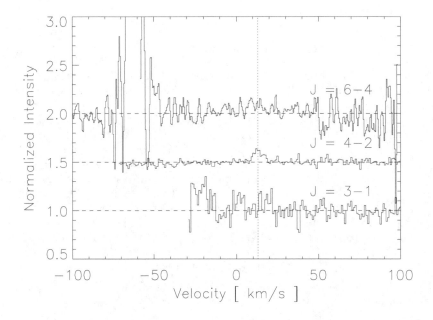

Figure 1. TEXES observations of the Herbig Ae star AB Aur in the three pure rotational lines available from the ground. Each of the spectra has been normalized to its median continuum value and offset by a constant. The $J = 4\text{--}2$ line is centered near the systemic velocity (10 km s^{-1}) and is fit by a Gaussian with FWHM = 7 km s^{-1} and a flux of 1.8×10^{-14} erg s-1 cm^{-2}.

We have begun constructing models of the molecular spectra arising from the disk around AB Aur. First, we compute the temperature and density structure of the disk using the best-fit parameters found by fitting the SED. AB Aur has a very well-observed SED and is well-fit by a model of a passive irradiated disk with a puffed-up inner rim (Dullemond, Dominik, & Natta 2001). We use the parameters of the best-fit SED disk model as input to the 1+1D vertical structure disk models by Dullemond, van Zadelhoff, & Natta (2002) to compute the radial and vertical temperature and density profile of the disk. Second, we input the resulting temperature and density structure into a three-dimensional, non-LTE radiative transfer code and compute the level populations of the lines of interest and the emergent spectrum. Due to the low A-values of the H$_2$ transitions and the large densities of these disks, we expect the H$_2$ level populations to be in local thermodynamic equilibrium. In addition to modeling the H$_2$ emission, we have CO M-band data for most of the sources in our sample that we would like to model simultaneously. Non-LTE effects like resonance flourescence likely play a key role as a mechanism for CO emission and so must be included in the radiative transfer. Finally, we take the emergent synthetic spectrum and compare it with the data. We can then iterate this procedure using different input disk parameters, such as dust settling, which still fit the SED but have a noticeable impact on the vertical structure and molecular emission to set constraints on the vertical disk structure.

4. Future Work: TEXES on Gemini N

We are currently modifying TEXES for use on the Gemini North telescope. With TEXES on Gemini, we will realize a factor of seven improvement in sensitivity and a factor of nearly three in spatial resolution. A fundamental limitation to the types of sources we can observe is the level of mid-infrared continuum as this affects our ability to guide on an object. On the IRTF, we require a continuum of 1.4 Jy at 12 microns. Using Gemini, this limit drops to 0.2 Jy, thus greatly expanding the number of sources in our survey as well as allowing us to observe disk sources with little dust continuum emission to investigate how much gas remains in the disk as the dust disappears. In addition to increasing the number of sources accessible to our survey, the increased sensitivity will allow us to get very high signal-to-noise spectra of the brighter sources we have observed so far. These improved spectra will greatly aid in our study of line profiles to infer radial information about the temperature and density in the disks. We expect TEXES to be available on Gemini beginning in Fall 2006.

References

Chen, C. H., & Jura, M. 2003, ApJ, 591, 267
Dullemond, C. P., Dominik, C., & Natta, A. 2001, ApJ, 560, 957
Dullemond, C. P., van Zadelhoff, G. J., & Natta, A. 2002, A&A, 389, 464
Eisner, J. A., Lane, B. F., Akeson, R. L., Hillenbrand, L. A., & Sargent, A. I. 2003, ApJ, 588, 360
Fukagawa, M., Hayashi, M., Tamura, M., Itoh, Y., Hayashi, S. S., Oasa, Y., Takeuchi, T., Morino, J.-i., Murakawa, K., Oya, S., Yamashita, T., Suto, H., Mayama, S., Naoi, T., Ishii, M., Pyo, T.-S., Nishikawa, T., Takato, N., Usuda, T., Ando, H., Iye, M., Miyama, S. M., & Kaifu, N. 2004, ApJ, 605, L53
Grady, C. A., Woodgate, B., Bruhweiler, F. C., Boggess, A., Plait, P., Lindler, D. J., Clampin, M., & Kalas, P. 1999, ApJ, 523, L151
Hernández, J., Calvet, N., Briceño, C., Hartmann, L., & Berlind, P. 2004, AJ, 127, 1682
Lacy, J. H., Richter, M. J., Greathouse, T. K., Jaffe, D. T., & Zhu, Q. 2002, PASP, 114, 153
Mannings, V., & Sargent, A. I. 1997, ApJ, 490, 792
Millan-Gabet, R., Schloerb, F. P., & Traub, W. A. 2001, ApJ, 546, 358
Najita, J., Carr, J. S., & Mathieu, R. D. 2003, ApJ, 589, 931
Semenov, D., Pavlyuchenkov, Y., Schreyer, K., Henning, T., Dullemond, C., & Bacmann, A. 2005, ApJ, 621, 853
van den Ancker, M. E., de Winter, D., & Tjin A Djie, H. R. E. 1998, A&A, 330, 145

Frank N. Bash Symposium 2005: New Horizons in Astronomy
ASP Conference Series, Vol. 352, 2006
S. J. Kannappan, S. Redfield, J. E. Kessler-Silacci, M. Landriau, and N. Drory

Spitzer Observations of WTTS Disks: New Constraints on the Timescale for Planet Building

Lucas A. Cieza[1] and the C2D Stars Team

[1]*Department of Astronomy, University of Texas, Austin, TX, USA*

Abstract. We report on the disk frequency of a sample of over 200 spectroscopically identified wTTs located in the IRAC and MIPS maps of the Ophiuchus, Lupus and Perseus Molecular Clouds from the *Spitzer* Legacy Project "From Molecular Cores to Planet-forming Disks" (c2d) Evans et al. (2003). We find that, overall, ~20% of the wTTs in the sample have noticeable IR-excesses indicating the presence of a circumstellar disk and that all the stars with disks are younger than ~10 Myr according to their position in the H-R diagram. Since *Spitzer* observations probe planet-forming regions of the disk ($r \sim 0.1 - 10 \, \mathrm{AU}$) and are capable of detecting IR excesses produced by very small amounts of dust, these results provide much stronger constraints on the time available for the formation of planets than those provided by previous studies based on detections of disks in the near-IR.

1. Introduction

Weak-lined T Tauri stars (wTTs) are low-mass pre-main sequence stars which occupy the same region of the H-R diagram as classical T Tauri stars (cTTs) but do not show clear evidence of accretion. Most wTTs show little or no near-IR excess. However, their mid- and far-IR properties remain unclear because past infrared (IR) telescopes such as *IRAS* and *ISO* lacked the sensitivity needed to detect mid- and far-IR excesses in low mass stars at the distance of nearest star-forming regions. Since cTTs are believed to become wTTs at the end of their accretion stage, wTTs are potentially extremely important in determining the timescale for planet building. One of the main goals of the *Spitzer* Legacy Project "From Molecular Cores to Planet-forming Disks" (c2d) Evans et al. (2003) is to determine whether or not most wTTs have circumstellar disks and study their properties and evolutionary status. Here we report on the disk frequency as a function of age of a sample of over 200 spectroscopically identified wTTs located in the *Spitzer* c2d Infrared Array Camera (IRAC) (3.6, 4.5, 4.8, and 8.0 μm) and Multiband Imaging Photometer (MIPS) (24 μm) maps of the Ophiuchus, Lupus, and Perseus Molecular Clouds. The main goal is to estimate the time available for the formation of planets in order to impose observational constraints on planet formation theories.

2. Sample Selection and *Spitzer* Observations

Our sample of over 200 spectroscopically identified wTTs was collected from the literature. The Lupus wTTs were taken from Hughes et al. (1994) and Wichmann et al. (1999), the Ophiuchus wTTs come from Bouvier & Appenzeller (1992) and Martin et al. (1998), and the Perseus objects were taken from Luhman et al. (2003). The Lupus and Ophiuchus objects are distributed across the cloud maps, while the targets in Perseus correspond exclusively to the IC 348 cluster. All the near-IR photometry (J, H, and K bands) comes from the Two Micron All Sky Survey (2MASS).

In order to identify the stars with disks, we place our sample on *Spitzer* color-magnitude diagrams (e.g., Fig. 1). Bare stellar photospheres have *Spitzer* colors ~0.0, while stars with IR-excess have positive colors. The broader the color baseline used, the larger the mean excess of the stars with disks; therefore, the color that provides the most clear separation between stars with and without disk is [3.6]-[24]. However, since c2d MIPS observations are not deep enough to reach the stellar photospheres of all the stars in the sample, we use the [3.6]-[8.0] color to identify stars with disks. In a few cases, the photosphere of substellar objects in IC 348 fall bellow the 8.0 μm (or 5.8 μm) c2d sensitivity limits, and we use [3.6]-[5.8] (or [3.6]-[4.5]) colors.

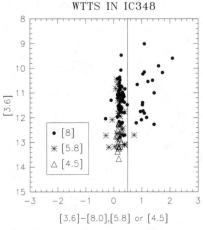

WTTS IN IC348

Figure 1.   [3.6] vs. [3.6]-[8.0] for the sample of IC 348 wTTs. We plot [5.8] or [4.5] whenever the objects are too faint to have a 8.0 μm from the c2d Survey. The vertical line denotes the IR color we use as the criterion to distinguish between stars with and without disks. Most of the stars (~80%) have ~0 color, characteristic of bare stellar photospheres.

3.   Age Estimation

In order to investigate the evolutionary status of wTTs with and without disks, we estimated the ages of our sample of the wTTs (Fig. 2) using the evolutionary tracks presented by D'Antona & Mazzitelli (1994, 1998, D98 hereafter) and Siess, Dufour, & Forestini (2000, S00 hereafter). To estimate the stellar effective temperatures ($T_{\rm eff}$), we adopt the spectral type-$T_{\rm eff}$ relations from Kenyon & Hartmann (1995). The stellar luminosities of the wTTs were obtained from the extinction corrected J-band magnitudes and the bolometric corrections, appropriate for the spectral type, from Hartigan, Strom, & Strom (1994). We

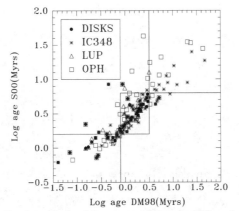

Figure 2.   Stellar ages derived for our sample of wTTs using two different sets of evolutionary tracks (D98 and S00). A clear decrease in the disk fraction is seen with increasing age. Approximately 50% of the targets that are 0.8-1.5 Myr old have disks. None of the stars that are older than $\sim$10 Myr according to either of the models have disks.

estimated the extinction from the *J-H* color excess assuming an extinction law with $R_V = 3.1$.

4.   Disk Mass Upper Limits Estimate

Based on the 24 μm flux sensitivities of the c2d survey ($4\sigma \sim 1$ mJy), we estimate that the wTTs in our sample without 24 μm excess (or detection) must be almost completely depleted of small grains ($r < 10$ μm). For these objects, we estimate mass upper limits for the warm dust ($T \sim 100$ K) of $M \sim 10^{-4} - 10^{-3}$ $M_\oplus$. If wTTs disks are depleted of small grains because they have been accreted into larger bodies (i.e., planetesimals and planets), a second generation of dust particles (e.g., debris disks) is expected to be generated by the mutual collision of these planetesimals. The size distribution of solid particles in collisional equilibrium is believed to quickly converge to a $dN \propto R^{-3.5} dR$ power law, were N is the number of particles of radius R. Combining our warm dust mass upper limits obtained above with the $dN \propto R^{-3.5} dR$ size distribution, we estimate that the size of the most massive objects must have reached the kilometer-size scale for the total mass of the solid material in the planet forming region to be of the order of several planetary masses. These are *very* rough mass estimates, but they suggest that the wTTs in our sample without 24 μm excesses (or detections) must be at a relatively advanced stage of the planet formation process if they are to form planets at all.

5.   Evidence for Very Early Disk Evolution

Fig. 2 suggests that the disks of some wTTs dissipate in timescales <1 Myr. Are these apparently very young diskless objects as young as the evolutionary tracks suggest or is their apparent youth just a product of the large age uncertainties?

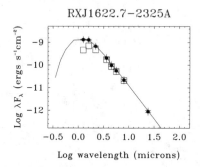

Figure 3.   The SED of one of the apparently youngest diskless wTTs in our
sample. Very young objects without 3.6-24 μm excess like this must be almost
completely depleted of small grains inward of $\sim$10 AU by an age of 0.3-1 Myr.

To address this question we plotted the spectral energy distributions (SEDs) of
some of the apparently youngest diskless wTTs in our sample (see Fig. 3 as an
example). The overall quality of the SED fits to stellar models strongly suggests
that both the spectral types and the luminosities are reasonably accurate. Still,
some of these objects are $\sim$15 times more luminous than main-sequence stars of
the corresponding spectral types, and therefore, probably *very* young.

6.   Conclusions

1) Overall, only $\sim$20% of the wTTs in the sample show significant IR-excess at
3.6-24 μm.
2) The primordial disk of $\sim$50% of the wTTs dissipates in the first 1-3 Myr.
Also, $\sim$100% of the wTTs lose their disks by an age of 10 Myr.
3) The wTTs in our sample without 24 μm excesses must be at a relatively
advanced stage of the planet formation process if they are to form planets at all.

References

Bouvier, J., & Appenzeller, I. 1992, A&AS, 92, 481
D'Antona, F., & Mazzitelli, I. 1994, ApJS, 90, 467
D'Antona, F., & Mazzitelli, I. 1998, in ASP Conf. Ser. 134: Brown Dwarfs and Extra-
        solar Planets, ed. R. Rebolo, E. L. Martin, & M. R. Z. Osorio (San Francisco:
        ASP), 442 (D98)
Evans, N. J., et al. 2003, PASP, 115, 965
Hartigan, P., Strom, K. M., & Strom, S. E. 1994, ApJ, 427, 961
Hughes, J., Hartigan, P., Krautter, J., & Kelemen, J. 1994, AJ, 108, 1071
Kenyon, S. J., & Hartmann, L. 1995, ApJS, 101, 117
Luhman, K. L., Stauffer, J. R., Muench, A. A., Rieke, G. H., Lada, E. A., Bouvier, J.,
        & Lada, C. J. 2003, ApJ, 593, 1093
Martin, E. L., Montmerle, T., Gregorio-Hetem, J., & Casanova, S. 1998, MNRAS, 300,
        733
Siess, L., Dufour, E., & Forestini, M. 2000, A&A, 358, 593 (S00)
Wichmann, R., Covino, E., Alcalá, J. M., Krautter, J., Allain, S., & Hauschildt, P. H.
        1999, MNRAS, 307, 909

Frank N. Bash Symposium 2005: New Horizons in Astronomy
ASP Conference Series, Vol. 352, 2006
S. J. Kannappan, S. Redfield, J. E. Kessler-Silacci, M. Landriau, and N. Drory

Is Sérsic Index a Good Detector of Pseudobulges?

David B. Fisher

Department of Astronomy, University of Texas, Austin, TX, USA

Abstract.   I discuss an effort to test the claim that Sérsic index of bulges can be used as a diagnostic of pseudobulges. For this, pseudobulges are identified using nuclear morphology (e.g., nuclear spirals, nuclear rings, nuclear bars). In addition, bulge-disk decompositions are carried out on composite surface brightness profiles. The result is thus far that there appear to be two distinct types of objects in parameter space. Classical (merger built) bulges have larger Sérsic index ($n > 2$) and pseudobulges have more nearly exponential surface brightness profiles ($n < 2$).

1.   Introduction

The dominant processes which governs galaxy evolution is changing. Previously hierarchical mergers created most of the structure in the universe. However, as galaxies are left untouched by major mergers, over time they are allowed to evolve via secular (internal) channels (bars, oval disks, and spiral structure). Kormendy & Kennicutt (2004) review these processes in detail, and provide an existence proof for this process. Here, I report on the efforts to make the classification of pseudobulges a more rigorous process.

2.   Secular Evolution and the Formation of Pseudobulges

Secular evolution of galaxies is actually a quite general process. The natural tendency of self-gravitating systems (in this case, disk galaxies) is the construction of a tightly bound central component. This can be illustrated by describing stellar dynamical kinetic energy as a thermodynamic property, $M\bar{v}^2 \propto k_B T$. For a self-gravitating system in equilibrium, the virial theorem states that total energy is equal to minus the kinetic energy. Therefore, the heat capacity of the system is negative. This means that if the system loses energy it becomes hotter ($dE \propto -dT$). If energy flows from the hotter inner disk to the colder outer of disk, the center loses energy and therefore gets hotter. Thus a disk tends to evolve such that the outer disks gains energy and spreads (in radius) while the center shrinks and becomes more tightly bound. Efficient drivers for this process are bars, ovals, spiral structure, etc.

The tendency of disks to spread leads to the construction of central component that looks like a classical bulge, however is different in origin and properties when compared to classical bulges. We call them "pseudobulges".

In early work on secular evolution Simkin, Su, & Schwarz (1980) simulate a galaxy in which gas clouds are represented by sticky particles. After several bar rotations, the gas clouds are rearranged into an outer ring, an inner ring (near

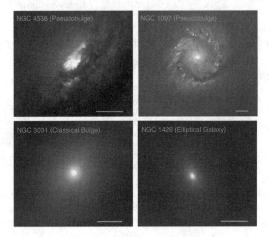

Figure 1.   Four *HST* images, shows a comparison of two "bulges" interpreted as being built by secular evolution (NGC 4526 and NGC 1097) to two
structures built by hierarchical mergers (NGC 3031 and NGC 1426). The bar
in each image represents 5 arcseconds. In all galaxies shown here the bulge
dominates the light of the galaxy at least beyond 16 arcseconds.

corotation of the bar) and a concentration of gas in the center of the galaxy.
We identify the simulated rings with inner and outer rings observed in barred
galaxies. In many barred galaxies, star forming nuclear rings are also present.
This gas forms stars and builds pseudobulges.

3.   Bulge Classification

For this study, I classify galaxies as having pseudobulges according to the prescription described in Kormendy & Kennicutt (2004), specifically the morphological indicators. Thus a bulge is called a pseudobulge if it shows spiral structure
in the center; its apparent flattening is similar to that of the outer disk; and/or
it is or contains a nuclear bar. To ensure that resolution is not a factor classification is done with *Hubble Space Telescope* (*HST*) archival images, and a distance
limit of 25 Mpc is placed on the sample.

Figure 1 shows an example of the classification. The central regions of four
galaxies are shown. Three of these galaxies are bulge-disk galaxies (NGC 4536,
NGC 1097, and NGC 3031) and for comparison an elliptical galaxy (NGC 1426)
is also shown. In all of these galaxies the bulge dominates the light in the galaxy
to at least 16 arcseconds. A white bar represents 5 arcseconds in each image.
NGC 4536 is an Sbc galaxy. I classify this as a pseudobulge due to the spiral
structure in the bulge. NGC 1097 is an SBb galaxy with a quite prominent
star bursting nuclear ring. Star bursting rings are disk phenomena, and thus I
classify this bulge as a pseudobulge. NGC 3031 an Sb galaxy with no signs of
disk-like morphology, thus it is classified as a classical bulge. Note the similarity
in morphology between NGC 3031 and NGC 1426, an elliptical galaxy.

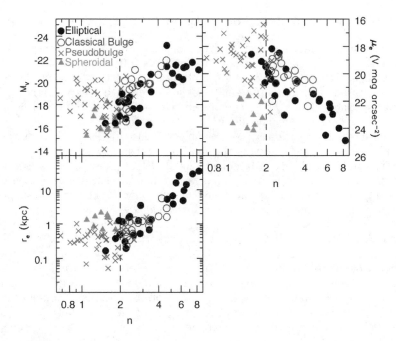

Figure 2.   Shown above are different photometric parameters plotted against the shape parameter n. It appears that the pseudobulges (crosses) do not follow the same correlations as objects known to have been built by mergers (solid and open circles).

To determine the Sérsic index I calculate composite *HST* plus wide field ground based, surface brightness profiles in the V band for ~60 bulge-disk galaxies varying in Hubble type from S0 to Sc. I also use composite surface photometry of elliptical galaxies in the Virgo cluster, as a control for objects known to have built through mergers. The resulting profiles are then fit with the Sérsic function (Sérsic 1968) plus an outer exponential disk (elliptical galaxies do not include the outer disk in fitting),

$$\mu(r) = \mu_e + b_n \left[\left(\frac{r}{r_e} \right)^{1/n} - 1 \right], \qquad (1)$$

where μ_e is the surface brightness at the half light radius (r_e), n is the Sérsic index, which gives information on the shape (smaller n means a steeper profile), and b_n is a constant function of n.

4.   Result: Pseudobulges Appear Distinct in Parameter Correlations

Many studies suggest that secular evolution creates bulges with nearly exponential surface brightness profiles (Carollo et al. 2001; Courteau, de Jong, & Broeils 1996; MacArthur, Courteau, & Holtzman 2003). These studies show that smaller bulges in later type galaxies are often nearly exponential. Simulations also show that a bulge component with a nearly exponential profile is expected from nonaxisymetric disturbance (i.e., secular drivers like bars) which can send gas inward within a disk (Pfenniger & Friedli 1991). However, a theoretical understanding as to why the central structure has such a profile shape is not in hand (Kormendy & Kennicutt 2004). What I test in this study is the prevalence of other, morphological, indicators that secular evolution is in place within a galaxy with the shape of the bulge component. The aim is to tie profile shape to formation mechanisms for particular galaxies.

I use well known scaling relations (e.g., Caon, Capaccioli, & D'Onofrio 1994; Graham & Colless 1997) between photometric parameters to compare the Sérsic indices of pseudobulges to those of classical bulges. If Sérsic index is a good diagnostic of secular evolution, the distribution of pseudobulges and objects built by mergers should appear disjoint in these parameter correlations. In Figure 2, one can easily see that multiple distributions of objects exist. Notice that the pseudobulges appear to have a flat distribution of Sérsic index, better resembling a scatter diagram than the correlated dependence of n on other parameters. Secondly, notice that the bulges classified as "classical bulge" obey the same correlations as elliptical galaxies. Spheroidal galaxies do not follow all trends with elliptical galaxies, especially in the $\mu_e - n$ plane. Like Kormendy (1985), I interpret spheroidal galaxies to be separate objects from elliptical galaxies.

The tentative result seems promising. Pseudobulges are a distinct sample of objects from elliptical galaxies and classical bulges. Further, it appears that the empirical evidence that $n = 2$ is a good dividing line between pseudobulges and classical bulges. And thus, is so far a robust detector of pseudobulges.

Acknowledgments.   I would like to thank John Kormendy for his helpful discussion. This paper is based partly on observations made with NASA/ESA *Hubble Space Telescope*, obtained from the data archive at the Space Telescope Science Institute. STScI is operated under NASA contract NAS 5-26555. We also used the NASA/IPAC Extragalactic Database (NED) which is operated by JPL and Caltech under contract with NASA.

References

Caon, N., Capaccioli, M., & D'Onofrio, M. 1994, A&AS, 106, 199
Carollo, C. M., Stiavelli, M., de Zeeuw, P. T., Seig ar, M., & Dejonghe, H. 2001, ApJ, 546, 216
Courteau, S., de Jong, R. S., & Broeils, A. H. 1996, ApJ, 457, L73
Graham, A., & Colless, M. 1997, MNRAS, 287, 221
Kormendy, J. 1985, ApJ, 295, 73
Kormendy, J., & Kennicutt, R. C. 2004, ARA&A, 42, 603
MacArthur, L. A., Courteau, S., & Holtzman, J. A. 2003, ApJ, 582, 689
Pfenniger, D., & Friedli, D. 1991, A&A, 252, 75
Sérsic, J. L. 1968, Atlas de Galaxias Australes, (Córdoba: Observatorio Astronomico)
Simkin, S. M., Su, H. J., & Schwarz, M. P. 1980, ApJ, 237, 404

Frank N. Bash Symposium 2005: New Horizons in Astronomy
ASP Conference Series, Vol. 352, 2006
S. J. Kannappan, S. Redfield, J. E. Kessler-Silacci, M. Landriau, and N. Drory

The Dark Halo in NGC 821

A. D. Forestell and K. Gebhardt

Department of Astronomy, University of Texas, Austin, TX, USA

Abstract.   The study of elliptical galaxy dark halos at large radii is important because it is the region where dark matter is thought to dominate the galaxy mass and can therefore provide the best constraints on observed dark halo properties. We present axisymmetric orbit-superposition models of the dark halo in the elliptical galaxy NGC 821 using line-of-sight velocity distributions obtained to $\sim 100''$ (over 2 effective radii) with long-slit spectroscopy from the Hobby-Eberly Telescope. We fit models with a range of dark halo density profiles and find that a power-law dark halo with a constant density of 0.0105 $M_\odot/\mathrm{pc}^3$ is the best-fitting model ruling out both the no dark halo and the Navarro, Frenk, & White (1996; NFW) models at a greater than 3σ confidence level. We show the internal moments σ_r, σ_θ, and σ_ϕ and find that the model with no dark halo is radially anisotropic at small radii and tangentially anisotropic at large radii while the best-fit halo models are slightly radially anisotropic at all radii. The dark halo we find is inconsistent with previous claims of little to no dark matter halo in this galaxy based on planetary nebula measurements.

1.   Observations

We obtained long-slit spectra with the Low-Resolution Spectrograph on the Hobby-Eberly Telescope using a 1 arcsec by 4 arcmin slit with a resolving power of 1300 over the wavelength range 4300–7300 Å. This setup gives a full-width half-maximum (FWHM) resolution of about 230 km/s and measurements of night sky line widths confirm that we can measure dispersions to about 110 km/s. NGC 821 was observed over eight nights in November 2003 for a total exposure time of approximately 5.5 and 2.3 hours on the major and minor axes respectively. The data reduction was performed with FORTRAN programs that we developed using standard techniques including overscan correction, flat correction, rectification, background subtraction, and image combination.

2.   Kinematics

We extracted the spectra in radial bins along the major and minor axes. Our farthest radial bin extends to $45''$ on the minor axis and $99''$ on the major axis. The measured R_e of NGC 821 varies throughout the literature from $50''$ (de Vaucouleurs et al. 1991, RC3) to $16.7''$ (Bender et al. 1988), thus our data extend to approximately 2 to 5 R_e.

We obtain a nonparametric line-of-sight velocity distribution (LOSVD) by deconvolving the galaxy spectrum with a set of stellar template spectra. Although we use the full nonparametric velocity profile in the dynamical modeling, in

Figure 1 we plot the first two Gauss-Hermite moments (mean velocity V and velocity dispersion σ).

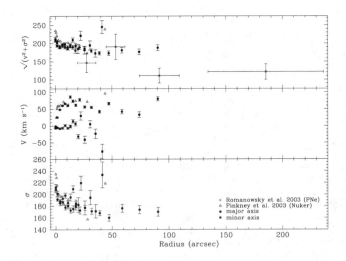

Figure 1.   Velocity and dispersion as a function of radius. Also plotted are comparison data from the literature. The rms line-of-sight velocity $(v^2+\sigma^2)^{1/2}$ is compared to the planetary nebula results of Romanowsky et al. (2003).

3.   Dynamical Models

We use axisymmetric orbit superposition models based on the method of Schwarzschild (1979). First, a surface brightness profile is converted to luminosity density using an assumed inclination. We assume an edge-on inclination for this analysis, which is reasonable given NGC 821's large ellipticity. The surface brightness profile is taken from Gebhardt et al. (2003). The luminosity density is converted to a mass density using a mass-to-light ratio (M/L_V) that is constant over the galaxy. A dark halo density profile is added to the stellar density and this total mass density gives the galaxy's graviational potential. Next, individual stellar orbits are sampled in energy (E), angular momentum (L_z), and the third integral (I_3), and these orbits are run for a significant number of orbital times in the specified potentials. The galaxy is divided spatially into cells both in 3-D and in projection, and the amount of time that an orbit spends in a cell represents the mass contributed by that orbit. Finally, the orbits are combined with nonnegative weights to find the best-fit superposition to match the data LOSVDs and light profile. This process is repeated for different dark halo density profiles and M/L_V values to find the halo potential that best fits the data, as determined by the χ^2 between the model and data LOSVDs. The change in χ^2 between different models determines the confidence levels of the fits.

4. Results

4.1. NFW Halo

We use the NFW (Navarro, Frenk, & White 1996) dark halo density profile, given as

$$\rho(r) = \frac{\rho_{crit}\,\delta_c}{(r/r_s)(1 + r/r_s)^2} \tag{1}$$

where r_s is the scale radius of the halo and $\rho_{crit} = 3H^2/8\pi G$ is the critical density. We use $H_0 = 70$ km s^{-1} Mpc^{-1}. The characteristic overdensity δ_c is related to a concentration parameter c by

$$\delta_c = \frac{200}{3}\frac{c^3}{\ln(1 + c) - c/(1 + c)}. \tag{2}$$

We refer to $\rho_{crit}\delta_c$ as the scale density. We freely vary both the concentration and scale radius, although there is a known correlation between them (Navarro, Frenk, & White 1996).

We find that the best-fit NFW dark halo density profile has scale radius 1050 kpc and scale density 1.14×10^{-4} $M_\odot$ pc^{-3}, corresponding to a c of 1.64. The no-halo model is ruled out with a change in χ^2 of 187 from the best-fit NFW halo. The scale radius is well beyond the radial extent of our modeling, so we do not claim to attach significance to this number. However, the large NFW scale radius is indicative of the need for a near power-law profile over the extent of our models.

4.2. Power-Law Halo

Because the best-fit NFW halo profiles are flat with the break radius beyond the extent of our modeling, we tried a simple power-law profile as well. We used power-law density profiles of the form

$$\rho(r) = \rho_o(\frac{r}{r_o})^{-n} \tag{3}$$

where n is the power-law slope, ρ_o is the characteristic density, and r_o is the characteristic radius such that $\rho(r=r_o) = \rho_o$. We use $r_o = 0.3'' = 35$ pc because it is the inner-most radial point calculated in the models.

The best-fit halo model has a constant density (slope $n = 0.0$) of $\rho_o = 0.0105$ $M_\odot$ pc^{-3}. This halo is a better fit to the data than the best NFW halo, with a $\Delta\chi^2 = 19$. This power-law slope is significantly more shallow than the 1.0 slope of an NFW profile.

Figure 2 shows the internal moments σ_r, σ_θ, and σ_ϕ, and the ratio of radial to tangential dispersion along the major axis for the model with no dark halo and the best-fit power-law halo model. The model without a dark halo shows radial anisotropy at small radii and tangential anisotropy at large radii. The best-fit power-law halo model has a slight radial anisotropy in the θ direction at all radii.

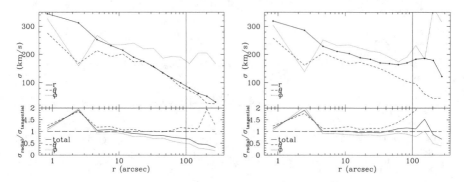

Figure 2.   Internal moments σ_r, σ_θ, and σ_ϕ and ratio of radial to tangential dispersion along the major axis for the model with no dark halo (left panel) and the best-fit power-law halo model (right panel).

5.   Conclusions

We find that the best-fit model of the dark halo in NGC 821 has a flat power-law density profile. This dark halo gives a better fit than both the NFW halo models and models without a dark halo at a greater than 3σ confidence level. This slope is shallower than the NFW profile, and may rule out halo profiles with inner slopes greater than one and support halos with a flat inner slope.

The models without a dark halo show tangential anisotropy at large radii. Tangential anisotropy at large radii could be an indication that a dark halo is necessary to explain the data because the observations better constrain both σ_r and σ_θ so σ_ϕ may be artificially increased to account for a large observed dispersion. The best-fit dark halo model shows a slight radial bias in the θ direction at all radii. This is in agreement with the hypothesis by Dekel et al. (2005) that the planetary nebulae analyses show little to no dark halo because isotropy is incorrectly assumed. In addition we find that our measured stellar velocity dispersions are larger than the reported planetary nebulae dispersions (Romanowsky et al. 2003) in the most radially extended regions.

In the future we will apply this method to several elliptical galaxies to learn about global elliptical galaxy dark halo properties. Improved correlations between the halo and galaxy properties, such as mass and environment, may give us information about the role dark halos play in galaxy formation and evolution.

References

Bender, R., Doebereiner, S., & Moellenhoff, C. 1988, A&AS, 74, 385
Dekel, A., et al. 2005, Nature, 437, 707
Gebhardt, K., et al. 2003, ApJ, 583, 92
Navarro, J. F., Frenk, C. S., & White, S. D. M. 1996, ApJ, 462, 563
Pinkney, J., et al. 2003, ApJ, 596, 903
Romanowsky, A. J., et al. 2003, Science, 301, 1696
Schwarzschild, M. 1979, ApJ, 232, 236
de Vaucouleurs, G., et al. 1991, Third Reference Catalogue of Bright Galaxies (New York: Springer) (RC3)

Frank N. Bash Symposium 2005: New Horizons in Astronomy
ASP Conference Series, Vol. 352, 2006
S. J. Kannappan, S. Redfield, J. E. Kessler-Silacci, M. Landriau, and N. Drory

Mid-Infrared Spectroscopy of Red 2MASS AGN

Lei Hao, D. Weedman, V. Charmandaris, M. S. Keremedjiev, S. J. U. Higdon, and J. R. Houck

Astronomy Department, Cornell University, Ithaca, NY, USA

Abstract. We observed the mid-IR spectra for a sample of Two Micron All Sky Survey selected red AGN using the Spitzer Space Telescope. Their mid-IR spectra show a great variety. The silicate features range from deep absorption to significant emission. The silicate features in general agree with the optical classification of the sources. The sources with the deepest silicate absorption are those that show Hα emission but not Hβ emission in their optical spectra. The average spectrum of the whole sample is flatter than for classical AGN, indicating that if extinction is the reason for our sources' red color, it is most likely due to warm dust.

1. Introduction

The Two Micron All Sky Survey (2MASS) has produced a sample of active galactic nuclei (AGN) that are not traditionally blue-selected objects (Cutri et al. 2002). The candidates are selected to have near-IR color $J - K_s > 2$ from the 2MASS Point Source Catalogue. Optical spectroscopic follow-up reveals 75% are previously unidentified emission-line AGN, with 80% of those showing broad optical emission lines. They have unusually high optical polarization levels P, with $\sim$10% showing $P > 3\%$ (Smith et al. 2002). But their host galaxy properties and environments are not very different from PG quasars, suggesting their red colors are simply caused by extinction of an otherwise typical QSO (Marble et al. 2003). X-ray observations of a well-defined subset (Wilkes et al. 2002) reveal that they are all X-ray faint, with the reddest being the faintest in X-rays. But the reddest objects do not have the hardest spectra, as might be expected in a simple optical and X-ray absorption scenario. We take their mid-IR spectra using the Spitzer Space Telescope (Werner et al. 2004) to directly observe their dust properties and better understand their red near-IR colors.

2. Our Sample and Their Mid-IR Spectra

Wilkes et al. (2002) have selected a well-defined, flux-limited, and color-selected subset of 26 2MASS AGN for X-ray observations. They satisfy $B - K_s > 4.3$ and $K_s < 13.8$, which include the brightest and reddest objects in the 2MASS AGN sample. All of the 26 sources are observed with the Spitzer Infrared Spectrograph (IRS; Houck et al. 2004) low-resolution modules covering a spectral range of 5 to 40 μm, and 20 of them are observed using the short wavelength high-resolution module (SH) with a wavelength coverage from 10 to 19 μm. The basic processing of the data, such as ramp fitting, dark sky subtraction and

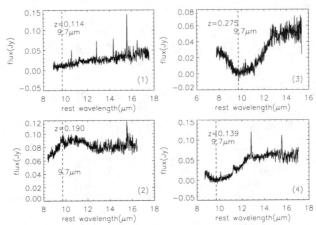

Figure 1.   High-resolution spectra of four sources after background subtraction.

removal of cosmic rays, was performed within the IRS pipeline at the Spitzer Science Center (version 11.0). The spectra are extracted with the reduction tool SMART (Higdon et al. 2004).

The background emission taken with the source observation can significantly change the shape of the spectrum and should be removed. For low-resolution spectra, the background can be easily removed by subtracting the spectral images of two nod positions before extracting the spectra. For hi-resolution spectra, the spectral images of the two nod positions are not sufficiently separated to use this technique. Since these sources are also observed by the IRS Blue Peak-up camera (BPU), we use their images in the BPU to estimate the flux of the sky adjacent to the source, and scale accordingly to the model background spectra obtained from SPOT. The final SH spectra are obtained by subtracting the background spectra from the original spectra. Figure 1 shows four examples of our high-resolution spectra after background subtraction.

3.   Results

The mid-IR spectra in our sample show a great variety, from having deep silicate absorption (Figure 1(3)) to significant silicate emission (Figure 1(2)). To measure the silicate strength at both 10 and 18 μm, we fit a continuum to the low-resolution spectra based on their spectral slope shortward of 8 μm and longward of 29 μm, where the spectra are least contaminated by features. We measure the ratio of the flux at the peak or depth of the features to the continuum flux to represent the silicate strengths. Figure 2 plots the 10 μm silicate strength compared with the 18 μm strength as measured above. There are 6 objects that show significant silicate emission at 10 μm. Based on their optical emission lines, four are classified as Type 1 and two as Type 1.5 (Smith et al. 2002). Two sources have very high linear polarization: $P = 9.37$ and 7.19. However, there is no correlation between the silicate strength and the linear polarization.

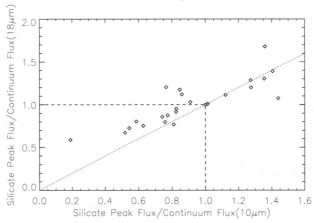

Figure 2.   The silicate strengths of the 10 μm feature compared with those of the 18 μm feature.

About 2/3 of the sample show silicate absorption features at 10 μm, but most of the absorption features are not as significant as 50% of the continuum ,except for one object. The two Type 2 AGN in the sample have 10 μm silicate depth as 0.77 and 0.59 of the continuum. They are not the most absorbed sources in the sample. Instead, most sources showing prominent silicate absorption are classified as Type 1.x by Smith et al. (2002), which means that they only show Hα emission lines, which are broad, but not the Hβ lines. However, careful examination of their optical spectra (Smith et al. 2002, and courtesy of Roc. Cutri) reveals that they are more likely to be Type 2 AGN. The limited spectral resolution does not separate the Hα line from the adjacent [NII] doublet and makes the Hα look broad. Their high Hα/Hβ ratio indicates that they are more extincted with their narrow-line region partly obscured. This is in agreement with their showing more prominent silicate absorption features.

Nine sources have silicate depth flux to continuum flux ratio less than 0.8. All of them except one are classified as Type 2 or Type 1.x by Smith et al. (2002), in agreement with the idea that Type 2 sources are more obscured than Type 1s. The exceptional one shows a clear quasar-like spectrum in the optical, with broad Hα and Hβ lines and a blue continuum. The mid-IR spectra looks very much like Mrk 231 (Weedman et al. 2005), with self-absorbed silicate features. Detailed analysis of this source in multiwavelength observations is needed to better understand its inner dust structure.

Half of the sample has detectable emission lines, including [SIV] 10.5 μm, [NeII] 12.8 μm, [NeV] 14.32 μm and [NeIII] 15.55 μm. The line ratios are not very different from classical AGN (Weedman et al. 2005).

In Figure 3, we plot the average spectrum of the SH spectra of the sample, compared with the classical Type 1 AGN NGC 4151 and Type 2 AGN Mrk 3. There is a weak PAH emission feature at 11.3 μm in the average spectrum and emission lines are generally weaker than those in NGC 4151 and Mrk 3. Furthermore, the slope of the average spectrum is flatter, indicating that on average the dust in 2MASS selected sources is warmer than in NGC 4151 and Mrk 3.

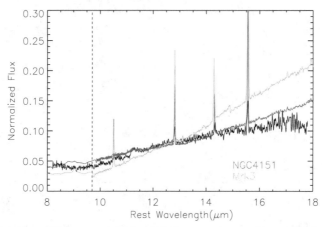

Figure 3.   The average high-resolution spectrum of our sample compared with the classical AGN NGC 4151 and Mrk 3.

4. Conclusion

The mid-IR spectra of the 2MASS selected red AGN show a great variety. The silicate features range from deep absorption to significant emission. The silicate features in general agree with their optical classifications: Type 1 AGN show silicate emission, and Type 2 AGN show silicate absorption. One source is exceptional, with a classical quasar spectrum in the optical but self-absorbed silicate features in the mid IR. The sources with the deepest silicate absorption are those that show Hα emission but not Hβ emission in their optical spectra.

The average spectrum of the whole sample is flatter than for classical AGN. But our mid-IR spectra show little other difference from blue-color selected AGN spectra. If extinction is the reason for our targets' red color, it is most likely due to warm dust.

Acknowledgments. This work is based on observations made with the Spitzer Space Telescope, which is operated by the JPL, Caltech under NASA contract 1407. Support for this work was provided by NASA through Contract Number 1257184 issued by JPL/Caltech.

References

Cutri, R. M., Nelson, B. O., Francis, P. J., & Smith, P. S. 2002, in ASP Conf. Ser. 284: AGN Surveys, ed. R. F. Green, E. Y. Khachikian, & D. B. Sanders, 127
Higdon, S. J. U., et al. 2004, PASP, 116, 975
Houck, J. R., et al. 2004, ApJS, 154, 18
Marble, A. R., Hines, D. C., Schmidt, G. D., Smith, P. S., Surace, J. A., Armus, L., Cutri, R. M., & Nelson, B. O. 2003, ApJ, 590, 707
Smith, P. S., Schmidt, G. D., Hines, D. C., Cutri, R. M., & Nelson, B. O. 2002, ApJ, 569, 23
Weedman, D. W., et al. 2005, ApJ, 633, 706
Werner, M. W., et al. 2004, ApJS, 154, 1
Wilkes, B. J., Schmidt, G. D., Cutri, R. M., Ghosh, H., Hines, D. C., Nelson, B., & Smith, P. S. 2002, ApJ, 564, L65

Frank N. Bash Symposium 2005: New Horizons in Astronomy
ASP Conference Series, Vol. 352, 2006
S. J. Kannappan, S. Redfield, J. E. Kessler-Silacci, M. Landriau, and N. Drory

White Dwarfs and Stellar Evolution

J. S. Kalirai

University of California Observatories/Lick Observatory,
University of California at Santa Cruz, Santa Cruz, CA, USA

Abstract. The initial-to-final mass relationship represents one of the most poorly understood aspects of fundamental stellar evolution. The relation connects the mass of the final products of stellar evolution for intermediate mass stars, white dwarfs, to their initial progenitor mass as main-sequence stars. It is a key input in several areas of astrophysics, including age determinations, distances of globular star clusters, constraining chemical evolution in galaxies, and understanding feedback processes and star formation in galaxies. The combination of current generation wide-field imaging cameras on 4-meter telescopes and multi-object spectrographs on 8–10-meter telescopes have recently provided us with an unprecedented view into this relation. By first imaging and then obtaining spectroscopy of white dwarf stars in rich, open star clusters we have now been able to put strong constraints on the amount of mass that stars lose through their evolution.

The relation between the initial mass of a main-sequence star and its final mass as a white dwarf (WD), for intermediate mass stars, is a fundamental input for several interesting astrophysical problems, for example, for the determination of the ages and distances of globular clusters from modeling their WD cooling sequences (Hansen et al. 2004), for constraining chemical evolution in galaxies, for determining supernova rates (van den Bergh & Tammann 1991) and for understanding feedback processes and star formation in galaxies (e.g., Somerville & Primack 1999). Yet, despite its fundamental importance, the initial-final mass relationship remains poorly constrained observationally.

Attempts to map the initial-final mass relationship have been made as early as 1977 (Weidemann 1977). The methods used require observing a sample of WDs in young star clusters and then spectroscopically measuring masses for the WDs through the modeling of their hydrogen Balmer absorption lines (Bergeron, Saffer, & Liebert 1992). These masses, combined with cooling models, directly yield WD cooling ages for the stars. As they are cluster members, the difference between the age of the cluster and the cooling ages yields the main-sequence lifetimes of the progenitor stars that created the WDs (and hence the progenitor masses). Therefore, a key input required for this study is an accurate age measurement for the cluster under investigation.

The Canada-France-Hawaii Telescope (*CFHT*) Open Star Cluster Survey (Kalirai et al. 2001a) is a large imaging project dedicated to observing many rich open star clusters. The primary goals of the survey are to better measure key parameters for the star clusters and to catalogue the entire WD populations within these systems for follow up spectroscopy. NGC 2099 is a prime example of a very rich, nearby cluster with an intermediate age. In Kalirai et al. (2001b), we identified several thousand cluster members in NGC 2099 and derived its age by comparing the observed main sequence, turn-off, and red giant clump with a set of theoretical isochrones (Ventura et al. 1998) and directly comparing post main-sequence evolutionary phases with synthetic color-magnitude diagrams (CMDs) based on Monte Carlo simulations (Kalirai & Tosi 2004). In this short paper, we describe the observations aimed at spectroscopically observing WDs in NGC 2099 and determining their masses.

1.   Observations

Spectroscopic observations of NGC 2099 were obtained with the *Gemini North* and *Keck I* telescopes. Wide-field *CFHT* imaging data was first used to select regions of the cluster with a large spatial density of WDs. With *Gemini*, we re-imaged three $5.'5 \times 5.'5$ fields using the GMOS multi-object imaging/spectroscopic instrument. With *Keck*, we re-imaged the same three fields with the LRIS imaging/spectroscopic instrument which has a $5' \times 7'$ field of view. These imaging data were not significantly deeper than the original *CFHT* data and were only used to ensure astrometric accuracy for the spectroscopy.

Multi-object spectra were obtained for a single *Gemini* field, and for two of the *Keck* fields. The *Gemini* observations used the B600 grating which simultaneously covers 2760 Å (centered at $\sim$4700 Å). The data were binned by a factor of four in the spectral direction to improve the signal-to-noise ratio (S/N). The *Keck* observations used the 600/4000 grism (blue side) which simultaneously covers 2580 Å (centered at 4590 Å). For the *Gemini* field, we obtained 22 individual 1-hour exposures spread over 22 days, taken mostly at low air-masses ($<$1.2) and good seeing ($\sim$0.''8). For the *Keck* fields, we obtained 4×2000 s exposures in each of the two fields, also at sub-arcsecond seeing.

The individual exposures were bias subtracted, flat-fielded, cleaned for cosmic rays, wavelength calibrated, sky subtracted, extracted, combined, and flux calibrated (using bright standard stars) as described in Kalirai et al. (2005a; 2005b). In total, we obtained spectroscopy of 24 individual WD candidates in the field of NGC 2099 (3 of these were found to not be WDs). Therefore, despite sampling only 14% of the total cluster area, we include almost 1/3 of the total WD population (cluster and field) given the careful positioning of the fields. This is therefore the largest individual star cluster WD sample that has ever been spectroscopically acquired. Individual spectra for 18 of these stars are shown in Figure 1.

2.   The Initial-Final Mass Relationship

Using the techniques described in Bergeron, Saffer, & Liebert (1992), we determine $T_{\rm eff}$ and $\log g$ for each WD by fitting the Balmer lines. The line profiles are first normalized using two points on the continuum on either side of each absorption line. The fitting of the line shapes uses the nonlinear least-squares method of Levenberg-Marquardt (Press et al. 1986). The χ^2 statistic is calculated and minimized for combinations of $T_{\rm eff}$ and $\log g$, using normalized model line profiles of all absorption lines simultaneously. The resulting 1-σ errors in $T_{\rm eff}$ and $\log g$ were tested by simulating synthetic spectra with the same number of absorption lines, and similar S/N, and measuring the output parameters from

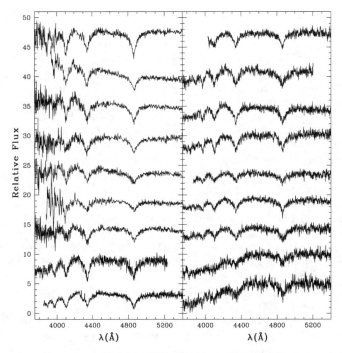

Figure 1.   Spectra of 18 faint WDs (22.46 ≤ V ≤ 23.66) observed in the direction of NGC 2099 with *Gemini* and *Keck*. All spectra have been flux calibrated except for bottom two in the right panel, which were secured at high air-masses so that the bluest flux has been lost due to atmospheric dispersion.

fitting these spectra. These results are found to have errors slightly less than those in the true spectra (as expected given small errors in flux calibration and other defects), and so we use the true spectra errors. Masses (M_f) and WD cooling ages (t_{cool}) are found by using the updated evolutionary models of Fontaine, Brassard, & Bergeron (2001) for thick hydrogen layers ($q(H) = M_H/M = 10^{-4}$) and helium layers of $q(He) = 10^{-2}$. The core is assumed to be a 50/50 C/O mix. The model atmosphere fits for each WD are given in Kalirai et al. (2005a).

The WD cooling age represents the time that each of these stars has spent traversing from the tip of the AGB down to its present WD luminosity. We can now calculate the progenitor main-sequence lifetime (t_{ms}, the total lifetime of the star up to the tip of the AGB) assuming an age for the cluster. The t_{ms} for each star is determined by subtracting the WD cooling ages (t_{cool}) from the cluster age. The initial progenitor masses (M_i) for the WDs are then calculated using the Hurley, Pols, & Tout (2000) models for $Z = 0.01$. For this metallicity, these models give very similar values to those derived from the Ventura et al. (1998) models. For these masses, a 100 Myr cluster age difference would only cause a ~0.3 $M_\odot$ initial mass difference (for 350 Myr lifetimes). This is a very small effect on our initial-final mass relationship, and therefore the results are not highly sensitive to the derived cluster age.

The initial-final mass relation for these stars is presented in Kalirai et al. (2005a). With the study of just one star cluster, the amount of data that constrains this relation has been nearly doubled. This is remarkable considering the time and effort required to establish the previous constraints (~30 years). Furthermore, Kalirai et al. (2005a) show that their initial-final mass relation is tightly constrained. Intermediate mass stars with initial masses in the range 2.8–3.4 $M_\odot$ are measured to end their lives with final masses (as WDs) in the range 0.7–0.9 $M_\odot$. This suggests that stars with masses between 2.8 and 3.4 $M_\odot$, and metallicities slightly less than Solar (NGC 2099 has $Z = 0.011 \pm 0.001$), will lose 70–75% of their mass through stellar evolution. This mass loss is found to be slightly less than that established from other clusters. However, this is consistent with expectations when considering the different metallicities of the clusters. Whereas most of the previous constraints established on the initial-final mass relation involve stars in clusters with Solar metallicities, NGC 2099 is metal-deficient. In fact, the amount of mass loss measured for NGC 2099 stars is consistent with theoretical expectations for this metallicity (Marigo 2001).

The spectroscopic measurement of masses of WDs with $V \simeq 23$ has not been previously accomplished. The number of targets that are accessible at these magnitudes is several orders of magnitudes larger than previously identified. The already published work in the *CFHT* Open Star Cluster Survey (see Kalirai et al. 2001a; 2001b; 2001c; 2003), as well as clusters currently being observed will form a key element of this study in the future. By observing both younger and older clusters, the entire initial progenitor mass range can be constrained and a detailed initial-final mass relation can be produced.

Acknowledgments. The author would like to thank Pierre Bergeron for providing us with his models and spectral fitting routines, as well as for numerous discussions. Many thanks to David Reitzel for helping reduce the *Keck* spectroscopic observations.

References

Bergeron, P., Saffer, R. A., & Liebert, J. 1992, ApJ, 394, 228
Fontaine, G., Brassard, P., & Bergeron, P. 2001, PASP, 113, 409
Hansen, B. M. S. et al. 2004, ApJS, 155, 551
Hurley, J. R., Pols, O. R., & Tout, C. A. 2000, MNRAS, 315, 543
Kalirai, J. S., Fahlman, G. G., Richer, H. B., & Ventura, P. 2003, AJ, 126, 1402
Kalirai, J. S., Richer, H. B., Reitzel, D., Hansen, B. M .S., Rich, R. M., Fahlman, G. G., Gibson, B. K., & von Hippel, T. 2005a, ApJ, 618, L123
Kalirai, J. S., Richer, H. B., Hansen, B. M .S., Reitzel, D., & Rich, R. M. 2005b, ApJ, 618, L129
Kalirai, J. S. & Tosi, M. 2004, MNRAS, 351, 649
Kalirai, J. S., Ventura, P., Richer, H. B., Fahlman, G. G., D'Antona, F., & Marconi, G. 2001b, AJ, 122, 3239
Kalirai, J. S., et al. 2001a, AJ, 122, 257
Kalirai, J. S., et al. 2001c, AJ, 122, 266
Marigo, P. 2001, A&A, 370, 194
Press, W. H. et al. 1986, Numerical Recipes (Cambridge: Cambridge Univ. Press)
Somerville, R. S. & Primack, J. R. 1999, MNRAS, 310, 1087
van den Bergh, S. & Tammann, G. 1991, ARA&A, 29, 363
Ventura, P., Zeppieri, A., Mazzitelli, I. & D'Antona, F. 1998, A&A, 334, 953
Weidemann, V. 1977, A&A, 59, 411

Frank N. Bash Symposium 2005: New Horizons in Astronomy
ASP Conference Series, Vol. 352, 2006
S. J. Kannappan, S. Redfield, J. E. Kessler-Silacci, M. Landriau, and N. Drory

Cosmological Implications of a Solid Upper Mass Limit Placed on DFSZ Axions Thanks to Pulsating White Dwarfs

A. Kim, M. H. Montgomery and D. E. Winget

Department of Astronomy, University of Texas, Austin, TX, USA

Abstract. Axions are theoretical particles that provide an elegant solution to the strong CP problem in the Standard Model of particle physics. In addition, they are potential dark matter candidates. The problem is that the theory of axions does not really constrain their mass, and axions can only account for dark matter if they have the right mass. Several searches are underway that try to find axions and/or constrain their mass. With pulsating white dwarfs, we can place a solid upper limit on the mass of DFSZ axions (a model for axions that interact with electrons, (Dine et al. 1981)). We present how this can be done and how it can be improved. We then try to see if we can draw any conclusion about axions as dark matter candidates by looking at the different cosmological models of their formation in the early universe.

1. Background

Axions were introduced to solve the strong CP problem in particle physics. In the Standard Model, there is no reason why Charge and Parity taken together (CP) should be conserved. Indeed, it has been shown that CP is not conserved in weak interactions (Christenson et al. 1964), but it appears to be precisely conserved in strong interactions. An elegant way to solve this problem is to introduce a single CP violating term in the Lagrangian of the fundamental interactions. This term gives rise to axions.

Like neutrinos, axions interact very little with baryonic matter, and in addition, they would constitute cold dark matter. Unfortunately, the theory of axions contains an essentially free parameter, the mass of the axion m_a. Axions can explain dark matter only if their mass is in the correct range, which is itself dependent on a cosmological model of how axions were formed.

2. Using Pulsating White Dwarfs to Place an Upper Limit on m_a

Axions, like neutrinos, would accelerate the cooling of white dwarfs, and therefore have an effect on the observed pulsation periods. As a white dwarf cools, the period of a given mode increases monotonously. The faster the cooling, the faster the period increases. We can make a prediction of how fast the star should be cooling through known sources, e.g. photons (Mestel 1952) and neutrinos (Itoh et al. 1996). A higher rate of period change ($\dot{P}$) than expected means that the star is cooling faster than expected, and indicates an extra source of energy loss (that is unless the change is periodical). $\dot{P}$ provides therefore a measure of the axion emission rate, which in turn depends on m_a.

We introduced axion energy losses in our White Dwarf Evolution Code and looked at the effects of different mass axions on $\dot{P}$ for different white dwarf models. The results for a 28,000 K model are shown in Figure 1. A 6 meV axion could hide in the uncertainties, but not an 8 meV axion. For cooler models, axions of higher mass can hide in the uncertainties.

The measured $\dot{P}$ for a cooler white dwarf, G117-B15A is consistent with simple Mestel cooling (Kepler et al. 2000). A rough analysis predicts that axions with a mass greater than about 10 meV would not be consistent with that measured value of $\dot{P}$, and that gives the best current upper limit on the mass of the axion. This limit can be improved by 1) a more careful analysis and 2) smaller uncertainties in $\log g$ and effective temperature. We have the tools for the first problem, and we will be able to address the second problem in the near future. With much smaller uncertainties in $\log g$ and effective temperature, G117-B15A could yield a limit as small as 4 meV (Corsico et al. 2001).

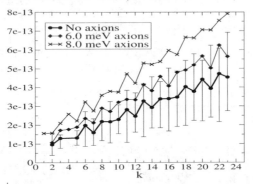

Figure 1. $\dot{P}$ versus k for $\ell = 1$ modes in a 28,000 K, $0.6M_\odot$ white dwarf model. The bold line includes neutrinos, but not axions. The error bars were obtained by running a grid of models differing from the fiducial model by 10% in mass and effective temperature (a reflection of the uncertainties in the observables) and calculating $\dot{P}$'s for those models.

3.   Cosmological Models for the Formation of the Axion

3.1.   Standard Thermal Scenario

Theory has it (Shellard & Battye 1998) that the axion field began to oscillate and axion strings formed when the universe cooled down to about f_a, which is an energy scale inversely proportional to m_a ($m_a < 10$ meV implies $f_a > 10^9$ GeV). At that time, axions were massless. At the much lower temperature $T \sim \Lambda_{QCD} \sim 200$ MeV axions acquired mass, and domain walls formed between the strings. Strings and domain walls went on to decay into axions.

In that scenario, axions may be produced thermally, result from the decay of axion strings, or result from the decay of the axion strings and domain walls later on. Thermally produced axions are viable dark matter candidates only if they are at least a few eVs. Axions this massive have been ruled out long ago in collider experiments (Raffelt 1990).

For the remaining two cases, the current axion density is given by an expression of the form

$$\Omega_a h^2 = \left(\frac{6\mu eV}{m_a} \right)^{7/6} Q \,, \tag{1}$$

where h is the dimensionless Hubble parameter ($H = h \times 100 \ km \ s^{-1} Mpc^{-1}$), and Q contains parameters not always very well constrained by the theory.

Decay of Axion Strings   In this scenario, finding the exact value of Q requires computationally tricky simulations. Estimates from various authors vary from $Q \sim 0.1$ (Battye & Shellard 1994) to $Q \sim 100,000$ (Frieman & Jaffe 1992), but most seem to believe that $Q \sim 1$–100 (Nagasawa & Kawasaki 1994, Shellard & Battye 1998, Yamaguchi et al. 1999). An upper limit on m_a of 10 meV leads to an *empirically* determined maximum possible value for Q of ~ 600.

Decay of Domain Walls Bounded by Strings   Most authors estimate that the contribution from the decay of domain walls formed when axions acquired mass is of the same order as the contribution from the decay of strings, or a little smaller. Q for domain wall decay ranges from 0.1 to ~ 3 (Chang et al. 1999, Shellard & Battye 1998).

The pulsating white dwarf upper limit on m_a does not help put any constraint on that model if one believes that indeed Q cannot be much greater than unity. Otherwise, the empirical limit is again $Q \sim 600$.

3.2.   Cosmological Lower Limit on m_a

Of the three contributions to the axion mass density, only the decay of strings and of domain walls matter for light axions. Adding the two together, we find that m_a^{min} satisfies

$$\Omega_a h^2 \sim 0.2 \left(\frac{6\mu eV}{m_a^{min}} \right)^{7/6} . \tag{2}$$

With the WMAP values $\Omega_{DM} \approx 0.2$ and $h \approx 0.7$ and assuming axions make up all of dark matter, we obtain $m_a^{min} \approx 11\mu eV$. Unfortunately, no current astrophysical observation can probe down to such low energies. Figure 2 summarizes the constraints on the axion mass for different axion strings models and different fractions of dark matter made up of axions.

3.3.   Inflationary Scenario

Under certain conditions, which have to do with the reheat temperature at the end of inflation, inflation can change the standard thermal scenario described above (Shellard & Battye 1998). In this scenario, the axion field starts out away from its equilibrium value and axions result from coherent oscillations of the axion field as it relaxes to its equilibrium value. This process is called "vacuum misalignment". Because we have no way of knowing what the initial value of the axion field was in our "corner" of the universe, the inflationary scenario is essentially non-constrainable (though limits to the madness do arise from quantum fluctuations as well as from temperature and density fluctuations in the CMB — see Shellard & Battye 1998 for an overview).

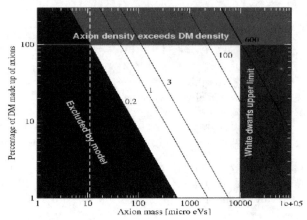

Figure 2.   The allowed axion mass range (white region) as a function of the fraction of dark matter made up of axions. The oblique lines are plots of equation (1) with varying values of Q, which are indicated next to the lines. The dark matter density which defines the horizontal light gray region was taken from WMAP results ($h = 0.71$ and $\Omega_{DM} \simeq 0.23$, (Bennett et al. 2003)). The dashed white line marks the minimum mass allowed by the cosmological models, 11 μeV.

4.   Conclusion

Given that there is more than one model of the axion and several cosmological models of how they were formed, it seems unlikely that any single measurement will make or break the axion as a dark matter candidate. Because they involve a solid measurement and fairly simple physics, pulsating white dwarfs place a strong upper limit on the mass of DFSZ axions. This limit of about 10 meV will improve in the near future and tighten constraints on some of the theories.

Acknowledgments.   This research was supported in part by NSF grant AST-0507639.

References

Battye, R. A., & Shellard, E. P. S. 1994, Phys.Rev.Lett, 73, 2954

Bennett, C. L., Halpern, M., Hinshaw, G., Jarosik, N. et al. 2003, ApJS, 148, 1

Christenson, J. K., Chronin, J. W., Fitch, V., & Turlay, R. 1964, Phys.Rev.Lett, 13, 138

Chang, S., Hagmann, C., & Sikivie, P. 1999, Phys.Rev.D, 59, 023505

Corsico, A. H., Benvenuto, O. G., Althaus, L. G., et al. 2001, New Astron., 6, 197

Dine, M., Fischler, W., & Srednicki, M. 1981, Phys.Lett.B, 104, 199

Frieman, J. A., & Jaffe, A. H. 1992, Phys.Rev.D, 45, 2674

Itoh, N., Hayashi, H., & Nishikawa, A. 1996, ApJS, 102, 411

Kepler, S. O., et al. 2000, ApJ, L534, 185

Mestel, L. 1952, MNRAS, 112, 583

Nagasawa, M., & Kawasaki, M. 1994, Phys.Rev.D, 50, 4821

Raffelt, G. G. 1990, Phys.Rep., 198, 1

Shellard, E. P. S, & Battye, R. A. 1998, in Proceedings of COSMO-97, ed. L. Roszkowski (Singapore: World Scientific), 233

Yamaguchi, M., Kawasaki, M., & Yokoyama, J. 1999, Phys.Rev.Lett, 82, 4578

Frank N. Bash Symposium 2005: New Horizons in Astronomy
ASP Conference Series, Vol. 352, 2006
S. J. Kannappan, S. Redfield, J. E. Kessler-Silacci, M. Landriau, and N. Drory

Hi-Res Spectroscopy of a Volume-Limited Hipparcos Sample within 100 parsec

P. L. Lim,[1] J. A. Holtzman,[1] V. V. Smith,[2] and K. Cunha[2]

[1]*Astronomy Department, New Mexico State University, Las Cruces, NM, USA*

[2]*NOAO Gemini Science Center, Tucson, AZ, USA*

Abstract. Accurate parallaxes from the Hipparcos catalog have enabled detailed studies of stellar properties. Subgiants are of particular interest because they lie in an area where isochrones are well separated, enabling dependable age determination. We have initiated a project to obtain hi-res spectra of Hipparcos subgiants in a volume-limited sample within 100 pc. We obtain stellar properties and abundances via a fully automatic analysis. We will use our results to constrain star formation history and chemical evolution in the solar neighborhood. We present initial results for our observed sample and discuss future development for this project.

1. Sample

We select Hipparcos subgiants in the Northern Hemisphere within 100 pc with $\frac{\sigma}{\pi_o} \leq 0.1$, where π_o is the stellar parallax, and σ is the 1σ parallax error, as shown in Fig. 1. We exclude stars with known multiplicity or variability to avoid ambiguity in abundance and age analysis. A clean sample is essential to test the quality of the automated method itself. We use the Astrophysical Research Consortium (ARC) 3.5-m telescope at the Apache Point Observatory (APO) to obtain high-resolution echelle spectra ($R \sim 37,500$) for our sample. We use standard Image Reduction and Analysis Facility (IRAF) techniques for spectrum extraction and calibration. By collecting our own data, we ensure a uniform sample with high signal-to-noise ratio (SNR) and consistency in data reduction.

2. Analysis

We use automated Gaussian fitting for clean absorption features to obtain equivalent widths. We employ MOOG (Sneden 1973) and Kurucz model atmospheres (Kurucz 1993) in a fully automated analysis to derive effective temperature (T_e), spectroscopic surface gravity ($\log g_s$), microturbulent velocity (ξ), and iron abundance. Values of T_e, $\log g_s$, and ξ are obtained by simultaneously solving for constant Fe I abundance across excitation potential and reduced equivalent width, and ionization balance between Fe I and Fe II. These solutions are used to obtain abundances for other elements. We use Bayesian estimation (Jørgensen & Lindegren 2005) of Padova isochrones (Girardi et al. 2002) to obtain age and mass.

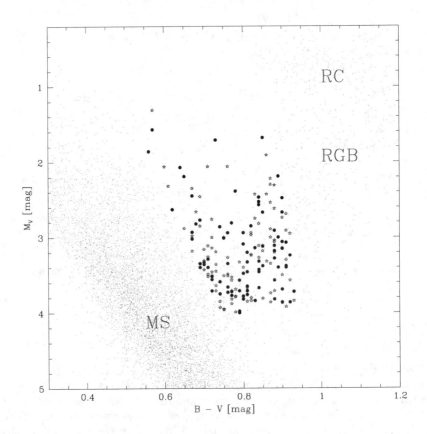

Figure 1.   Clean subgiants in our sample are shown as large dots (observed as of 18 July 2005) and stars (unobserved as of 18 July 2005). Small points are all Hipparcos stars within 100 pc. Other stellar populations seen here are the main sequence (MS), the red giant branch (RGB), and the red clump (RC).

3.   Initial Results — Stellar Parameters & Abundances

We use hi-res spectral atlases of the Sun and Arcturus to test the validity of our method. For Arcturus, we also have observed data to test the stability of our method at our echelle resolution. The results for these two stars agree with current standard values within errors, which implies that the method should work for our subgiants.

Disagreement between spectroscopic and trigonometric surface gravities has been reported by Allende Prieto et al. (1999) and Thorén, Edvardsson, & Gustafsson (2004), but is not well understood. In our results, we see an average scatter of ~0.2 dex in the difference between spectroscopic surface gravity from our automatic method and its trigonometric counterpart from Bayesian estimation.

The scatter is slightly larger at lower metallicity, which could be due to non-local thermodynamic equilibrium (NLTE) effects (Allende Prieto et al. 1999).

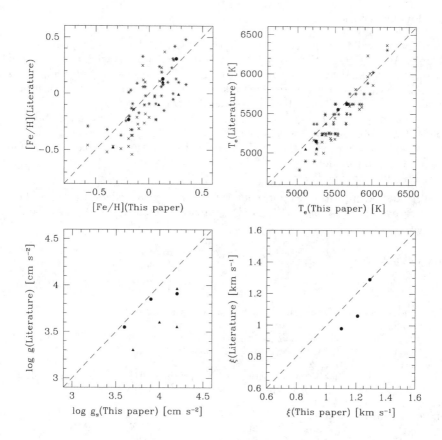

Figure 2.   Comparison of metallicity and stellar parameters between our method and literature. Filled circles, filled triangles, crosses, and asterisks represent values from Takeda et al. (2005), Cayrel de Strobel, Soubiran, & Ralite (2001), Nordström et al. (2004), and Feltzing, Holmberg, & Hurley (2001), respectively. All filled symbols are values from spectroscopic studies. Crosses and asterisks are values from photometric studies.

Comparison of our results with literature values are shown in Fig. 2. Our values are consistent with those from Takeda et al. (2005), who also used a fully automated analysis but with a different algorithm. It is difficult to determine the significance of our disagreement with Cayrel de Strobel et al. (2001) because their catalog is a compilation of various papers.

We have considerable scatter and apparent offsets of ~0.1 dex and ~100 K for [Fe/H] and T_e respectively compared to results from photometric studies (Feltzing et al. 2001; Nordström et al. 2004). However these offsets are within the errors. Several stars that are in both photometric studies demonstrate [Fe/H] discrepancies as large as ~0.5 dex between the papers, which suggests inconsistency in photometric calibrations.

4. Future Work

More spectra will be acquired until we have covered all the clean subgiants in our sample. We are currently investigating the discrepancy between our data and literature values. We will improve the accuracy of our abundances by enhancing the robustness of our method. Our results will be available in the form of a catalog that is easily accessible to the scientific community.

The ultimate scientific goal of the project is to use the observed metallicity to constrain solar neighborhood star formation history (SFH) and chemical evolution. By incorporating observables into SFH derivation via color magnitude diagram (CMD) modeling, we will have results that are not only qualitatively accurate but also quantitatively.

Acknowledgments. Support for this work is provided by the NSF. This research has made use of the SIMBAD database, operated at CDS, Strasbourg, France.

References

Allende Prieto, C., García López, R. J., Lambert, D. L., & Gustafsson, B. 1999, ApJ, 527, 879
Cayrel de Strobel, G., Soubiran, C., & Ralite, N. 2001, A&A, 373, 159
Feltzing, S., Holmberg, J., & Hurley, J. R. 2001, A&A, 377, 911
Girardi, L., Bertelli, G., Bressan, A., Chiosi, C., Groenewegen, M. A. T., Marigo, P., Salasnich, B., & Weiss, A. 2002, A&A, 391, 195
Jørgensen, B. R., & Lindegren, L. 2005, A&A, 436, 127
Kurucz, R. 1993, ATLAS9 Stellar Atmosphere Programs and 2 km/s grid, CD-ROM No. 13. (Cambridge: Smithsonian Astrophysical Observatory)
Nordström, B., et al. 2004, A&A, 418, 989
Sneden, C. 1973, ApJ, 184, 839
Takeda, Y., Ohkubo, M., Sato, B., Kambe, E., & Sadakane, K. 2005, PASJ, 57, 27
Thorén, P., Edvardsson, B., & Gustafsson, B. 2004, A&A, 425, 187

Frank N. Bash Symposium 2005: New Horizons in Astronomy
ASP Conference Series, Vol. 352, 2006
S. J. Kannappan, S. Redfield, J. E. Kessler-Silacci, M. Landriau, and N. Drory

Mapping Convection Using Pulsating White Dwarf Stars

M. H. Montgomery

Department of Astronomy, University of Texas, Austin, TX, USA

Abstract.   We demonstrate how pulsating white dwarfs can be used as an astrophysical laboratory for empirically constraining convection in these stars. We do this using a technique for fitting observed non-sinusoidal light curves, which allows us to recover the thermal response timescale of the convection zone (its "depth") as well as how this timescale changes as a function of effective temperature. We also obtain mode identifications for the pulsation modes, allowing us to use asteroseismology to study the interior structure of these stars.

1.   The Model

Our approach for deriving information on the depth of the convection zone and its temperature sensitivity is based on the seminal numerical work of Brickhill (1992) and on the complementary analytical treatment of Goldreich & Wu (1999) and Wu (1997, 2001); the details of the implementation are described in Montgomery (2005a,b). In terms of the fitting procedure, the parameters to be derived are

θ_i inclination angle
τ_0 average convective response timescale
N "convective exponent"
A intrinsic amplitude of mode
ℓ angular quantum number
m azimuthal quantum number

Over the temperature range of the ZZ Ceti (11000–12000 K) and DBV (22000–28000 K) instability strips, we parameterize the instantaneous convective response timescale as

$$\tau_C = \tau_0 \left(\frac{T_{\text{eff}}}{T_{\text{eff},0}} \right)^{-N},$$

where τ_0 , N, and $T_{\text{eff},0}$ are constants. From theoretical models using mixing-length theory, we expect that $N \sim 95$ for the ZZ Cetis and ~ 25 for the DBVs.

2.   Fits to the Observations

In Figure 1, we show our fits to the observed light curves (lower panels) and the thermal response timescale of the convection zone, τ_C, which we derive from these fits (upper panels). The dashed line in the upper panels indicates the "prediction" of mixing length theory for the variation with T_{eff} of τ_C. While

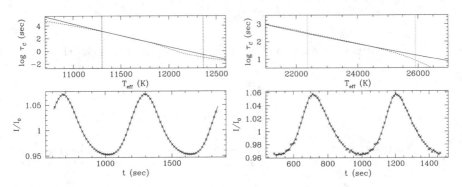

Figure 1.   Light curve fits for G29-38 (left) and PG1351+489 (right)

Table 1.   Derived parameters for G29-38

θ_i (deg)	τ_0 (sec)	N	Amp	ℓ	m	MSD[b]
65.5[a]± 3.4	187.4 ± 20.3	95.0 ± 7.7	0.259 ± 0.011	1	1	0.160 ± 0.045
73.9 ± 0.7	150.1 ± 6.5	7.1 ± 0.6	0.417 ± 0.025	1	0	0.178 ± 0.035

[a]preferred fit
[b]MSD $\equiv \frac{1}{N} \sum_{i=1}^{N} [I_{\mathrm{obs}}(i) - I_{\mathrm{calc}}(i)]^2$

Table 2.   Derived parameters for PG1351+489

epoch	θ_i (deg)	τ_0 (sec)	N	Amp	ℓ	m	MSD[a]
1995	57.8 ± 1.6	86.7 ± 8.3	22.7 ± 1.3	0.328 ± 0.018	1	0	4.15 ± 0.65
2004	58.9 ± 3.1	89.9 ± 3.6	19.2 ± 2.1	0.257 ± 0.021	1	0	0.95 ± 0.25
1995	0.0 ± 5.9	85.1 ± 8.8	18.1 ± 1.4	0.305 ± 0.014	2	0	4.04 ± 0.74
2004	0.0 ± 6.1	89.0 ± 6.1	16.0 ± 1.1	0.233 ± 0.011	2	0	0.94 ± 0.15

this theoretical line may be moved up or down by a different choice of the mixing length to scale height ratio (α), its slope is relatively independent of α, so these fits indicate that parameterizing the convective response through an appropriate choice of α may be meaningful.

For G29-38, two fits yielded comparable values of MSD, the "mean squared deviation" (see Table 1 for a definition of MSD). However, the fit with $\ell = 1$, $m = 0$ yields an improbably low value for the exponent N: the value $N \sim 7$ is not only inconsistent with any version of mixing length theory, it also disagrees with the results of numerical hydrodynamical simulations (Freytag et al. 1995; Kupka & Montgomery 2002). Therefore, the only viable solution is the $m = 1$ solution, which has a value of N exactly in the range expected from mixing length theory.

For PG1351+489, data from two different epochs were analyzed, and as was the case for G29-38, there are two fits with comparable residuals to choose from. Concentrating on the $\ell = 1$ fit, we see that all the parameters are approximately the same in the 1995 data set as they were in the 2004 data set, with the exception of the amplitude. Thus, the background state of the star has remained

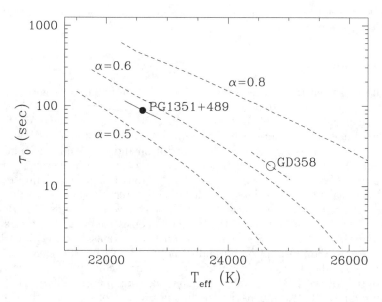

Figure 2.   Mapping convection across the DB instability strip.

the same while the pulsation mode has decreased in amplitude by about 25%. These conclusions remain unchanged if the $\ell = 2$ fit is used. We note that the value of τ_0 is statistically the same for both solutions, and that the only derived parameter which is markedly different is the inclination angle, θ_i, which is $0.0°$ (i.e., "pole on") for the $\ell = 2$ solution. Although such a viewing angle is unlikely, and the value of N is somewhat lower than we theoretically expect, we are not yet in a position to rule out this solution.

3.   Mapping the Instability Strips

The next goal of this research is to map how the convective properties change as one crosses the DAV and DBV instability strips. In particular, we are proposing to observe the DBV GD 358 with the Whole Earth Telescope (WET) in May/June 2006, and this should allow us to perform a similar analysis of its convective properties.

In the Figure 2, we show our previous determination (Montgomery 2005b) of the position and slope of PG1351+489 in the $\log \tau_0 - T_{\text{eff}}$ plane, and we show the predictions of mixing length theory for various values of α (dashed curves). The open circle with dashed line indicates the added constraints which a determination for GD 358 could provide.

In addition to this, we have begun a research program at the University of Texas and McDonald Observatory to observe and analyze suitable stars throughout both the DAV and DBV instability strips. The Argos instrument (Nather & Mukadam 2004), which is placed at prime focus of the McDonald 2.1m telescope, is ideal for this work and has revolutionized the time series observation of faint stellar objects.

4. Conclusions

The solutions shown in the preceding sections provide remarkably good fits to the observed light curves. In addition, the fact that the derived values of N are in the range expected ($N \sim 95$ for the ZZ Cetis and $N \sim 25$ for the DBVs) lends further legitimacy to this approach. From a theoretical standpoint, although N is relatively insensitive to the convective prescription and/or the mixing length α, τ_0 is a strong function of α. Using values for T_{eff} and $\log g$ (surface gravity) from the literature (G29-38: Bergeron et al. 2004; PG1351+489: Beauchamp et al. 1999), we can calculate the value of α which best matches our derived values of τ_0. Using ML2 to denote the "efficient" version of convection proposed by Böhm & Cassinelli (1971), we find that ML2/α=0.6 is consistent with the results for G29-38 and that ML2/α=0.5 is consistent with the results for PG1351+489. This agrees with the prescription normally adopted for fitting the spectra of ZZ Cetis (ML2/α = 0.6, Bergeron et al. 2004) but is markedly different from that often adopted for the DBVs (ML2/α = 1.25, Beauchamp et al. 1999).

In future analyses we will consider the fitting process for the case when there are two or more modes simultaneously present. This is necessary since most pulsating white dwarfs are multi-periodic. The ultimate goal of this research is to apply this method to all suitable stars across both instability strips and to map out the function $\tau(T_{\mathrm{eff}}, \log g)$. This will provide a valuable contact point for future theories of convection.

Acknowledgments. The author would like to thank D. Koester, D. O. Gough, D. E. Winget, and T. S. Metcalfe for useful discussions, and F. Mullally, S. J. Kleinman, and S. O. Kepler for providing some of the data analyzed in section 2. This research was supported in part by NSF grant AST-0507639.

References

Beauchamp, A., Wesemael, F., Bergeron, P., Fontaine, G., Saffer, R. A., Liebert, J., & Brassard, P. 1999, ApJ, 516, 887

Bergeron, P., Fontaine, G., Billères, M., Boudreault, S., & Green, E. M. 2004, ApJ, 600, 404

Böhm, K. H., Cassinelli, J., 1971, A&A, 12, 21

Brickhill, A. J. 1992, MNRAS, 259, 519

Freytag, B., Steffen, M., & Ludwig, H.-G. 1995, in Proceedings of the 9th European Workshop on White Dwarfs, ed. D. Koester & K. Werner (Berlin: Springer-Verlag), 88

Goldreich, P., & Wu, Y. 1999, ApJ, 511, 904

Kupka F., Montgomery M. H., 2002, MNRAS, 330, L6

Montgomery, M. H. 2005a, ApJ, 633, 1142

Montgomery, M. H. 2005b, in Proceedings of the 14th European Workshop on White Dwarfs, ed. D. Koester & S. Moehler (San Francisco: ASP), 483

Nather, R. E., & Mukadam, A. S. 2004, ApJ, 605, 846

Wu, Y. 1997, PhD thesis, California Institute of Technology

Wu, Y. 2001, MNRAS, 323, 248

Frank N. Bash Symposium 2005: New Horizons in Astronomy
ASP Conference Series, Vol. 352, 2006
S. J. Kannappan, S. Redfield, J. E. Kessler-Silacci, M. Landriau, and N. Drory

Searching for Planets around Pulsating White Dwarf Stars

F. Mullally,[1] D. E. Winget,[1] and S. O. Kepler[2]

[1] *Department of Astronomy, University of Texas, Austin, TX, USA*
[2] *Instituto de Física, Universidade Federal do Rio Grande do Sul, RS, Brazil*

Abstract. We report on our continuing search for planets around pulsating white dwarf stars. All stars with a main sequence mass less than $\sim$1 $M_\odot$ are expected to become white dwarf stars (WDs), so a search for planets around WDs is a search around the evolutionary end point of a wide range of main sequence spectral types. The range of spectral types also means a broad range of main-sequence lifetimes enabling us to probe planet formation rates as a function both of spectral type and Galactic history. We present data for three stars in our sample and predict the region of mass-orbital separation space explored by monitoring these stars for just three years.

1. Introduction

White dwarf stars (WDs) of spectral type DA (hydrogen atmosphere) pulsate in an instability strip between 11,000 K and 12,500 K (see review by Winget 1998) and are called ZZ Ceti stars. High frequency modes in the hot ZZ Ceti stars, also known as hDAV (hot DA Variable) stars, exhibit extreme amplitude and frequency stability. Well monitored hDAVs show a drift rate less than a few times 10^{-15} for the most stable modes (Kepler et al. 2005; Mukadam et al. 2003), making them more stable than most pulsars and comparable to atomic clocks. The drift in period is caused primarily by the cooling of the star and therefore the period increases monotonically.

When such stable clocks possess an orbital companion, their reflex motion around the center of mass of the system becomes measurable, providing a means to detect a planet. The change in position changes the light travel time of the pulses from the star to the Earth and hence the measured phase of the light curve. The change in pulse arrival times $(O - C)$ due to the presence of a planet is given by

$$O - C = \frac{a m_p \sin(i)}{M_{wd}}, \tag{1}$$

where a is the semi-major axis of the orbit, m_p is the planet mass, i is the angle of inclination, and M_{wd} is the mass of the white dwarf. This is similar to the way in which planetary mass objects have been detected around pulsars (Wolszczan & Frail 1992). Their method differs from ours only in the nature (and history) of the clock, and the periods of the pulses (Figures 1a, 2a).

As the amplitude of the change in light travel time is determined by the distance from the star to the center of mass of the system, our planet detection sensitivity increases linearly with both the mass of the planet and its distance

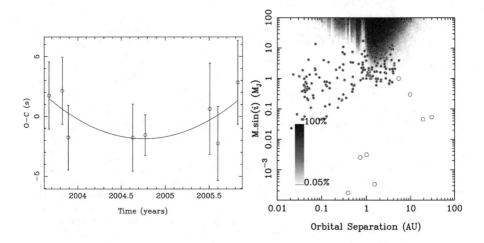

Figure 1.   *Left:* O-C diagram for a sample star from this survey (GD244) with data spanning approximately two years. A positive value of O-C means the pulsations were observed later than expected. The solid line shows the best fit parabola to the data which indicates the presence of a planet through Equation 2. Although this curvature looks impressive, it is consistent within the measurement uncertainties with a straight line. There is as yet no evidence of the presence of a planetary companion around either star.
Right: Region of detection space where the presence of planets has been constrained for GD244 thus far. The open circles are planets from our own solar system. The filled dots are known extra solar planets, mostly discovered by the radial velocity method. The shaded area indicates the region where planets would have been detected had they been present, with black indicating unambiguous detection and white for regions where planets would remain undetected.

away from the star. We can therefore probe as yet unexplored regions of the mass - orbital separation plane (Figures 1b, 2b).

Planets in orbits very much longer than our baseline of observations will be observable as an acceleration of the star, causing a change in pulsational period ($\dot{P}$). By constraining $\dot{P}$ we can place an upper limit on the region of detectability of planets. This limit can rule out the presence of a Jupiter mass planet at 5 AU after 3 years. $\dot{P}$ is given by

$$\dot{P} = \frac{P_{puls}}{c} \frac{Gm_p}{a^2} \tag{2}$$

(Kepler et al. 1991) where P_{puls} is the pulsation period, c is the speed of light, and G is Newton's constant of gravitation. Our measurement or constraint of $\dot{P}$ constrains the presence of planets in orbit around the star.

With a sufficiently long baseline of observations ($\sim$10 years) our technique becomes sensitive to pulsational period drift due to cooling of the star. Cooling places the ultimate limit on our ability to constrain planets in long orbital periods. A typical $\dot{P}_{cooling} \sim 10^{-15}$ has the equivalent effect on arrival times as a 1 M_J companion with an orbital separation of approximately 50 AU.

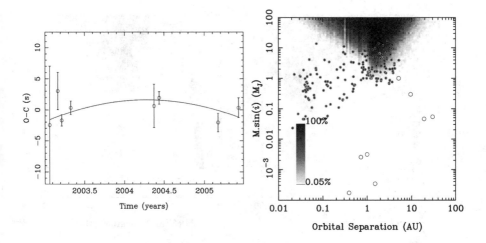

Figure 2. Same as Figure 1 except for a different star, WD1354+0108. There is no evidence in this data of a companion to this star.

1.1. Will They Survive?

Theoretical work indicates outer terrestrial planets and gas giants will survive the red giant phase (e.g., Rasio et al. 1996), and be stable on timescales longer than a billion years, long enough for the star to cool to the DAV pulsational instability strip (Duncan & Lissauer 1998). A sub-stellar companion, recently discovered around the K giant ι Dra (Frink et al. 2002), the first planet or brown dwarf found orbiting a red giant star, lends observational support to these calculations. However a null result will have implications for the survival of planets during the red giant phase, and will still be scientifically interesting. Non detection of planets has implications for our understanding of the red giant phase and long term planetary stability.

2. Our Project

Since beginning this project in late 2001, we have increased the number of known DAVs suitable for long term monitoring (the hDAVs) to over 30 (Mukadam et al. 2004; Mullally et al. 2005) using the Argos prime focus CCD photometer (Nather & Mukadam 2004) on the 2.1 m Otto Struve Telescope at McDonald Observatory. This is the ideal instrument for our search, with a minimum of optical surfaces to maximize light gathering capability and a frame transfer read out which eliminates time lost to chip readout.

We have been monitoring each star since discovery. Data on one previously known star (Figure 3 left) extends back as far as the 1970s, however the majority of our sample have baselines that start in 2002 or 2003. In Figure 3 (right), we predict the limits we will be able to place on stars based on three years of data.

Acknowledgments. This work is supported by a grant from the NASA Origins program, NAG5-13094 and performed in part under contract with the

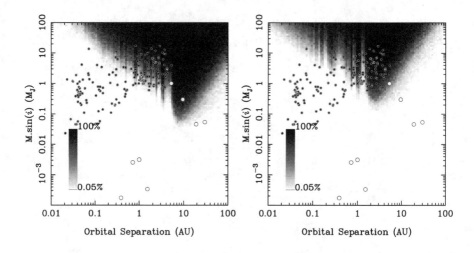

Figure 3.   *Left*: Detection limits for planetary companions around G117-
B15A based on data spanning 30 years. With this dataset, the presence of a
planet with $m_p \sin i$ and orbital separation similar to Saturn is ruled out.
Right: The same plot indicating projected detection efficiency for our survey
after 3 years. Planets with orbital properties similar to many known extrasolar planets as well as Jupiter analogues will be detectable.

Jet Propulsion Laboratory (JPL) funded by NASA through the Michelson Fellowship Program. JPL is managed for NASA by the California Instituute of
Technology.

References

Duncan, M. J. & Lissauer, J. J. 1998, Icarus, 134, 303
Frink, S., Mitchell, D. S., Quirrenbach, A., Fischer, D. A., Marcy, G. W., & Butler,
        R. P. 2002, ApJ, 576, 478
Kepler, S. O., et al. 2005, ApJ, 634, 1311
Kepler, S. O., et al. 1991, ApJ, 378, L45
Mukadam, A. S., et al. 2004, ApJ, 607, 982
Mukadam, A. S., et al. 2003, Baltic Astronomy, 12, 71
Mullally, F., et al. 2005, ApJ, 625, 966
Nather, R. E. & Mukadam, A. S. 2004, ApJ, 605, 846
Rasio, F. A., Tout, C. A., Lubow, S. H., & Livio, M. 1996, ApJ, 470, 1187
Winget, D. E. 1998, Journal of the Physics of Condensed Matter, 10, 11247
Wolszczan, A. & Frail, D. A. 1992, Nat, 355, 145

Frank N. Bash Symposium 2005: New Horizons in Astronomy
ASP Conference Series, Vol. 352, 2006
S. J. Kannappan, S. Redfield, J. E. Kessler-Silacci, M. Landriau, and N. Drory

Evidence for an Intermediate Mass Black Hole in Omega Centauri

Eva Noyola,[1] Karl Gebhardt,[1] and Marcel Bergmann[2]

[1] *Department of Astronomy, University of Texas, Austin, TX, USA*
[2] *NOAO Gemini Science Center (Chile), Tucson, AZ, USA*

Abstract.
    The globular cluster ω Cen is one of the largest and most massive members of the galactic system. However, its classification as a globular cluster has been challenged making it a candidate for being the stripped core of an accreted dwarf galaxy; this together with the fact that it has one of the largest velocity dispersions for star clusters in our galaxy makes it an interesting candidate for harboring an intermediate mass black hole. We measure the surface brightness profile, and we find a central power-law cusp of slope -0.08. We analyze Gemini GMOS-IFU kinematic data for a $5'' \times 5''$ field centered on the cluster, as well as for a field $14''$ away. We detect a clear rise in the velocity dispersion from 20 km/s at $14''$ to 25 km/s in the center. An isotropic, spherical dynamical model requires a rise in the central M/L of a factor of three. Assuming a constant M/L for the stars, this rise implies a black hole of mass of $5 \times 10^4 \ M_\odot$.

1. Photometry

We measure integrated light from an ACS F435W image applying the technique in Noyola & Gebhardt (2005), which uses a robust statistical estimator, the biweight, to calculate the number of counts per pixel on a given annulus around the center of the cluster. As a test, we also measure the profile from a Fabry-Perot narrow band Hα image with lower spatial resolution. Since both images have a limited radial extent, we need to complete the surface brightness profile to cover the entire radial extent of the cluster; for this we use the Chebyshev fit of Trager et al. (1995). This profile also provides the means to normalize our photometry, so in effect, we are updating the profile for the inner $40''$. The measured profiles from the two images are consistent as can be seen in Figure 1. The solid line is a smooth fit made to the combination of the photometric points from ACS inside $40''$ and Trager's Chebyshev fit outside $40''$. For comparison, we include Trager et al.'s photometric points in the plot. The surface brightness profile shows a continuous rise toward the center with a logarithmic slope of -0.08, which is in contrast to the common notion that ω Cen has a flat core. Baumgardt et al. (2005) performed N-body models of star clusters modeled with King profiles and containing a central black hole. They predict the formation of a shallow cusp of -0.2 logarithmic slope.

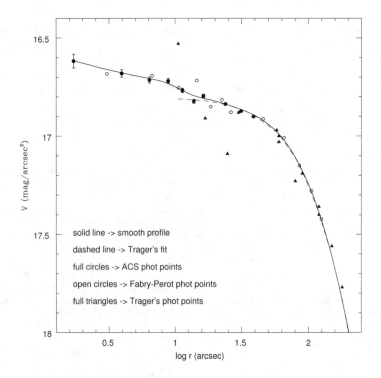

Figure 1.   Surface brightness profiles for ω Cen. The circles show our measured photometric points from the ACS (full) and Fabry-Perot (open) images. The triangles show photometric points obtained from ground based images by Trager et al. (1995). The dashed line is Trager's Chebyshev fit. The solid line is our smooth fit to the combination of the ACS points inside 40″ and Trager's Chebyshev fit outside.

2.   Kinematics

We obtained nod-and-shuffle observations using the Gemini GMOS integral field unit (IFU) with a 5″×5″ field of view. Two fields were observed, one around the cluster center and one more centered 14″ away from it. We obtain a spectrum for every fiber on each field and analyze the CaII triplet region (8490 to 8680 Å). We compare the reconstructed image from the IFU fibers for the central frame with the acquisition image as well as the same region on the ACS image. We also create a convolved image of the ACS frame. The same match is performed for the field 14″ away. Both ACS fields contain ∼100 resolved stars. A combination of the integrated background spectra can measure the velocity dispersion of the ∼40 brightest unresolved stars in those fibers. We use the brightest star on the central field as a template to measure the velocity dispersion of the combined spectra for each field. We combine individual spectra using a biweight estimator for each frame. Since we do not want to be dominated by the brightest stars, a cut

n5139

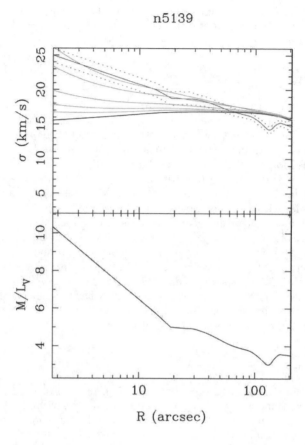

Figure 2. Our measured velocity dispersion profile in comparison with a set of isotropic spherical models of varying black hole masses. In the top panel, the line within dotted lines shows our measured velocity dispersion, the bottom line is the model without a black hole, and the lines above it represent models with central black holes ranging in mass from 5×10^3 $M_\odot$ to 6×10^4 $M_\odot$. The lower panel shows the inferred mass to light ratio (M/L) rise due to the velocity dispersion increase.

in brightness is chosen in order to exclude the fibers for which the contribution of a single bright star is large. Once the bright fibers have been excluded, we combine the spectra of all remaining fibers into one and we use it to measure the line of sight velocity dispersion. Our analysis technique uses a maximum penalized likelihood estimate to obtain a nonparametric line of sight velocity distribution, as described in detail in Gebhardt et al. (2000). We measure a

25 km/sec dispersion for the central field and 20 km/sec for the field $14''$ away. The middle measurement coincides with the central velocity dispersion value measured for ω Cen by various authors. The two measurements provide us with the two innermost points of a velocity dispersion profile completed with Fabry-Perot radial velocity measurements of individual stars at larger radii.

3. Modeling

In order to explore the effect of a central black hole inside ω Cen, we create a set of isotropic models using the non-parametric method described in Gebhardt et al. (1996). We deproject the surface brightness profile to obtain a luminosity density profile. Assuming a constant mass to light ratio (M/L) of 4 we obtain a velocity dispersion profile. We then add a central point mass of various masses ranging from $5 \times 10^3\ M_\odot$ to $6 \times 10^4\ M_\odot$. Figure 2 shows the comparison of the different models and the measured dispersion profile. As can be seen, an isotropic model with no black hole present predicts a small decline in velocity dispersion, but we instead observe a clear rise toward the center. The shape of the outer part of the profile does not match very well and this can be due to the fact that the M/L profile is most likely not constant around the core. Despite the shallow cusp, ω Cen has a close to flat central density profile, which implies that the potential is very shallow and therefore mass segregation cannot be an important effect. There is no reason to expect a large variation of M/L inside the core due to stellar content, so a detected rise in M/L is likely to come from the presence of a concentrated massive object.

4. Summary

We measure the surface brightness profile for ω Cen in the central $40''$. The profile shows a continuous rise toward the center with a logarithmic slope of -0.08, in contrast with previous measurements that found a flat core. We measure the velocity dispersion for two $5'' \times 5''$ regions, one at the center of the cluster and the other $14''$ away. We detect a rise in the velocity dispersion from 20 km/sec for the outer field to 25 km/sec for the central one. Comparing the velocity dispersion profile with a series of isotropic models we conclude that a black hole of $5 \times 10^4\ M_\odot$ is necessary to match the observations.

References

Baumgardt, H., Makino, J., & Hut, P. 2005, ApJ, 620, 238
Gebhardt, K., et al. 1996, AJ, 112, 105
Gebhardt, K., et al. 2000, AJ, 119, 1157
Noyola, E. & Gebhardt, K. 2005, AJ, submitted
Trager, S. C., King, I. R., & Djorgovski, S. 1995, AJ, 109, 218

Frank N. Bash Symposium 2005: New Horizons in Astronomy
ASP Conference Series, Vol. 352, 2006
S. J. Kannappan, S. Redfield, J. E. Kessler-Silacci, M. Landriau, and N. Drory

Photometry of Near Earth Asteroids at McDonald Observatory

J. Györgyey Ries,[1] F. Varadi,[2] E. S. Barker,[3] and P. J. Shelus[4]

[1] *McDonald Observatory, University of Texas, Austin, TX, USA*
[2] *Institute of Geophysics and Planetary Physics, UCLA, Los Angeles, CA, USA*
[3] *Johnson Space Center, Houston, TX, USA*
[4] *Center for Space Research, University of Texas, Austin, TX, USA*

Abstract. The McDonald Observatory Near Earth Object (NEO) group has been involved in confirmation and follow-up efforts since 1995. Expanding this program from astrometry to astrophysics, we are attempting to derive refined absolute magnitudes and rotational periods for Near Earth Asteroids including potentially hazardous objects. We have obtained lightcurves for 2002 EZ11, 2001 CC21, 2003 UV11, 65803 (1996 GT), and 2003 SS84. We were able to determine rotational periods for 2002 EZ11 and 2001 CC21 and identified a short period brightness variation superimposed on a longer trend for the other minor planets. The rotational periods were determined using Singular Spectrum Analysis, which has proven to perform quite well on short, noisy time series. 2003 SS84 and 65803 (1996 GT) were also observed at Arecibo and Goldstone providing a comparison for our results. This research is funded by NASA's NEO Observation Program Grants NAG5-10183 and NAG5-1330.

1. Introduction

Asteroid rotation rate statistics are essential in understanding the dynamical and physical evolution of a planetary system. To correctly deduce the frequency and severity of possible collisions with Earth, we need to know their physical properties. We have rotational periods for about 300 Near Earth Asteroids (NEAs), while the number of known NEAs is now about 3600. The rotation rate distribution of NEAs is non-Maxwellian, and it is not consistent with collisional processes (Harris 2002). The lightcurves also provide information on the size and shape of the asteroids, which in combination with the rotation rates can constrain the bulk density. The variation in brightness is mostly due to rotation for single asteroids, but it can be complex if the asteroid is tumbling. For binaries, it is a combination of rotation and orbital geometry.

2. Observational Aspects

We use differential CCD photometry in the R band with the Prime Focus Camera on the 30 inch telescope at McDonald Observatory. Our limiting magnitude is near $M_V = 22$, using a 15 min exposure. Exposure times are limited by the lack of telescope tracking in declination, or by trailing stellar images in right ascension for fast moving objects. We restrict our exposures to less than 300 seconds in order to collect sufficient numbers of points for our time series analysis. We also assume that the sky is uniform enough over our field of view to carry out relative photometry, or, in other words, that conditions are stable during the exposure.

3.   Time Series Analysis

We use Singular Spectrum Analysis (SSA, Ghil et al. 2002), which was originally developed to deal with short and noisy time series in geophysics. SSA decomposes the time series using the eigenvalues and eigenvectors of a matrix, with elements that are lagged autocorrelations. The largest eigenvalues correspond to the largest oscillatory components in the time series, the smaller ones to noise. A single dominant oscillation is usually represented by a pair of eigenvalues of similar magnitudes (Fig. 1a).

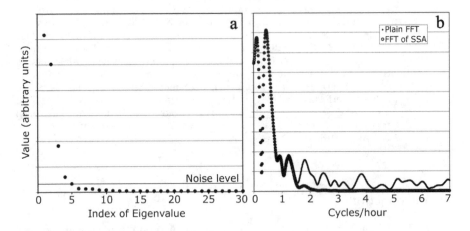

Figure 1.   SSA determined eigenvalues for (65803) 1996 GT (a) and a comparison of the Fourier spectrum of the original and the filtered data (b).

The eigenvectors associated with the eigenvalues provide moving average filters. Filtering with the eigenvectors of the largest eigenvalues reconstructs the strongest signals above the noise (Fig. 1b). The filters are data-adaptive, not necessarily sinusoidal, and can accommodate modulations in the amplitudes of the oscillations.

4.   Results

In 2003, we observed 2002 EZ11 during bright time, through the R filter, spanning 6.5 hours. We found a 0.10 ± 0.02 magnitude peak-to-peak variation in instrumental (relative) magnitudes (Ries et al. 2003). The large field of view of the prime focus corrector (PFC) insured that we had enough comparison stars for a well-determined relative photometric measurement, although the object moved more than 2 degrees on the sky during the night.

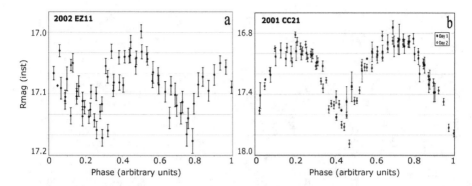

Figure 2.   Composite lightcurves created by folding the time series onto one full phase with the periods determined from SSA.

The lightcurve of 2002 EZ11 (Fig. 2a) is simple. We assumed, that the variations are due to rotation, and that the shape of the asteroid can be approximated by a triaxial ellipsoid, giving two brightness peaks during one revolution. The SSA decomposition indicated one dominant period at 1.16 hours. If our assumption is correct, 2002 EZ11 is not an unusually fast rotator. However, the 2.32 hour period combined with the small amplitude and its **H** (absolute magnitude, defined as the magnitude of an asteroid at zero phase angle and at unit heliocentric and geocentric distances), indicates that it is close to the upper limit of spin rates for bodies held together by self gravity only. We observed (98943) 2001 CC21 on two consecutive nights, and had enough data to obtain separate period estimates. For the composite lightcurve (Fig. 2b) we used 5.02 hours, which is between the two separate estimates. (We did correct for brightness change due to changing distance, but not for light travel time difference.) This rotation rate is not unusual, but the amplitude of the lightcurve is 2 or 3 times larger than the average (0.2 mag) for a typical NEA. Table 1 summarizes the asteroids for which we have obtained lightcurves. (**H** is adopted from the Minor Planet Center (MPC) database and M(V) is the estimated visual magnitude at the time of the observations.)

Table 1.   NEAs observed at McDonald Observatory

Designation	Date	Phase(deg)	**H**	M_V	ΔM_R	Period(h)
65803	11/27/2003	8.3 − 7.6	18.4	13.4	≈ 0.25	2.29*
98943	12/1, 2/2003	23 − 24	18.6	17.5	≈ 0.8	5.02
2002 EZ11	2/17/2003	38 − 35	18.1	15.6	0.10	2.32
2003 SS84	9/28/2003	24.8	21.8	18.1	≥ 0.3	TBD
2003 UV11	10/28/2003	20 − 22	19.4	15.5	≥ 0.3	TBD

Our quick-look photometry program calculates catalog magnitude and instrumental magnitude averages after removing the largest reference star outliers. The differences of the averages are used to correct for changing sky conditions. Although these values have some error due to scatter in the catalog averages, by

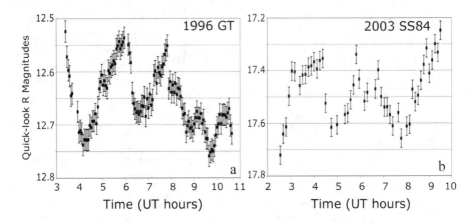

Figure 3.   Quick-look lightcurves.   1996 GT, a binary, has a complex lightcurve (a). While the quick-look data is noisy, we were able to recover the 2.29 hr short period variations reported by Pravec et al. (2005).

the end of the night we can determine whether the target merits follow-up and rigorous calibration (Fig. 3).

5. Summary

Due to the nature of NEA orbits, the viewing geometry changes rapidly, so lightcurves from different observing runs cannot be combined without performing a complete analysis of the orbital position and rotational state of the asteroid. However, with lightcurves obtained at different orbital phases, we can determine **H** and can attempt to determine the orientation of the rotation axis, the general size, and the aspect ratio of the asteroid (Kaasalainen & Torppa 2001). We need more observations for a period determination for 2003 UV11, which shows short period variations superimposed on a longer trend. The source 2003 SS84 also requires detailed observations since it is a binary, and its lightcurve is made more complex by the orbital motion of its components.

References

Ghil, M., et al. 2002, Reviews of Geophysics, 40, 3
Harris, A. W. 2002, Icarus, 156, 184
Kaasalainen, M. & Torppa, J. 2001, Icarus, 153, 24
Pravec, P., et al. 2005, Icarus, 173, 108
Ries, J. G., Barker, E. S., Shelus, P. J., & Ricklefs, R. L. 2003, AAS/Division of
    Dynamical Astronomy Meeting, 34

Frank N. Bash Symposium 2005: New Horizons in Astronomy
ASP Conference Series, Vol. 352, 2006
S. J. Kannappan, S. Redfield, J. E. Kessler-Silacci, M. Landriau, and N. Drory

The Black Hole Mass – Galaxy Bulge Relationship for QSOs in the SDSS DR3

S. Salviander,[1] G. A. Shields,[1] K. Gebhardt,[1] and E. W. Bonning[2]

[1]*Department of Astronomy, University of Texas, Austin, Texas, USA*
[2]*Laboratoire de l'Univers et de ses Théories, Observatoire de Paris, Meudon, France*

Abstract.   We investigate the relationship between black hole mass, M_{BH}, and host galaxy velocity dispersion, σ_*, for QSOs in Data Release 3 of the Sloan Digital Sky Survey. We derive M_{BH} from the broad $H\beta$ line width and continuum luminosity, and the bulge stellar velocity dispersion from the [O III] narrow line width ($\sigma_{[O\ III]}$). Use of Mg II and [O II] in place of $H\beta$ and [O III] allows work at higher redshifts. For redshifts $z < 0.5$, our results agree with the $M_{BH} - \sigma_*$ relationship for nearby galaxies. For $0.5 < z < 1.19$, the $M_{BH} - \sigma_*$ relationship shows nominal evolution with redshift in the sense that, in the mean, the bulges are too small for their black holes. However, we find that approximately half of this trend can be attributed to observational biases. Accounting for these biases, we find an upper limit of 0.3 dex on any evolution in the $M_{BH} - \sigma_*$ relationship between now and redshift $z \approx 1$.

1.   Introduction and Methodology

Recent work has established that the mass of a supermassive black hole, M_{BH}, correlates with properties of the host galaxy's bulge, especially the stellar velocity dispersion, σ_* (Gebhardt et al. 2000; Ferrarese & Merritt 2000; Kormendy & Gebhardt 2001). Tremaine et al. (2002) give this relationship as

$$M_{BH} = (10^{8.13} M_\odot)(\sigma_*/200 \text{ km s}^{-1})^{4.02}. \tag{1}$$

Studying the evolution of this relationship over cosmic time may yield clues as to its origin.

Shields et al. (2003, "S03") used QSO emission lines to study the $M_{BH} - \sigma_*$ relationship for large lookback times. The black hole "photoionization mass" (based on Kaspi et al. 2000) was derived from

$$M_{BH} = (10^{7.69} M_\odot)v_{3000}^2 L_{44}^{0.5}, \tag{2}$$

where $v_{3000} \equiv \text{FWHM}(H\beta)/3000 \text{ km s}^{-1}$ and $L_{44} \equiv \nu L_\nu/(10^{44} \text{ erg s}^{-1})$, the continuum luminosity at 5100 Å. Following Nelson (2000), σ_* was estimated as $\sigma_{[O\ III]} = \text{FWHM}([O\ III])/2.35$. S03 found little change in the $M_{BH} - \sigma_*$ relationship from redshifts $z \approx 2$ to today. However, they had only a small sample of high redshift objects and none in the range $0.3 < z < 1.1$. Here we use the Sloan Digital Sky Survey[1] Data Release 3 (SDSS DR3; Abazajian et al. 2005) to study the $M_{BH} - \sigma_*$ relationship at redshifts up to $z \simeq 1.2$.

We have analyzed the DR3 QSO spectra with the aid of an automated spectrum fitting program. A lower-redshift sample ($0.10 < z < 0.80$) was used for study of the $M_{BH} - \sigma_*$ relationship using Hβ and [O III] (the "HO3" sample). A higher redshift sample ($0.46 < z < 1.19$) was chosen to study the $M_{BH} - \sigma_*$ relationship using Mg II and [O II] in place of Hβ and [O III], and the continuum luminosity at 4000 Å scaled to 5100 Å by assuming a power law function fitted by Vanden Berk et al. (2001) for SDSS quasar composite spectra, $F_\nu \propto \nu^{-0.44}$ (the "MO2" sample). The algorithm fit the selected lines and continuum wavelengths, and the line FWHM was determined from the fits. Data with poor error bars were discarded, and the remaining fits were inspected visually for reliability. The widths of Mg II and Hβ agree closely in the mean, as do those of [O II] and [O III], supporting the use of Mg II and [O II] at higher redshifts (see also McLure & Jarvis 2002). This material will be presented in more detail in Salviander et al. 2005.

2.   The $M_{BH} - \sigma_*$ Relationship

The $M_{BH} - \sigma_*$ relationship for both the HO3 and MO2 samples is shown in Figure 1. For low redshifts, the results are consistent with the findings of S03 based on Hβ and [O III]. For higher redshifts, the mean $\sigma_{[O\ III]}$ and $\sigma_{[O\ II]}$ depart from the σ_* expected from the $M_{BH} - \sigma_*$ relationship for nearby galaxies, in the sense that $\sigma_{[O\ III]}$ and $\sigma_{[O\ II]}$ are too small.

3.   Redshift Dependence

A diagnostic of the evolution of the $M_{BH} - \sigma_*$ relationship with lookback time is shown in Figure 2. We compare M_{BH} calculated with Equation 2 to the "[O III] mass" of S03—that is, M_{BH} calculated with Equation 1 using $\sigma_{[O\ III]}$ in place of σ_*. The mean $\Delta \log M_{BH} \equiv \log M(H\beta) - \log M([O\ III])$ is +0.14 for the HO3 sample, but this offset is comparable to the calibration uncertainties of Equation 2. The MO2 sample shows a mean $\Delta \log M_{BH}$ of +0.40 dex.

At face value, Figure 2 indicates that $\Delta \log M_{BH}$ evolves slightly for $z > 0.5$, both for the MO2 sample and the higher redshift objects in the HO3 sample. Is this a real dependence of $M_{BH} - \sigma_*$ relationship on cosmic time? One concern is the effect of the limiting magnitude of the sample survey along with our own selection for good-quality data, which favors brighter objects. This leads to a correlation between L, M_{BH}, and z. For nearby galaxies, Tremaine et al. (2002) find an rms dispersion of 0.3 dex in $\log M_{BH}$ at fixed σ_*. For a given galaxy mass, M_{gal}, galaxies with larger M_{BH} tend to have higher L and will thus be over-represented in a magnitude-limited sample. This introduces a positive bias in the mean $\Delta \log M_{BH}$ for the sample. To estimate this bias we conducted simple Monte Carlo trials. Using simple assumptions about the distribution of M_{gal} and $L/L_{Eddington}$ we modeled a distribution of black hole masses and

---

[1]The SDSS Web site is http://www.sdss.org/

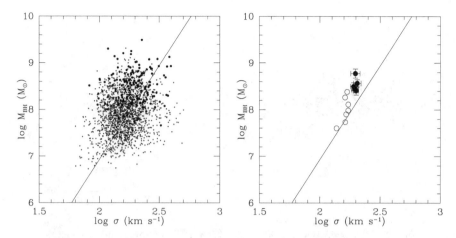

Figure 1. The $M_{BH} - \sigma_*$ relationship. M_{BH} is calculated from Equation 2; σ_* is derived from the FWHM of [O III] or [O II]. Left: Small and large circles show data for the HO3 and MO2 samples. The solid line is the Tremaine relationship for nearby galaxies (Equation 1) and is not a fit to the data. Right: Mean M_{BH} and σ_* for redshift bins $\Delta z = 0.1$. Open and filled circles show data for the HO3 and MO2 samples. Error bars show the standard error of the mean. HO3 sample error bars are smaller than the data points.

luminosities, and discarded all QSOs below a threshold luminosity, L_{cut}, which increases with redshift. We then computed mean values of $\log M_{gal}$, $\log M_{BH}$, $\log L$, and $\Delta \log M_{BH}$. As expected, we found a positive bias in $\Delta \log M_{BH}$ that increases with increasing L_{cut}. The result of our simulations suggest that, due to the scatter in the $M_{BH} - \sigma_*$ relationship, observational selection favors the brighter QSOs, and hence the bigger black holes, for a given galaxy mass. This bias can account for ~ 0.1 dex of the trend in $\Delta \log M_{BH}$ in Figure 2.

Another issue is the possible exclusion of objects with larger $\sigma_{[O III]}$ or $\sigma_{[O II]}$. The tendency for spectra to become noisier with redshift means that for a given equivalent width, wider [O III] and [O II] lines are more difficult to measure. This increases the chance that higher-redshift QSOs with large $\sigma_{[O III]}$ or $\sigma_{[O II]}$ will be rejected from the sample due to large error bars. To test the degree of bias introduced by this effect, we generated mock QSO spectra with noise to mimic our HO3 sample. For a given Hβ FWHM and continuum luminosity, we calculated M_{BH} and inferred $\sigma_{[O III]}$ using Equation 1 with a dispersion $\delta \sigma_{[O III]} = 0.13$ dex from Bonning et al. (2005). We then used our algorithm to fit Hβ, [O III], and the continuum flux at 5100 Å in the mock spectra as though they were real data, and we computed mean values for $\Delta \log M_{BH}$ in redshift bins as in Figure 2. We found a positive bias in $\Delta \log M_{BH}$ that increases with redshift. The result of this simulation suggests that due to increasingly noisy spectra with redshift, our sample favors narrower $\sigma_{[O III]}$ and $\sigma_{[O II]}$ for a given M_{BH}. This bias can account for ~ 0.1–0.2 dex of the trend in $\Delta \log M_{BH}$ in Figure 2.

Since these biases arise from two unrelated phenomena—the scatter in the $M_{BH} - \sigma_*$ relationship and the scatter in the $\sigma_{[O III]} - \sigma_*$ relationship—we may add

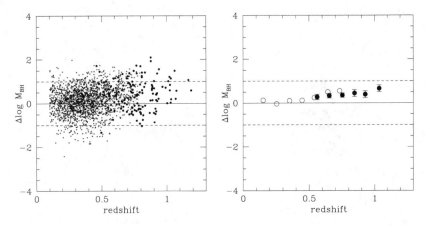

Figure 2.   Redshift dependence of $\Delta \log M_{\rm BH}$. Left: Small and large circles show data for the HO3 and MO2 samples. Right: Mean $\Delta \log M_{\rm BH}$ as a function of redshift. Open and filled circles show the average $\Delta \log M_{\rm BH}$ in redshift bins $\Delta z = 0.1$ for the HO3 and MO2 samples. Error bars show the standard error of the mean. HO3 sample error bars are smaller than the data points.

them linearly to estimate their cumulative effect on the mean $\Delta \log M_{\rm BH}$. We find that half of the overall trend in Figure 2 can be attributed to observational selection effects. Given the uncertainties in our estimates, we therefore place an upper limit of 0.3 dex on the evolution of $\Delta \log M_{\rm BH}$ with redshift between now and redshift $z \approx 1$. This suggests, within a factor of two, either contemporaneous growth of black holes and bulges or that black holes and bulges have completed their growth by $z \approx 1$.

Acknowledgments.   This work was supported by Texas Advanced Research Program grant 003658-0177-2001, by NSF grant AST-0098594, and by Space Telescope Science Institute grant GO-09498.04-A.

References

Abazajian, K., et al. 2005, AJ, 129, 1755
Bonning, E. W., et al. 2005, ApJ, 626, 89
Ferrarese, L. & Merritt, D. 2000, ApJ, 539, L9
Gebhardt, K., et al. 2000, ApJ, 539, L13
Kaspi, S., et al. 2000, ApJ, 533, 631
Kormendy, J. & Gebhardt, K. 2001, in AIP Conf. Proc. 586: 20th Texas Symposium on Relativistic Astrophysics, ed. J. C. Wheeler & H. Martel, 363
McLure, R. J., & Jarvis, M. J. 2002, MNRAS, 337, 109
Nelson, C. H. 2000, ApJ, 544, L91
Salviander, S., et al. 2005, ApJ, in prep.
Shields, G. A., et al. 2003, ApJ, 583, 124
Tremaine, S., et al. 2002, ApJ, 574, 740
Vanden Berk, D. E., et al. 2001, AJ, 122, 549

Frank N. Bash Symposium 2005: New Horizons in Astronomy
ASP Conference Series, Vol. 352, 2006
S. J. Kannappan, S. Redfield, J. E. Kessler-Silacci, M. Landriau, and N. Drory

Double Barred Galaxies in N-Body Simulations

Juntai Shen[1,2] and Victor P. Debattista[3,4]

[1] *McDonald Observatory, University of Texas, Austin, TX, USA*
[3] *Astronomy Department, University of Washington, Seattle, WA, USA*

Abstract.   At least one quarter of early-type barred galaxies have secondary bars, which are embedded in their large-scale primary counterparts. The dynamics of such double barred galaxies are still not well understood. We can form long-lived double-barred systems in collisionless N-body simulations. In this paper we focus on a simulated double-barred galaxy and study its dynamical properties. We find the secondary bar rotates faster than its primary counterpart. The amplitude of the secondary bar pulsates and its pattern speed varies regularly, as it rotates through the primary. When the two bars align, the amplitude of the secondary bar is weaker, and it rotates more rapidly than when they are perpendicular. The pulsating nature of secondary bars could have important implications for understanding the central region of such galaxies.

1.   Introduction

Recent imaging surveys have revealed the frequent existence of nuclear bars in a large number of barred galaxies, e.g., Erwin & Sparke (2002) found that double-barred galaxies are surprisingly common: at least one quarter of their sample of 38 early-type optically-barred galaxies harbor small-scale secondary bars. They found that a typical secondary bar is about 12% the size of its primary counterpart. The facts that inner bars are also seen in near-infrared (e.g. Mulchaey et al. 1997; Laine et al. 2002), and they are often found in gas-poor S0s indicate that most of them are stellar structures. Results from these surveys also show inner bars are at a random angle relative to the primary bars, implying that they are probably dynamically independent structures. Shlosman et al. (1989) invoked multiple nested bars to channel gas inflow into galactic centers to feed AGN, in a similar fashion as the primary bar drives gas inward. However, recent work suggests that this mechanism may not be as efficient as originally hoped (e.g. Maciejewski et al. 2002).

Recently Maciejewski & Sparke (1997, 2000) considered the potential of two rigid bars rotating at constant pattern speeds, and found regular orbits in which particles return to a *loop*, instead of the same point, after two bars return to the same relative orientation. Particles trapped around these stable loops could be used to build a long-lived double-barred system. On the simulational side, most of numerical studies have required gas to form directly-rotating secondary bars (Friedli & Martinet 1993; Shlosman & Heller 2002). Since most observed secondary bars are made of old stars, it is more likely that they form out of stars,

---

[2] Harlan Smith Fellow.
[4] Brooks Fellow.

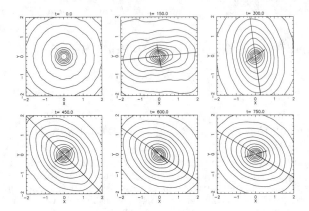

Figure 1.   Snapshots of the iso-density contours of stars at various times.
The short and long straight lines mark the major axes of the secondary and
primary, respectively. The panels are equally spaced in time by about 1.8 Gyr
and are 4 × 4 in units of the scale-length of the initial disk.

instead of gas. Rautiainen et al. (2002) can form long-lived stellar secondary
bars as they simulate central structure of barred galaxies. Their nuclear bars
often have "a vaguely spiral shape" in direct images and do "not reach the
nucleus", so it is unclear how closely they resemble the observed secondary bars.
Kormendy & Kennicutt (2004) pointed out that seeing an inner bar is strong
evidence of the dominance of a pseudobulge in a galactic center. In this work,
we investigate whether or not a rotating bulge can give rise to, or support, the
formation of a long-lived secondary bar in collisionless N-body simulations.

2.   Model Setup

We focus on one example of long-lasting double-barred systems formed in our
simulations. Our high-resolution simulation consists of a live disk and bulge
component. We have not yet included a halo component for simplicity, also
because secondary bars are very small-scale phenomena in galactic centers where
visible matter is dominant. The initial disk has the exponential surface density
profile and Toomre's $Q \sim 2.0$. The bulge was generated using the method
of Prendergast & Tomer (1970) as described in Debattista & Sellwood (2000),
where a distribution function is integrated iteratively in the global potential,
until convergence. The bulge has the mass of $M_b = 0.2M_d$ and is truncated at
$0.7R_d$. The bulge set up this way is un-rotating. As a crude first step, we give
the rotation of the bulge by simply reversing the negative azimuthal velocities of
all bulge particles into the same positive values. We have checked that systems
set up this way are in very good virial equilibrium.

3.   Results

Figure 1 illustrates how iso-density contours of stars evolve. The small-scale bar
forms very early on ($t < 10$), as the dynamical times in the inner galaxy are

much shorter than in the outer part. The pattern rotation of the secondary is very rapid at this stage. The large-scale primary bar starts to develop between $t = 100$ and 200. Due to the formation of the primary bar, the secondary bar is quickly slowed down, and its amplitude drops rapidly. The quantitative evolution of the amplitudes of both bars is shown in Figure 2.

After $t > 250$, the properties of the secondary bar becomes more or less settled. The amplitude of the secondary bar displays regular, yet significant, oscillations. Its amplitude is greater when the bars are perpendicular, and weaker when the two are parallel to each other (Figure 2). On the contrary, the amplitude of the primary bar varies in the opposite sense to the secondary with respect to the relative phase of the two bars. The variation of the primary is, however, much weaker than that of the secondary counterpart. The amplitude of the

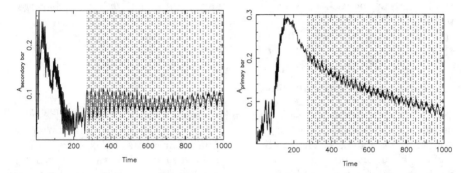

Figure 2.   Time evolution of the bar amplitude of the secondary bar (*left*) and the primary bar (*right*). Dashed lines mark time when the two bars align, while dotted lines mark the time when they are perpendicular to each other.

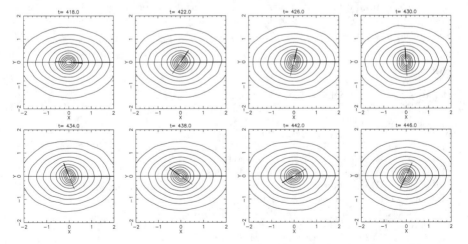

Figure 3.   Non-uniform relative rotation of the secondary bar for roughly half of a period, in the rest frame of the primary bar which remains horizontal. The panels are equally spaced in time. The straight line marks the major axis of the secondary bar.

primary continues to decay (Figure 2) on a timescale of almost a Hubble time, possibly because orbits supporting the primary bar are gradually disrupted by the relatively strong inner bar (Maciejewski & Sparke 2000). The extent of the variation in the secondary bar amplitude also decreases as the primary weakens.

The long-lived secondary bar rotates faster than the primary: between $t=400$ and 500, the average pattern rotation period of the secondary is about $T_p \sim 17.9$, and for the primary $T_p \sim 29.7$. Intriguingly, the pattern speed of the secondary also varies with the relative phase of the two bars: it rotates faster than average when the two bars align, and slower when the bars are orthogonal (Figure 3). The variation of the pattern speed can be more than 20% for the secondary, while much less significant for the primary.

The orbital analysis by Maciejewski & Sparke (2000) found that x_2 loops, which can support the secondary bar, change axis ratios and lead or trail the rigid figure of the secondary bar, as the bars rotate. The pulsating character of the self-consistent secondary bar in our work provides strong evidence that the stars supporting this secondary may be trapped around these x_2 loops, as expected from Maciejewski & Sparke (2000).

4.   Conclusions and Summary

We are able to form a long-lived secondary bar in pure stellar N-body simulations. Bulge rotation may be an important ingredient for the formation of double-barred galaxies. The resultant secondary bar rotates faster than the primary. The amplitude and pattern speed of the secondary bar vary significantly depending on the relative phase of the two bars: the secondary is stronger and rotates more slowly when the bars are perpendicular than when they align. This finding supports that x_2 loops found in Maciejewski & Sparke (2000) may be the backbone of a self-consistent secondary bar. So the assumption of rigidly rotating secondary bars in many previous studies is too strong, and may give misleading results. The implications of the pulsating nature of secondary bars on central gas inflow are unclear and to be studied.

References

Debattista, V. P., & Sellwood, J. A. 2000, ApJ, 543, 704
Erwin, P., & Sparke, L. S. 2002, AJ, 124, 65
Friedli, D., & Martinet, L. 1993, A&A, 277, 27
Kormendy, J., & Kennicutt, R. C. 2004, ARA&A, 42, 603
Laine, S., Shlosman, I., Knapen, J. H., & Peletier, R. F. 2002, ApJ, 567, 97
Maciejewski, W., & Sparke, L. S. 1997, ApJ, 484, L117
Maciejewski, W., & Sparke, L. S. 2000, MNRAS, 313, 745
Maciejewski, W., Teuben, P. J., Sparke, L. S., & Stone, J. M. 2002, MNRAS, 329, 502
Mulchaey, J. S., Regan, M. W., & Kundu, A. 1997, ApJS, 110, 299
Prendergast, K. H., & Tomer, E. 1970, AJ, 75, 674
Rautiainen, P., Salo, H., & Laurikainen, E. 2002, MNRAS, 337, 1233
Shlosman, I., Frank, J., & Begelman, M. C. 1989, Nat, 338, 45
Shlosman, I., & Heller, C. H. 2002, ApJ, 565, 921

Frank N. Bash Symposium 2005: New Horizons in Astronomy
ASP Conference Series, Vol. 352, 2006
S. J. Kannappan, S. Redfield, J. E. Kessler-Silacci, M. Landriau, and N. Drory

Washington Photometry and Hectochelle Spectroscopy of Giant Stars in the Leo II dSph Galaxy

Michael H. Siegel[1] and Steven R. Majewski[2]

[1] *McDonald Observatory, University of Texas, Austin, TX, USA*

[2] *Department of Astronomy, University of Virginia, Charlottesville, VA, USA*

Abstract.   We have used the KPNO 4-meter telescope to obtain high-precision Washington+$DDO51$ photometry over a five square degree region centered on the Leo II dSph galaxy. We identify Leo II red giant branch (RGB) star candidates using two-color and color-magnitude diagrams (CMDs), a very reliable method of identifying giant stars amongst the overwhelming foreground of Galactic dwarfs. The spatial distribution of the giant candidates follows the canonical single-mass King profile in the inner regions of Leo II, but shows faint extended structure beyond the nominal limiting radius. MMT Hectochelle spectra demonstrate that the majority of our Leo II giant candidates have radial velocities near that of Leo II, including 3-5 stars beyond Leo II's King limiting radius. The observed extended structure of Leo II either is the result of tidal disruption during a past perigalacticon or traces an extended halo of dark matter. With high precision ($\sim$1 km s^{-1}) radial velocities, we hope to clarify these issues.

1.   Introduction

Initial surveys of the density profiles of the dSph satellite galaxies of the Milky Way indicated that they conformed closely to a canonical single-mass King (1962) profile (Irwin & Hatzidimitriou 1995). However, recent high-precision wide-field surveys have revealed significant low surface brightness departures from King profiles at large radii (Majewski et al. 2000a; Palma et al. 2003; Majewski et al. 2003; Westfall et al. 2006; Sohn et al. 2006). The extended structures could represent extended bound halos or unbound tidal debris. N-body simulations of satellite galaxy evolution predict tidal disruption leading to the formation of long-lived streams along the dSph's orbit (Johnston et al. 1995, Majewski et al. 2003). If this were the case, the velocity distribution of stars just beyond the dSph limiting radius would constrain the disposition of tidal streams and the orbit of the satellite (Sohn et al. 2006).

Extended structure has been detailed in the Ursa Minor, Draco, Sculptor, Leo I, Carina, Sextans and Sagittarius dSph galaxies. While the nature of the extended structure remains controversial (see Munoz et al. 2005; Wilkinson et al. 2004), it is hoped that high-precision radial velocities will clarify the dynamical origin of the extended structures.

In this contribution, we detail an extensive survey of the Leo II dSph. We use a reliable photometric method to survey a large area around Leo for member stars and use echelle spectra to confirm any extended structure. However, work remains to be done before the dynamical history of Leo II can be revealed.

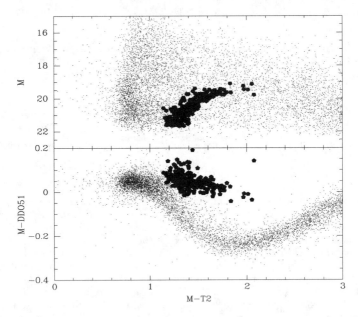

Figure 1.   Selection of Leo II giant star candidates.   Stars with
$\sigma_{M,T_2,DDO51} < 0.05$ are shown. Large dots are probable Leo II RGB stars.

2.   Observations

We observed the Leo II field in November 1998, May 1999 and May 2002 using
the KPNO 4-meter telescope with the MOSAIC camera and Washington M,
T_2 and $DDO51$ filters. Photometry was measured with DAOPHOT/ALLSTAR
(Stetson 1987) and calibrated to the standards of Geisler (1990, 1996).

We used the MMT Hectochelle spectrograph in March 2005 to obtain high-
resolution ($R \sim 30,000$) spectra of targets extracted from the photometric catalog. We integrated for six hours in a single fiber arrangement with the spectrograph at 31st order, centered at 5150 Å. The spectra were extracted using the
IRAF task DOHYDRA and radial velocities were measured by cross-correlation
with standards in the SA 57 field kindly provided by D. Latham.

3.   Results and Conclusions

The Washington+$DDO51$ system of filters provides measures of both stellar effective temperature and the strength of the gravity-sensitive magnesium MgH+b
stellar absorption lines.   Two-color diagrams using this filter system can effectively distinguish low surface-gravity red giant stars from foreground, high
surface-gravity dwarfs (Majewski et al. 2000b), and the method becomes more
powerful (with 90+% efficiency) when combined with a further selection of stars
with the proper color-magnitude combination for a dSph RGB locus.

Figure 1 shows this photometric system applied to Leo II. The arc of stars
at negative $M - DDO51$ color are the foreground of Galactic dwarf stars. The

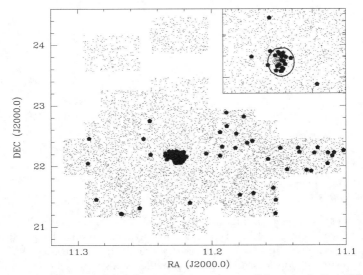

Figure 2.   The spatial distribution of giant star candidates in the Leo II field. Heavy dots are Leo II giants selected using color-color and color-magnitude criteria. The inset shows an approximately one square degree close-up of the core of Leo II, with spectroscopically confirmed members marked with heavy dots.

stars clustered at positive $M - DDO51$ are primarily Leo II giant stars, with a small contribution of compact galaxies, metal-poor subdwarfs and photometric contaminants mixed in (see Siegel et al. 2005). Leo II stars were selected as being along both the giant locus and Leo II's prominent RGB in the CMD.

Figure 2 shows the spatial distribution of Leo II giant candidates across the survey field. While most of the candidates are concentrated within the King limiting radius of Leo II, there is a faint curtain of Leo II giant candidates across the field. This is highly reminiscent of previous results from other dSph galaxies. Determining whether the candidates outside the King limiting radius are *actual* Leo II RGB stars requires high-resolution spectroscopy.

Figure 3 shows the preliminary radial velocities measured for Leo II stars with the Hectochelle spectrograph. Many of the target stars were too faint against the bright moonlit sky for reliable velocities. However, about two dozen of the brightest targets are clustered near a radial velocity of 80.8 km s^{-1}. This clump has a velocity dispersion of 5.3 km s^{-1}. Both numbers are very close to those derived by Vogt et al. (1995; 76 and 6.7 km s^{-1}), indicating that these are, in fact, *bona fide* Leo II members.

The inset of Figure 2 shows the spatial distribution of confirmed Leo II members. Depending on the assumed King radius, 3-5 Leo II stars are outside the canonical boundary of the dwarf, confirming the extended structure.

These widely separated Leo II members either represent the debris of a past interaction with the Milky Way or they trace an extended bound halo of the Leo II dSph. Of course, three extra-tidal stars do not hold enough information to determine Leo II's dynamical history. We are presently engaged in a more

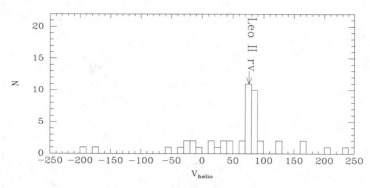

Figure 3.   Velocity histogram of stars in the Leo II field.

extensive spectroscopic campaign to obtain ~ 1 km s^{-1} precision radial velocities for a much larger sample of Leo II stars. Such precise measures have shown that the extended structure in Leo I – a dSph at a similar distance to Leo II – appears to be the result of tidal disruption and provided insight into Leo I's orbit and dynamical history.

Acknowledgments.   This research has been supported by NSF grants AST-0306884, AST-9702521 and AST-0307851.

References

Geisler, D. 1990, PASP, 102, 344

Geisler, D. 1996, PASP, 111, 480

Irwin, M. & Hatzidimitriou, D. 1995, MNRAS, 277, 1354

Johnston, K. V., Spergel, D. N. & Hernquist, L. 1995, ApJ, 451, 598

King, I. 1962, AJ, 67, 471

Majewski, S. R., Ostheimer, J. C., Kunkel, W. E. & Patterson, R. J. 2000b, AJ, 120, 2550

Majewski, S. R., Ostheimer, J. C., Patterson, R. J., Kunkel, W. E., Johnston, K. V. & Geisler, D. 2000a, AJ, 119, 760

Majewski, S. R., Skrutski, M. F., Weinberg, M. D. & Ostheimer, J. C. 2003, ApJ, 599, 1082

Munoz, R. R., et al. 2005, ApJ, 631, L137

Palma, C., Majewski, S. R., Siegel, M. H., Patterson, R. J., Ostheimer, J. C. & Link, R. 2003, AJ, 125, 1352

Siegel, M. H., Majewski, S. R., Gallart, C. Sohn, S. T., Kunkel, W. E. & Braun, R. 2005, ApJ, 623, 181

Sohn, S. T., Majewski, S. R., Munoz, R. M., Kunkel, W. E., Johnston, K. V., Ostheimer, J. C., Guhathakurta, P., Patterson, R. J., Siegel, M. H. & Cooper, M. 2006, in preparation

Stetson, P. B. 1987, PASP, 99, 191

Westfall, K. B., Majewski, S. R., Ostheimer, J. C., Frinchaboy, P. M., Kunkel, W. E., Patterson, R. J. & Link, R. 2006, AJ, 131, 375

Wilkinson, M. I., Kleyna, J. T., Evans, N. W., Gilmore, G. F. Irwin, M. J. & Grebel, E. K. 2004, ApJ, 611, 21

Vogt, S. S., Mateo, M, Olszewski, E. W. & Keane, M. J. 1995, AJ, 109, 151

Frank N. Bash Symposium 2005: New Horizons in Astronomy
ASP Conference Series, Vol. 352, 2006
S. J. Kannappan, S. Redfield, J. E. Kessler-Silacci, M. Landriau, and N. Drory

Time Series Spectroscopy: Mode Identification of White Dwarf Stars

Susan E. Thompson

The Colorado College, Colorado Springs, CO, USA

Abstract.   I show the preliminary results of optical time series spectroscopy taken of the pulsating white dwarf star G 29-38. With these spectra, I identify the spherical degree of the star's largest pulsation modes; a necessary quantity to accurately probe the stellar interior with asteroseismology. Because limb-darkening is wavelength dependent, the observed line-shape variations of the broad Hydrogen Balmer lines depend on the spherical degree of the mode. Thompson et al. (2004) discovered that constrained fitting of the spectral lines enables mode identification where direct measurements of amplitude from the spectra failed. I use the free parameters from fitting the spectra to quantify the variations in the line shape. To identify the spherical degree, the periodic change in the parameter values are compared to fits of spectra created from model atmospheres of white dwarf stars. Simulations of noisy model spectra demonstrate that the scatter around the model limits identification to the the largest amplitude modes.

1.   Introduction

The Hydrogen atmosphere white dwarf pulsators, called DAVs or ZZ Cetis, have evaded unambiguous asteroseismological modeling in part because of the difficulty in measuring the spherical degree (ℓ) of their pulsation modes. Robinson et al. (1995) showed that ℓ can be obtained from time series optical and UV spectroscopy. The effects of limb darkening increase in the blue wavelengths and through the spectral lines, changing the observed surface area of the star and the amplitude of the pulsation mode at that wavelength. Thompson et al. (2004) improved the technique used to analyze the optical spectra by fitting the spectral lines. The chromatic amplitudes (measured mode amplitude at each wavelength bin) created from these fits provides mode identifications with data that otherwise would be unusable, revealing the first $\ell = 4$ mode on a DAV.

I fit each Balmer line to quantify the changing line shapes during pulsation. By creating a time series from the free parameters of the fits, I can easily characterize how the line shape changes during pulsation and compare that with model pulsations of different ℓ. I also run simulations of noisy model spectral series to evaluate the scatter associated with the measurements. I use this method to confirm the mode identifications of G 185-32, established with chromatic amplitudes (Thompson et al. 2004), and then apply the method observations of G 29-38 taken with the Very Large Telescope (VLT).

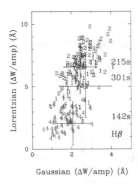 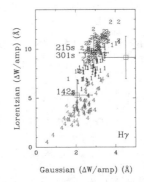

Figure 1.   Mode Identification of G 185-32 using the free parameters of the fits to the spectra. Series of simulated spectra with noise were fit to create the expected scatter around each model spherical degree ($\ell = 1$, 2 and 4). The pulsated model atmosphere has $T = 11,750$ and $log(g) = 8.25$ (cgs units).

2.   Line Shape Variations and G 185-32

The Hβ and Hγ lines of the DAVs are well matched by a linear continuum plus a Gaussian and a Lorentzian function. I fit the average spectrum to establish the basic shape of the line and use these parameters as the initial conditions to fit each individual spectrum. The slope and flux of a linear continuum along with the area of the Gaussian and the Lorentzian vary with each fit. The continuum varies with the brightness variations, while the two areas measure the shape of the line. I use the time series of the free parameters to quantify the changing line shape. Essentially, I normalize the fitted equivalent width of the Gaussian and Lorentzian components by the fractional amplitude of the pulsation mode.

With this quantitative analysis of the changing line shape, simulations of noisy spectra are fit in the same manner to see how noise scatters the ideal measurement. Many series of spectra with the same noise level, time spacing, and time span as the data are created and fit in the same manner described above. This gives a measure for the expected scatter due to noise around the model for each ℓ.

To demonstrate this technique, I analyze the spectroscopy presented by Thompson et al. (2004) of G 185-32. They showed that all data regarding the 142 s mode are consistent with $\ell = 4$, the only high ℓ identified on a DAV. The free parameter fits to the data in Figure 1 agree with the chromatic amplitude determinations of ℓ. The 142 s mode most closely matches the $\ell = 4$ model, while the two largest modes agree with $\ell = 1$ or 2.

3.   G 29-38

G 29-38 is one of the brightest DAVs with large amplitude pulsations. Clemens et al. (2000) first observed chromatic amplitudes with spectra of this DAV and found a potential $\ell = 2$ mode at 776 s. Van Kerkwijk et al. (2000) successfully measured the periodic line-of-sight velocities associated with the pulsations, providing further evidence of the $\ell = 2$ mode.

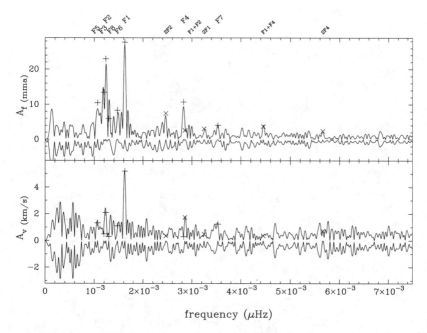

Figure 2.   The Fourier transforms of the light (top) and velocity (bottom) curves of G 29-38. The modes found in the data are labeled at the top. Sinusoidal fits to each curve at these frequencies are marked on the FT. The residuals are shown as reflections beneath each FT.

I obtained a 6 hour series of spectra (604 images) from the VLT, observed with the FORS1 spectrograph on August 27, 1999. Each reduced spectrum spans 3300 Å to 5700 Å, has a dispersion of 2.35 Å per binned pixel, and a signal-to-noise of 80. I created a light curve by averaging the counts collected in the continuum redward of the Hβ line. The velocity curve is a weighted average of the Doppler shifts found by fitting the Hβ–Hδ individually. The largest frequency peak of the velocity curve corresponds to the F1 mode found in the light curve. I found eight individual frequencies and five combination frequencies in the light curve. The amplitude and frequency found by fitting the velocity and light curve at the same frequencies are marked on the Fourier transforms (FTs) in Figure 2. Comparing the flux and velocity measurements can give insight to the convective driving theory (Goldreich & Wu 1999), but is beyond the scope of this article (see van Kerkwijk et al. 2000 and Thompson et al. 2003).

3.1.   Mode Identification of G 29-38

I measure the Gaussian and Lorentzian widths for each mode and compare them to that obtained from model spectra (Figure 3). Error bars are measured from fits at the stated period to the time series of the free parameters. Only the largest amplitude modes result in unambiguous mode identifications. The 810 s and 615 s mode are consistent with $\ell = 1$. The 835 s mode is either $\ell = 1$ or 2 and

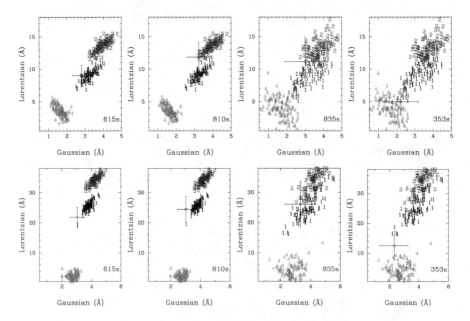

Figure 3.   Variations of the Hβ (top) and Hγ (bottom) line shape quantified by the normalized Gaussian and Lorentzian parameters of the 4 largest observed modes on G 29-38 (labeled F1–F4 on Figure 2). Model simulations, representative of the noise and mode amplitude, are shown for $\ell = 1, 2$ and 4. The pulsated model atmosphere has $T = 11,750$ and $log(g) = 8.25$ (cgs units).

the 353 s mode is $\ell = 1$ or 4. Higher quality data or more nights are necessary to make more definitive statements about the low amplitude pulsations on G 29-38.

These preliminary results on G 29-38 show that time-series spectroscopy is effective for mode identification of DAVs. Current telescopes and instruments, along with this analysis technique, will make identification possible and provide necessary constraints on asteroseismological models of white dwarf stars.

Acknowledgments.   Special thanks are given to Marten van Kerkwijk and J. Christopher Clemens for obtaining the data. I thank F. Mullally for useful discussions about the simulations. Funds to complete this project were provided by NASA through a AAS small research grant.

References

Clemens, J. C., van Kerkwijk, M. H., & Wu, Y. 2000, MNRAS, 314, 220

Goldreich, P., & Wu, Y. 1999, ApJ, 511, 904

Robinson, E. L., et al. 1995, ApJ, 438, 908

Thompson, S. E., Clemens, J. C., van Kerkwijk, M. H., & Koester, D. 2003, ApJ, 589, 921

Thompson, S. E., Clemens, J. C., van Kerkwijk, M. H., O'Brien, M. S., & Koester, D. 2004, ApJ, 610, 1001

van Kerkwijk, M. H., Clemens, J. C., & Wu, Y. 2000, MNRAS, 314, 209

Frank N. Bash Symposium 2005: New Horizons in Astronomy
ASP Conference Series, Vol. 352, 2006
S. J. Kannappan, S. Redfield, J. E. Kessler-Silacci, M. Landriau, and N. Drory

Modeling Star Formation with Dust

Andrea Urban,[1] Neal J. Evans II,[1] Steven D. Doty,[2] and Hugo Martel[3]

[1] *Department of Astronomy, University of Texas, Austin, TX, USA*
[2] *Department of Physics and Astronomy, Denison University, Granville, OH, USA*
[3] *Département de physique, de génie physique et d'optique, Université Laval, Québec, QC, Canada*

Abstract.   Recent simulations of clustered star formation have incorporated many physical processes, yet coupling radiative transfer to the hydrodynamics has proven to be a difficult problem. We attempt to simplify this problem by making approximations to the heating and cooling terms in the energy equations, specifically gas-dust grain collisions, CO cooling rates, and cosmic-ray heating. Combining a recently developed hydrodynamics algorithm, SPH with particle-splitting, with our simplified heating-cooling algorithm, we plan to study the formation of young stars in a cluster while stars form nearby and heat the surrounding medium.

1.   Introduction

Modeling present-day star formation is a very difficult problem. This is due to the large range of densities involved and the various physical processes that occur at different scales. On the largest scale, molecular clouds are heated by the interstellar radiation field and cosmic rays. They can cool through molecular, primarily CO, rotational transitions. The heating and cooling terms balance and the molecular cloud maintains a temperature of 8-12 K. At larger densities, the interstellar radiation field is blocked, but the gas can still be heated by cosmic rays. As the density increases and stars form in dense clumps within the molecular cloud, dust is heated by the newly-formed stars and begins to contribute to the heating term in the molecular cloud via gas-dust grain collisions. Although there is more gas by volume than dust in the interstellar medium, dust is a better absorber of photons due to its broadband absorption properties.

In this study of star formation, we are interested in the effect dust has on heating the surrounding molecular cloud gas and how its presence and temperature affect future star formation in the cloud. Therefore, we start our simulation in the isothermal regime at $T = 10$ K and study the effects of stellar heating of dust on the molecular cloud. We plan to run two simulations with the same initial conditions. One simulation will not include heating by the dust and the other one will. For more details on the simulation see Fig. 1.

2.   Simulation Details

We combine a Smoothed Particle Hydrodynamics (SPH) algorithm with particle splitting for the hydrodynamics and a Particle-Particle/Particle-Mesh (P3M)

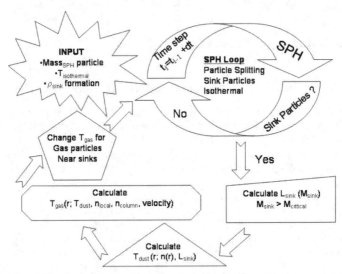

Figure 1.   Simulation Loop: Density contrast, isothermal temperature, and minimum sink mass define the input parameters of our cluster formation simulation. Before sink particles form, only gravity and gas dynamics are modeled. When the first sink particle forms with a mass greater than $M_{critical}$, we assume its luminosity can be related to the its mass and mass accretion rate as described by Wuchterl & Tscharnuter (2003). Then the dust temperature distribution around it is calculated, as described in Sec. 3. Finally, the gas temperature is calculated, as described in Sec. 4. With a non-isothermal temperature near the sink particle, the simulation moves forward in time and starts the loop over gravity and gas dynamics again.

algorithm for the gravity (see Martel, Evans, & Shapiro 2006). We simulate a cubic volume with periodic boundary conditions, representing a small part of a giant molecular cloud. Our simulation starts with an initial density of $n = 1.0 \times 10^5$ cm^{-3}. We set the mass resolution of the simulation to be $0.008\ M_\odot$, which is 10% of the minimum mass for hydrogen burning. Therefore, we can resolve objects of $M = 0.008\ M_\odot$, which sets the minimum Jeans mass of our isothermal simulation at $T = 10$ K. This corresponds to a density of $n = 5.8 \times 10^8$ cm^{-3}. The density contrast of our simulation is then $\approx$5800. Using two generations of particle splitting we are able to simulate a region with $M = 671\ M_\odot$ and size of 0.492 pc containing 1088 M_{Jeans}.

3.   Dust Temperature Calculation

In order to calculate the dust temperature distribution near protostellar sources we use the spherical radiative transfer code, DUSTY (Nenkova, Ivezić, & Elitzur 2000) using dust opacities from the fifth column of Table 1 in Ossenkopf & Henning (1994). Using DUSTY to calculate the dust temperature distribution around every sink/star that forms is impractical due to computational limitations. Therefore, we have developed an algorithm to calculate the dust temperature distribution based on a grid of DUSTY models. We first estimate the dust temperature distribution for a given model using a power law for which we define two parameters, K and β

$$T_d(\Delta r) = K\left[\frac{L}{(\Delta r)^2}\right]^{\frac{1}{4+\beta}}, \tag{1}$$

where L is the luminosity of the source and Δr is the distance from the source. Fig. 2 compares the dust temperature distribution calculated using Eq. 1 to the distribution calculated by DUSTY. The percentage error between the DUSTY solution and the fit to the DUSTY solution using Eq. 1 is low at most radii except at the center where the dust becomes optically thick. Therefore, we also model the dust temperature distribution with a power-law profile in $\log T - \log r$ space. With these two descriptions of the dust temperature profile for a variety of models in our grid, we are able to interpolate over any combination of the parameters that arise in the cluster simulation in order to determine the dust temperature near a sink particle.

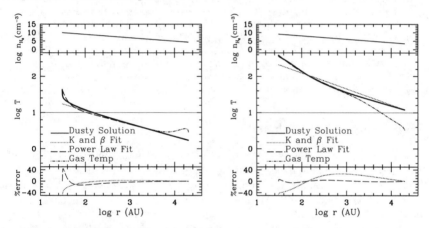

Figure 2. Compares dust temperature distribution derived by different methods. $L = 0.01\ L_\odot$ for left figure and $L = 100\ L_\odot$ for right figure. For both figures, $\tau = 2.89$. The top panel shows the density distribution. The large middle panel shows the dust temperature distribution. Various fits to the DUSTY solution have been attempted here. The bottom panel shows the error for various methods of fitting the DUSTY derived temperature distribution.

When the dust temperature profiles of sources overlap, we can no longer use Eq. 1 or the power-law profile. Therefore we use the following algorithm to calculate dust temperature distribution. First, we calculate the value of the dust temperature using a modified form of Eq. 1 in which we sum over luminosity sources and replace K and β with the average of the sources' K's and β's. Then we calculate the temperature at the same location in the simulation volume due to each source using the individual power-law fits. Finally, we choose the highest of the two temperatures as the value of the dust temperature at that point in the simulation. Fig. 3 shows the dust temperature distribution between two sources compared to the dust temperature distribution around the sources individually.

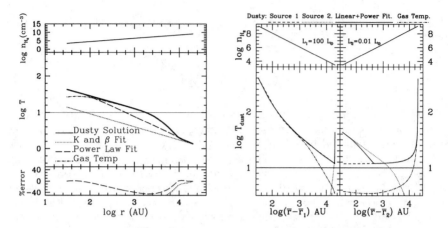

Figure 3. Left: This model of the dust temperature distribution has the same parameters as the model in the left panel of Fig. 2, except the density profile is inverted. This dust temperature profile was used to calculate the dust temperature between this low luminosity source and the bright source shown in the right panel of Fig. 2. Right: The right and left panel in this figure show the same data, except that the x-axis is reversed in order to show the dust temperature profile in detail close to each source individually. The density profile is plotted at the top of the figure. The key at the top of the figure describes the lines used in the figure.

4. Gas Temperature Calculation

We use an energy rate balance code to calculate the gas temperature. This code includes gas-dust collisional temperature coupling, cosmic-ray heating, and CO cooling (see Young et al. 2004). We input a cosmic-ray rate of 3×10^{-17} s^{-1} (van der Tak & van Dishoeck 2000). We assume a CO fractional abundance relative to H$_2$ of 1.58×10^{-4}. From the SPH/gravity code, we will input the local density, the weighted mean value of the column density, and a velocity parameter to derive the gas temperature, which we will feed back to the simulation. In Figs. 2 and 3, we have also plotted the gas temperature distribution.

Acknowledgments. We would like to thank NASA for providing AU with a Graduate Student Researchers Program Fellowship.

References

Martel, H., Evans, N. J., & Shapiro, P. R. 2006, ApJS, in press (astro-ph/0505008)

Nenkova, M., Ivezić, Ž., & Elitzur, M. 2000, in ASP Conf. Ser. 196: Thermal Emission Spectroscopy and Analysis of Dust, Disks, and Regoliths, ed. M. L. Sitko, A. L. Sprague, & D. K. Lynch (San Francisco: ASP), 77

Ossenkopf, V., & Henning, T. 1994, A&A, 291, 943

van der Tak, F. F. S., & van Dishoeck, E. F. 2000, A&A, 358, L79

Wuchterl, G., & Tscharnuter, W. M. 2003, A&A, 398, 1081

Young, K. E., Lee, J.-E., Evans, N. J., Goldsmith, P. F., & Doty, S. D. 2004, ApJ, 614, 252

Eiichiro Komatsu, Frank Bash, and Andrea Urban sample appetizers at the opening reception.

Paul Strycker, Ryan Campbell, and Pey-Lian Lim unwind at the reception after a long drive from New Mexico State University.

Doug Mar chats with *Physics Today* editor Toni Feder, who later wrote an article about the Bash Symposium that appeared in the December 2005 issue of *Physics Today*.

300

Lei Hao, Juntai Shen, and Pey-Lian Lim contribute to the symposium's diversity, representing three different institutions and three different subfields of astronomy.

Frank Bash thanks conference participants during the banquet at Fonda San Miguel.

Kaethe Podgorski and Seth Redfield enjoy the conversation at the conference dinner.

Author Index

ASTRONOMICAL SOCIETY OF THE PACIFIC
(ASP)

An international, nonprofit, scientific and educational organization
founded in 1889, established the
ASP CONFERENCE SERIES
in 1988, to publish books on recent developments in
astronomy and astrophysics.

All book orders or inquiries concerning

ASTRONOMICAL SOCIETY OF THE PACIFIC
CONFERENCE SERIES
(ASP - CS)

and

INTERNATIONAL ASTRONOMICAL UNION VOLUMES
(IAU)

should be directed to the:

Astronomical Society of the Pacific Conference Series
390 Ashton Avenue
San Francisco CA 94112-1722 USA

Phone:	**800-335-2624**	**(within USA)**
Phone:	**415-337-2126**	
Fax:	**415-337-5205**	

E-mail: **service@astrosociety.org**
Web Site: **http://www.astrosociety.org**

E-book Site: http://www.aspbooks.org

Complete lists of proceedings of past IAU Meetings are maintained at the
IAU Web site at the URL: http://www.iau.org/publicat.html

Volumes 32 - 189 in the IAU Symposia Series may be ordered from:

Kluwer Academic Publishers
P. O. Box 117
NL 3300 AA Dordrecht
The Netherlands

Kluwer@wKap.com

ASP CONFERENCE SERIES VOLUMES
Published by the Astronomical Society of the Pacific

ASP CONFERENCE SERIES VOLUMES
Published by the Astronomical Society of the Pacific

ASP CONFERENCE SERIES VOLUMES
Published by the Astronomical Society of the Pacific

PUBLISHED: 2004 (* asterisk means OUT OF PRINT)

Vol. CS-312   Third Rome Workshop on GAMMA-RAY BURSTS IN THE AFTERGLOW
ERA
eds. Marco Feroci, Filippo Frontera, Nicola Masetti and Luigi Piro
ISBN: 1-58381-165-6

Vol. CS-313   ASYMMETRICAL PLANETARY NEBULAE III: WINDS, STRUCTURE AND
THE THUNDERBIRD
eds. Margaret Meixner, Joel H. Kastner, Bruce Balick and Noam Soker
ISBN: 1-58381-168-0

Vol. CS 314   ASTRONOMICAL DATA ANALYSIS SOFTWARE AND SYSTEMS
(ADASS) XIII
eds. Francois Ochsenbein, Mark G. Allen and Daniel Egret
ISBN: 1-58381-169-9   ISSN: 1080-7926

Vol. CS 315   MAGNETIC CATACLYSMIC VARIABLES, IAU Colloquium 190
eds. Sonja Vrielmann and Mark Cropper
ISBN: 1-58381-170-2

Vol. CS 316   ORDER AND CHAOS IN STELLAR AND PLANETARY SYSTEMS
eds. Gene G. Byrd, Konstantin V. Kholshevnikov, Aleksandr A. Mylläri,
Igor' I. Nikiforov and Victor V. Orlov
ISBN: 1-58381-172-9

Vol. CS 317   MILKY WAY SURVEYS: THE STRUCTURE AND EVOLUTION OF OUR
GALAXY, The 5th Boston University Astrophysics Conference
eds: Dan Clemens, Ronak Shah and Tereasa Brainerd
ISBN: 1-58381-177-X

Vol. CS 318   SPECTROSCOPICALLY AND SPATIALLY RESOLVING THE
COMPONENTS OF CLOSE BINARY STARS
eds. Ronald W. Hilditch, Herman Hensberge and Krešimir Pavlovski
ISBN: 1-58381-179-6

Vol. CS-319   NASA OFFICE OF SPACE SCIENCE EDUCATION AND PUBLIC
OUTREACH CONFERENCE
eds. Carolyn Narasimhan, Bernhard Beck-Winchatz, Isabel Hawkins and
Cassandra Runyon
ISBN: 1-58381-181-8

Vol. CS-320   THE NEUTRAL ISM OF STARBURST GALAXIES
eds. Susanne Aalto, Susanne Hüttemeister and Alan Pedlar
ISBN: 1-58381-182-6

Vol. CS-321   EXTRASOLAR PLANETS: TODAY AND TOMORROW
eds. J. P. Beaulieu, A. Lecavelier des Etangs and C. Terquem
ISBN: 1-58381-183-4

Vol. CS-322   THE FORMATION AND EVOLUTION OF MASSIVE YOUNG STAR
CLUSTERS
eds. Henny J. G. L. M. Lamers, Linda J. Smith and Antonella Nota
ISBN: 1-58381-184-2

Vol. CS-323   STAR FORMATION IN THE INTERSTELLAR MEDIUM:
In Honor of David Hollenbach, Chris McKee and Frank Shu
eds. D. Johnstone, F. C. Adams, D. N. C. Lin, D. A. Neufeld and
E. C. Ostriker
ISBN: 1-58381-185-0

PUBLISHED: 2004 (* asterisk means OUT OF PRINT)

Vol. CS-324   DEBRIS DISKS AND THE FORMATION OF PLANETS:
A Symposium in Memory of Fred Gillett
eds. Larry Caroff, L. Juleen Moon, Dana Backman and Elizabeth Praton
ISBN: 1-58381-186-9

Vol. CS-325   THE SOLAR-B MISSION AND THE FOREFRONT OF SOLAR PHYSICS,
The Fifth Solar-B Science Meeting
eds. Takashi Sakurai and Takashi Sekii
ISBN: 1-58381-187-7

Vol. CS-326   GRAVITATIONAL LENSING: A UNIQUE TOOL FOR COSMOLOGY
eds. David Valls-Gabaud and Jean-Paul Kneib
ISBN: 1-58381-188-5

Vol. CS-327   SATELLITES AND TIDAL STREAMS
eds. F. Prada, D. Martínez Delgado and T. J. Mahoney
ISBN: 1-58381-190-7

PUBLISHED: 2005

Vol. CS-328   BINARY RADIO PULSARS
eds. F. A. Rasio and I. H. Stairs
ISBN: 1-58381-191-5

Vol. CS-329   NEARBY LARGE-SCALE STRUCTURES AND THE ZONE OF
AVOIDANCE
eds. A. P. Fairall and P. A. Woudt
ISBN: 1-58381-192-3

Vol. CS-330   THE ASTROPHYSICS OF CATACLYSMIC VARIABLES AND RELATED
OBJECTS
eds. J.-M. Hameury and J.-P. Lasota
ISBN: 1-58381-193-1

Vol. CS-331   EXTRA-PLANAR GAS
ed. Robert Braun
ISBN: 1-58381-194-X

Vol. CS-332   THE FATE OF THE MOST MASSIVE STARS
eds. Roberta M. Humphreys and Krzysztof Z. Stanek
ISBN: 1-58381-195-8

Vol. CS-333   TIDAL EVOLUTION AND OSCILLATIONS IN BINARY STARS:
THIRD GRANADA WORKSHOP ON STELLAR STRUCTURE
eds. Antonio Claret, Alvaro Giménez and Jean-Paul Zahn
ISBN: 1-58381-196-6

Vol. CS-334   14TH EUROPEAN WORKSHOP ON WHITE DWARFS
eds. D. Koester and S. Moehler
ISBN: 1-58381-197-4

Vol. CS-335   THE LIGHT-TIME EFFECT IN ASTROPHYSICS
Causes and Cures of the $O - C$ Diagram
ed. Christiaan Sterken
ISBN: 1-58381-200-8

Vol. CS-336   COSMIC ABUNDANCES as Records of Stellar Evolution and
Nucleosynthesis, in honor of Dr. David Lambert
eds. Thomas G. Barnes, III and Frank N. Bash
ISBN: 1-58381-201-6

ASP CONFERENCE SERIES VOLUMES
Published by the Astronomical Society of the Pacific

ASP CONFERENCE SERIES VOLUMES
Published by the Astronomical Society of the Pacific

PUBLISHED: 2006 (* asterisk means OUT OF PRINT)

Vol. CS 350   BLAZAR VARIABILITY WORKSHOP II: ENTERING THE GLAST ERA
eds. H. R. Miller, K. Marshall, J. R. Webb and M. F. Aller
ISBN: 1-58381-218-0

Vol. CS 351   ASTRONOMICAL DATA ANALYSIS SOFTWARE AND SYSTEMS
(ADASS) XV
eds. Carlos Gabriel, Christophe Arviset, Daniel Ponz and Enrique Solano
ISBN: 1-58381-219-9   ISSN: 1080-7926

Vol. CS 352   New Horizons in Astronomy: Frank N. Bash Symposium 2005
eds. Sheila J. Kannappan, Seth Redfield, Jacqueline E. Kessler-Silacci,
Martin Landriau, and Niv Drory
1-58381-220-2

A listing of IAU Volumes publish by the ASP Conference Series follows

INTERNATIONAL ASTRONOMICAL UNION VOLUMES
Published by the Astronomical Society of the Pacific

PUBLISHED: 1999 (* asterisk means OUT OF PRINT)

Vol. No. 190   NEW VIEWS OF THE MAGELLANIC CLOUDS
eds. You-Hua Chu, Nicholas B. Suntzeff, James E. Hesser
and David A. Bohlender
ISBN: 1-58381-021-8

Vol. No. 191   ASYMPTOTIC GIANT BRANCH STARS
eds. T. Le Bertre, A. Lèbre and C. Waelkens
ISBN: 1-886733-90-2

Vol. No. 192   THE STELLAR CONTENT OF LOCAL GROUP GALAXIES
eds. Patricia Whitelock and Russell Cannon
ISBN: 1-886733-82-1

Vol. No. 193   WOLF-RAYET PHENOMENA IN MASSIVE STARS AND STARBURST
GALAXIES
eds. Karel A. van der Hucht, Gloria Koenigsberger and
Philippe R. J. Eenens
ISBN: 1-58381-004-8

Vol. No. 194   ACTIVE GALACTIC NUCLEI AND RELATED PHENOMENA
eds. Yervant Terzian, Daniel Weedman and Edward Khachikian
ISBN: 1-58381-008-0

PUBLISHED: 2000

Vol. XXIVA   TRANSACTIONS OF THE INTERNATIONAL ASTRONOMICAL UNION
REPORTS ON ASTRONOMY 1996–1999
ed. Johannes Andersen
ISBN: 1-58381-035-8

Vol. No. 195   HIGHLY ENERGETIC PHYSICAL PROCESSES AND MECHANISMS FOR
EMISSION FROM ASTROPHYSICAL PLASMAS
eds. P. C. H. Martens, S. Tsuruta, and M. A. Weber
ISBN: 1-58381-038-2

Vol. No. 197 *   ASTROCHEMISTRY: FROM MOLECULAR CLOUDS TO PLANETARY
SYSTEMS
eds. Y. C. Minh and E. F. van Dishoeck
ISBN: 1-58381-034-X

Vol. No. 198   THE LIGHT ELEMENTS AND THEIR EVOLUTION
eds. L. da Silva, M. Spite and J. R. de Medeiros
ISBN: 1-58381-048-X

PUBLISHED: 2001

IAU SPS   ASTRONOMY FOR DEVELOPING COUNTRIES
Special Session of the XXIV General Assembly of the IAU
ed. Alan H. Batten
ISBN: 1-58381-067-6

Vol. No. 196   PRESERVING THE ASTRONOMICAL SKY
eds. R. J. Cohen and W. T. Sullivan, III
ISBN: 1-58381-078-1

Vol. No. 200 *   THE FORMATION OF BINARY STARS
eds. Hans Zinnecker and Robert D. Mathieu
ISBN: 1-58381-068-4

INTERNATIONAL ASTRONOMICAL UNION VOLUMES

Published by the Astronomical Society of the Pacific

INTERNATIONAL ASTRONOMICAL UNION VOLUMES

Published by the Astronomical Society of the Pacific